Water Resources Engineering and Management

Water Resources Engineering and Management

Edited by **Herbert Lotus**

R CALLISTO
REFERENCE

New York

Published by Callisto Reference,
106 Park Avenue, Suite 200,
New York, NY 10016, USA
www.callistoreference.com

Water Resources Engineering and Management
Edited by Herbert Lotus

International Standard Book Number: 978-1-63239-618-1 (Hardback)

Printed in the United States of America.

Contents

Preface

The engineering as well as management of water resources are described in this book. Hydrology is a scientific field which deals with the processes governing the exhaustion and renewal of water resources of the earth's land areas. The aim of this book is to combine the latest advancements in hydrology and the engineering of water resources. It discusses surface water and groundwater modeling, and covers various topics related to the management of surface water and groundwater resources. Overcoming the impact of climate change on water resources is also presented in the book. Other topics discussed are the interpretation of field knowledge, advancement of models, the usage of computational models based upon analytical and numerical techniques, evaluation of model performance and their usage for predictive purposes. This book will be beneficial for students and professionals dealing with this field.

The information contained in this book is the result of intensive hard work done by researchers in this field. All due efforts have been made to make this book serve as a complete guiding source for students and researchers. The topics in this book have been comprehensively explained to help readers understand the growing trends in the field.

I would like to thank the entire group of writers who made sincere efforts in this book and my family who supported me in my efforts of working on this book. I take this opportunity to thank all those who have been a guiding force throughout my life.

Editor

Part 1

Surface Water Modeling

Strengths, Weaknesses, Opportunities and Threats of Catchment Modelling with Soil and Water Assessment Tool (SWAT) Model

Matjaž Glavan and Marina Pintar
University of Ljubljana, Biotechnical Faculty, Department of Agronomy,
Chair for Agrometeorology, Agricultural Land Management,
Economics and Rural Development,
Slovenia

1. Introduction

Changed water regime in watercourses and high loads of sediment as a product of surface flow soil erosion can cause reductions in biodiversity, which is becoming one of the main indicators of environmental quality. Especially in the light of the European Union Water Framework Directive (WFD) (2000/60/EC) which requires new approaches, methods and tools for improvement, protection and prevention of further decreasing water quality. The main aim of the WFD is to achieve good quality status of water bodies in Europe by 2015 (Volk et al., 2009). Changes in social system and climate may, regardless environmental legislation restrictions, lead to changes in land use and hence in quantity of water flow and sediment concentrations in waters. Merging the different spatial and environmental data is time consuming; therefore, the use of computer modelling tools is necessary.

Models for catchments modelling can be divided into empirical-statistical (GLEAMS, MONERIS, N-LES), physical (WEPP, SA) and conceptual (distributed or semi-distributed - SWAT, NL-CAT, TRK, EveNFlow, NOPOLU, REALTA) (Kronvang et al., 2009; Hejzlar et al., 2009). Distributed and semi-distributed models have the best features for catchment modelling, as they can divide catchments into several smaller subcatchments and hydrological response units (HRUs) with its unique properties. This allows you to explore the responses of the catchments at different spatial and temporal dimensions. In contrast to the distributed models, semi-distributed models have limitation in simulating sediment and chemicals transport processes between HRUs. Generally have catchment models gained new value after they were placed in the geographical information system (GIS) environment. With the use of different cartographic databases it became easier to work with and spatially oriented results are more understandable to different target groups.

It is often claimed that models with diffused sources of pollution are not designed to simulate individual events but are primarily designed to assess long-term average values (Garen & Moore, 2005, Arnold et al., 1998). Current models for simulating water quality from diffuse sources have a lot of uncertainty in the projections (Garen & Moore, 2005) due to the influence of the study areas - heterogeneity, imperfect algorithms and scarce monitoring network. Modellers, regulators and policy makers must keep limitations of the models in their mind while they evaluate the results of the model simulations.

The Soil and Water Assessment Tool (SWAT) model was developed to help water resource managers in evaluating the impact of agricultural activities on waters and diffuse pollution in medium and large river catchments. The core of the model was developed on the basis of 30 years of experiences with modelling, as a continuation of Department of Agriculture of the United States of America (USDA) Agricultural Research Service (ARS) work (Gassman et al., 2007). The SWAT model is continuous-time, semi-distributed, process based catchment model, which was developed by merging three existing models - CREAMS, EPIC and GLEAMS (Gassman et al, 2007).

The model is extensively used in the world for modelling hydrology, water quality and climatic change (Krysanova & Arnold, 2008, Gassman et al., 2007). Applications of the model in the European Union (EU) were driven by Water Framework Directive enforced in 2000, which encourage the use of tools for assessment of diffused pollution, such as SWAT model (Barlund et al., 2007; Gassman et al., 2007). European Commission, for the purposes of ensuring adequate tools for the end-user that could satisfy the European need for harmonization of quantitative estimates of chemical losses from diffuse sources, has facilitated enforcement of the Water Framework Directive and the Nitrates Directive, funded EUROHARP project (Kronvang et al., 2009a, Kronvang et al., 2009b).

According to the Special Report on Emissions Scenarios (SRES) of the Intergovernmental Panel on Climate Change (IPCC) greenhouse gas emissions on a global scale is expected to increase the temperatures by 2100 (°C) for six SRES scenarios B1: 1.8 (1.1- 2.9), B2: 4.2 (1.4-3.8), A1T: 2.4 (1.4-3.8), A1B: 2.8 (1.7-4.4), A2: 3.4 (2.0-5.4) and A1FI: 4.0 (2.4-6.4) (Meehl et al ., 2005; Knutti et al., 2008). The possibility of realization of all scenarios in the future is equally probable, as they are based on different assumptions about the probable socio-economic development in the future. There is no doubt that air temperature will rise, however reliability of precipitation projections is lower, although there prevails a negative trend in the changes. Modelling of the IPCC scenarios have shown their impact on the catchments hydrology, on the infrastructure, on accumulation of carbon, on nutrient availability and plant growth (Sardans & Penuelas, 2004; Purkey et al., 2007).

General circulation models (GCM) well describe the process at a global level, but it is less reliable at the regional scale, since they do not include regional surface details. Therefore, the direct application of the results of GCM simulations in local and regional climate change negatively impact studies on agriculture, forestry, energy and water management (Bergant & Kajfež Bogataj, 2004). To bridge the gap between global and regional scale it is essential to couple GCM with nested regional model and provide potentially consistent geographical and physical estimates of regional climate change. To present local climate change features further empirical downscaling and various mathematical models are needed (Bergant & Kajfež Bogataj, 2005). However, projections of future climate changes, particularly at the local level, are accompanied by a number of uncertainties, which must be included in the analysis at the time of interpretation (Bergant & Kajfež Bogataj, 2004).

The aim of this chapter is to present strengths, weaknesses, opportunities and threats of the catchment model, Soil and Water Assessment Tool (SWAT), through two case studies catchments in Slovenia (Dragonja in Istria and Reka in Goriška Brda area).

Strengths, Weaknesses, Opportunities and Threats of Catchment Modelling with Soil and Water Assessment Tool (SWAT) Model

5

2. Materials and methods

2.1 Study areas characteristics

The river Reka catchment spreads over 30 km² and is located in the north-western part of the Slovenia in Goriška Brda region (Fig. 1). Altitude ranges between 75 m and 789 m above sea level (a.s.l.). Very steep ridges of numerous hills, which are directed towards the southwest, characterizes the area. The catchment landscape is agricultural with large percentages of forest (56 %) and vineyards (23 %).

The river Dragonja catchment spreads over 100 km² and is located in far south-west of the Slovenia in Istria region (Fig. 1). This is a coastal catchment (Adriatic Sea), with an altitude ranging between 0 and 487 m above sea level (a.s.l). The ridges of the hills are designed as a plateau with flat tops and steep slopes. The landscape is largely overgrown with forest (63 %) and grassland (18 %). Terracing is typical for both areas and depends on natural conditions, steepness of the slopes (erosion), geological structure (sliding) and climatic conditions (rainfall).

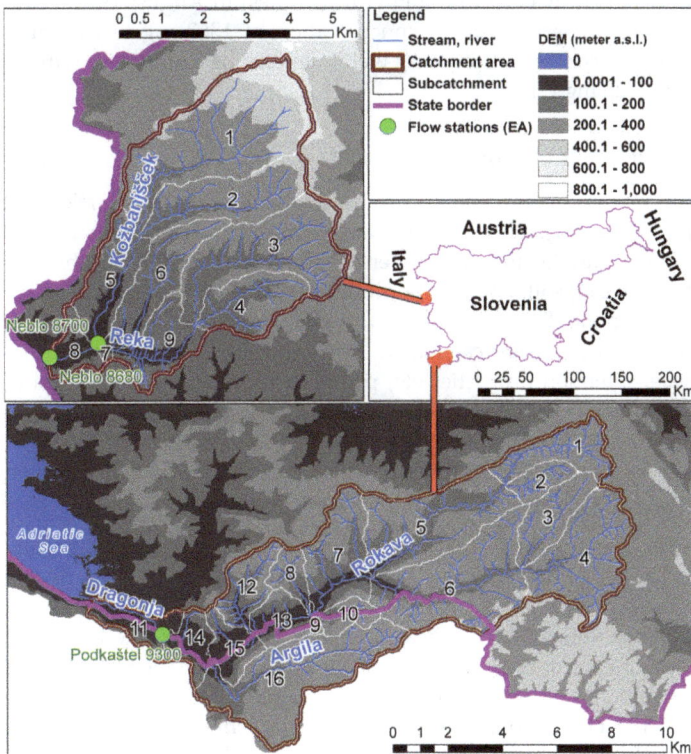

Fig. 1. The river Reka and Dragonja catchment case areas divided in sub-catchments

The bedrock of the study areas is Eocene flysch which consists of repeated sedimentary layers of sandstone and claystone. Soils are shallow brown euthric with silt-loam-clay texture that is difficult for tillage. With appropriate agro-technical measures (deep ploughing, organic fertilisers) they obtain properties for vine or olive production. In case of inappropriate agricultural activities and land management, we can witness very strong erosion processes.

Both areas are characterized by sub-Mediterranean climate (N Mediterranean – Adriatic Sea) with south-western winds and warm and moist air. Average annual temperature at the station Bilje (the Reka catchment), for the period 1991–2009, was 13.3 °C, and average annual rainfall in the period 1992 – 2008, was 1397 mm. Average annual temperature at Portorož station (the Dragonja catchment), for the period 1991 – 2009, was 14.1 °C, and average annual rainfall in the period 1993 – 2008, was 930 mm.

River network of the two areas is very extensive. Rivers are torrential and mediterranean. The average annual flow (1993-2008) of the rivers Reka and Dragonja is 0.59 m^3 s^{-1} and 0.87 m^3 s^{-1}, respectively, respectively, however summers have dry periods.

The annual average concentration of sediment in Reka catchment for year July 2008 – June 2009 was 32.60 mg l^{-1} (365 samples). In the river Dragonja catchment, the average annual concentration of sediment was 29.10 mg l^{-1} (107 samples) in August 1989 – December 2008. In January 2007, the highest sediment concentration of 1362 mg l^{-1} was measured. Data shows that sediment concentrations are wellin excess as compared with Environment Agency guide level (25 mg l^{-1}).

2.2 Database development for the model build

Field tour to the research areas and review of available data was carried out before the modelling started (Table 1). Since the available data was insufficient for modelling, we performed additional monitoring of sediment at the Reka tributary Kožbanjšček. Soil data was gathered from digital soil map and additional excavation of soil profiles and laboratory measurements were done. Using established model standards curve number (CN), albedo and organic carbon values were calculated. For water-physical soil properties (hydraulic conductivity, water-retention properties etc) pedotransfer functions were used (Saxton et al., 1986; Neisch et al., 2005; Pedosphere, 2009). For the purpose of this study we used the SWAT 2005 model, Geographic Information System (GIS) 9.1 software and ArcSWAT interface. Extensions necessary for SWAT functioning in GIS environment are Spatial Analyst, Project Manager and SWAT Watershed Delineator, which enables visualisation of the results.

SWAT is capable of simulating a single catchment or a system of hydrological linked subcatchments. The model of GIS based interface ArcSWAT defines the river network, the main point of outflow from the catchment and the distribution of subcatchments and hydrological response units (HRU). Subcatchments are spatially related to each other. HRUs are basically parts of each subcatchment with a unique combination of land use, soil, slope and land management and those are not spatially related. This allows modelling of different evapotranspiration (ET), erosion, plant growth, surface flow, water balance, etc for each subcatchment or HRU, thus increases accuracy of the simulations (Di Luzio et al., 2005). The number of HRU-s in each subcatchment was set by a minimum threshold area of 5%:5%:5% for land use, soil and slope classes, respectively. Classes that cover less than 5% of area are eliminated in order to minimise the number of HRU-s whilst not overly compromising on model accuracy. The river Reka catchment was delineated on 9 subcatchment and 291 HRUs while the river Dragonja catchment on 16 subcatchments and 602 HRUs.

Data type	Scale	Source	Description/properties
Topography	25m×25m	The Surveying and Mapping Authority of the Republic of Slovenia	Elevation, slope, channel lengths
Soils	Slovenia: 1:25,000 Croatia: 1:50,000	Ministry of Agriculture, Forestry and Food of the Republic of Slovenia (MAFFRS); University of Ljubljana; University of Zagreb	Spatial soil variability. Soil types and properties.
Land use	Slovenia: 1m×1m (Graphical Units of Agricultural Land) Croatia: 100m×100m (CORINE)	MAFFRS; European Environment Agency	Land cover classification and spatial representation
Land management	/	Chamber of Agriculture and Forestry of Slovenia; MAFFRS (Mihelič et al., 2009); Interviews with farmers	Crop rotations: planting, management, harvesting. Fertiliser application (rates and time)
Weather stations	Reka: 2 rainfall, 1 meteo (wind, temp., rain, humidity, solar) Dragonja: 3 rainfall, 1 meteo	Environment Agency of the Republic of Slovenia (EARS)	Daily precipitation, temperature (max., min.), relative humidity, wind, solar radiation.
Water abstraction	46 permits (136 points)	EARS	From surface and groundwater.
Waste water discharges	Reka: 2 points Dragonja: 1 point	EARS	Registered domestic, Industrial discharge
River discharge	Reka: 2 stations Dragonja: 1 station	EARS	Daily flow data (m³ day⁻¹)
River quality	Reka: 0 monitoring station; Dragonja: 1 monitoring station	EARS	Sediment (mg l⁻¹)

Table 1. Model input data sources for the Reka and Dragonja catchments

2.3 Model performance objective functions

The Pearson coefficient of correlation (R^2) (unit less) (1) describes the portion of total variance in the measured data as can be explained by the model. The range is from 0.0 (poor model) to 1.0 (perfect model). A value of 0 for R^2 means that none of the variance in the measured data is replicated by the model, and value 1 means that all of the variance in the measured data is replicated by the model. The fact that only the spread of data is quantified which is a major drawback if only R^2 is considered. A model that systematically over or under-predicts all the time will still result in "good" R^2 (close to 1), even if all predictions are wrong (Krause et al., 2005).

$$R^2 = \left(\frac{\sum_{i=1}^{n}(simulated_i - simulated_{average})(measured_i - measured_{average})}{\sqrt{\sum_{i=1}^{n}(simulated_i - simulated_{average})^2}\sqrt{\sum_{i=1}^{n}(measured_i - mesured_{average})^2}} \right)^2 \qquad (1)$$

Function bR^2 is defined as coefficient of determination R^2 multiplied by the absolute value of the coefficient (slope) of the regression line (b) between simulated (S) and measured (M), with treatment of missing values (2). It allows accounting for the discrepancy in the magnitude of two signals (depicted by b) and their dynamics (depicted by R^2). The slope b is computed as the coefficient of the linear regression between simulated and measured, forcing the intercept to be equal to zero. The objective function is expressed as:

$$bR^2 = \begin{cases} |b| \times R^2 \ for\ b \leq 1 \\ |b| \div R^2 \ for\ b > 1 \end{cases} \qquad (2)$$

The Nash-Sutcliffe simulation efficiency index (E_{NS}) (unit less) (3) is widely used to evaluate the performance of hydrological models. It measures how well the simulated results predict the measured data. Values for E_{NS} range from negative infinity (poor model) to 1.0 (perfect model). A value of 0.0 means, that the model predictions are just as accurate as measured data (minimally acceptable performance). The E_{NS} index is an improvement over R^2 for model evaluation purposes because it is sensitive to differences in the measured and model-estimated means and variance (Nash & Sutcliffe, 1970). A disadvantage of Nash-Sutcliffe is that the differences between the measured and simulated values are calculated as squared values and this places emphasis on peak flows. As a result the impact of larger values in a time series is strongly overestimated whereas lower values are neglected.

$$E_{NS} = 1 - \left(\frac{\sum_{i=1}^{n}(measured_i - simulated_i)^2}{\sum_{i=1}^{n}(measured_i - measured_{average})^2} \right) \qquad (3)$$

Root Mean Square Error – RMSE (4) is determined by calculating the standard deviation of the points from their true position, summing up the measurements, and then taking the square root of the sum. RMSE is used to measure the difference between flow ($q^{simulated}$) values simulated by a model and actual measured flow ($q^{measured}$) values. Smaller values indicate a better model performance. The range is between 0 (optimal) and infinity.

$$RMSE = \sqrt{\frac{\sum_{i=1}^{n}(q_t^{measured} - q_t^{simulated})^2}{n}} \qquad (4)$$

Percentage bias – PBIAS (%) (5) measures the average tendency of the simulated flows ($q^{simulated}$) to be larger or smaller than their measured ($q^{measured}$) counter parts (Moriasi et al., 2007). The optimal value is 0, and positive values indicate a model bias towards underestimation and vice versa.

$$PBIAS = \left(\frac{\sum_{i=1}^{n}(q_t^{measured} - q_t^{simulated})}{\sum_{i=1}^{n}(q_t^{measured})} \right) \cdot 100\% \qquad (5)$$

Model calibration criteria can be further based on recommended percentages of error for annual water yields suggested from the Montana Department of Environment Quality (2005) that generalised information related to model calibration criteria (Table 2) based on a number of research papers.

Errors (Simulated-Measured)	Recommended Criteria
Error in total volume	10%
Error in 50% of lowest flows	10%
Error in 10% of highest flows	15%
Seasonal volume error (summer)	30%
Seasonal volume error (autumn)	30%
Seasonal volume error (winter)	30%
Seasonal volume error (spring)	30%

Table 2. Model calibration hydrology criteria by Montana Department of Environment Quality (2005)

For the detection of statistical differences between the two base scenarios and alternative scenarios Student t-test statistics should be used ($\alpha = 0.025$, degrees of freedom (SP = n-1)), for comparing average annual value of two dependent samples at level of significance 0.05 (6). Where $\bar{x}$ is a sample arithmetic mean (alternative scenario), μ an average of the corresponding random variables (base scenario), s is sample standard deviation (alternative scenario) and n number of pairs (alternative scenario). Variable, which has approximately symmetrical frequency distribution with one modus class, is in the interval $\bar{x} \pm s$ expected 2/3 of the variables and in $\bar{x} \pm 2s$ approximately 95% of the variables and in $\bar{x} \pm 3s$ almost all variables. Confidence interval ($l_{1, 2}$) for Student distribution ($t_{\alpha/2}$) for all sample arithmetic means can be calculated (7).

$$t = \frac{\bar{x} - \mu}{s / \sqrt{n}} \tag{6}$$

$$l_{1,2} = \bar{x} \pm t_{\frac{\alpha}{2}}(n-1) \cdot \frac{s}{\sqrt{n}} \tag{7}$$

3. Sensitivity analysis

Sensitivity analysis limits the number of parameters that need optimization to achieve good correlation between simulated and measured data. The method of analysis in the SWAT model called ParaSol is based on the method of Latin Hypercube One-factor-at-a-Time (LH-OAT). ParaSol method combines the objective functions into a global optimization criterion and minimizes both of them by using the Shuffled Complex (SCE-UA) algorithm (van Griensven et al., 2006). The new scheme allows the LH-OAT to unmistakably link the changes in the output data of each model to the modified parameter (van Griensven et al., 2006).

Tool within the SWAT model can automatically carry out the sensitivity analysis without the measured data or with the measured data. The tool varies values of each model parameter within a range of (MIN, MAX). Parameters can be multiplied by a value (%), part of the value can be added to the base value, or the parameter value can be replaced by a new value. The final result of the sensitivity analysis are parameters arranged in the ranks, where the parameter with a maximum effect obtains rank 1, and parameter with a minimum effect obtains rank which corresponds to the number of all analyzed parameters. Parameter that has a global rank 1, is categorized as "very important", rank 2 – 6 as "important", rank 7 – 41 (i.e. the number of parameters in the analysis – i.e. flow 7 - 26) as "slightly important" and rank 42 (i.e. flow 27) as "not important" because the model is not sensitive to change in parameter (van Griensven et al., 2006).

Beside in the model build tool for the sensitivity analysis and calibration a special standalone tool called SWAT-CUP is available which includes all important algorithms (GLUE, PSO, MCMC, PARASOL and SUFI-2) of which Sequential Uncertainty Method (SUFI-2) has shown to be very effective in identifying sensitive parameters and calibration procedures (Abbaspour et al., 2007). With the right choice of a method and tool we can substantially shorten the process of parameter sensitivity identification.

Sensitivity analysis was performed for subcatchment 5 on the river Reka tributary Kožbanjšček and subcatchment 14 on the river Dragonja. This were the only points in both catchments where were alongside the river flow also sediment and nutrients concentrations measured. The presented analysis was performed for an average daily flow and sediments. Table 3 represents for each model the first 10 parameters that have the greatest impact on the model when they are changed. The sensitivity analyses demonstrated great importance of the hydrological parameters that are associated with surface and subsurface runoff. Parameter sensitivity ranking and value range is greatly dependent on the uncertainties originating in model simplification, in processes not included, unknown or unaccountable and in measured data errors or in time step of the measured data (daily, monthly, yearly).

Surlag represents the surface runoff velocity of the river and **Alpha_Bf** factor determines the share between the base and surface flow contribution to the total river flow. **Cn2** curve runoff determines the ratio between the water drained by the surface and subsurface runoff in moist conditions. **Ch_K2** describes the effective hydraulic conductivity of the alluvial river bottom (water losing and gaining). **Esco** describes evaporation from the soil and **Rchrg_Dp** fraction of groundwater recharge to deep aquifer. For the sediment modelling the most important parameters are **Spcon** and **Spexp** that affect the movement and separation of the sediment fractions in the channel. **Ch_N** – Manning coefficient for channel, determines the sediment transport based on the shape of the channel and type of the river bed material. **Ch_Erod** – Channel erodibility factor and **Ch_Cov** – Channel cover factor that has proved to be important for the Dragonja catchment. Soil erosion is closely related to the surface runoff hydrological processes (Surlag, Cn2). The analysis showed importance of the hydrological parameters that are associated with surface and subsurface runoff (Cn2, Surlag) and base flow (Alpha_Bf), suggesting numerous routes by which sediment is transported (Table 3).

| Base model | Sensitivity Analysis Objective function (SSQR) | | Category |
	Flow	Sediment	
Reka	Surlag	Spcon	Very important
	Alpha_Bf	Ch_N	Important (2-6)
	Cn2	Surlag	
	Ch_K2	Spexp	
	Esco	Cn2	
	Ch_N	Alpha_Bf	
Dragonja	Cn2	Spcon	Very important
	AlphaBf	Ch_Erod	Important (2-6)
	Ch_K2	Ch_Cov	
	RchrgDp	Ch_N	
	Esco	Spexp	
	Surlag	Surlag	

Table 3. SWAT parameters ranked by the sensitivity analysis for the Reka subcatchment 5 and the Dragonja subcatchment 14 (1998 - 2005)

4. Calibration and validation

Many of the model's input parameters cannot be measured for different reasons, such as high cost of equipment or lack of time or personnel which means that the model must be calibrated. During the model calibration parameters are varied within an acceptable range, until a satisfactory correlation is achieved between measured and simulated data. Usually, the parameters values are changed uniformly on the catchment level. However, certain parameters (Cn2, Canmx, Sol_Awc) are exceptions, because of the spatial heterogeneity. The variable to which the model is most sensitive should be calibrated first, that is usually hydrology. Firstly manual calibration, parameter by parameter, should be carried out with gradual adjustments of the parameter values until a satisfactory output results (E_{NS} and $R^2 >$ 0.5) (Moriasi et al., 2007, Henriksen et al., 2003). This procedure may be time consuming for inexperienced modellers. In the process of autocalibration only the most sensitive parameters are listed that showed the greatest effect on the model outputs. For each of the parameter a limit range (max, min) has to be assigned.

Validation is the assessment of accuracy and precision, and a thorough test of whether a previously calibrated parameter set is generally valid. Validation is performed with parameter values from the calibrated model (Table 4) and with the measured data from another time period. Due to the data scarcity, the model was validated only for the hydrological part (flow). The river Reka sediment data covers only one year of daily observations, which was only enough for the calibration. For the river Dragonja a 14 years long data series of sediment concentrations were available, but the data was scarce (92 measurements). It should be pointed out that samples taken during monitoring represents only the current condition of the river in a certain part of the day (concentration in mg l[-1]) which has to be recalculated to load, while the simulated value is a total daily transported load (kg day[-1]) in a river.

Calibration of the daily flow for the rivers Reka (subcatchment 8) and Dragonja (subcatchment 14) catchments was performed for the period from 1998 to 2005. According to the availability of data we selected different periods for the daily flow validation of the Reka (1993–1997, 2006–2008) and Dragonja (1994–1996, 2006–2008). Due to the lack of data we performed the sediment calibration for the river Reka tributary Kožbanjšček (subcatchment 5) daily between 1. 7. 2008 – 30. 6. 2009 and for the river Dragonja (subcatchment 14) daily calibration (montly sampling frequency) for the period 1994 – 2008.

	Parameter	Default value	Range	Calibrated values	
				Reka	Dragonja
1	Alpha_Bf	0.048	0–1	0.30058	0.45923
2	Canmx[1]	0	0–20	8, 4, 2	8, 4, 2
3	Ch_K2	D	0–150	7.0653	3.7212
4	Ch_N	D	0–1	0.038981	0.04363
5	Cn2	D*	–25/+25%	–8, –15 [2]	+14
6	Esco	0.95	0–1	0.8	0.75
7	Gw_Delay	31	0–160	131.1	60.684
8	Gw_Revap	0.02	0–0.2	0.19876	0.069222
9	Gwqmn	0	0–100	100	0.79193
11	Surlag	4	0.01–4	0.28814	0.13984
E_{NS}				**0.61**	**0.57**

Legend: [1] - forest, permanent crops, grassland + arable; [2] - subcatchment 1-2-5, subcatchment 3-4-6-7-8-9; D - depends on river channel characteristics; D*- depends on soil type, land use and modeller set up

Table 4. Hydrological parameters, ranges and final values selected for the model calibration for the rivers Reka and Dragonja catchments

4.1 Hydrology

Base flow simulations are an important part of the catchment model calibration (Moriasi et al., 2007). The total flow consists of a base and surface flow, where base flow constitutes a major proportion of the measured total flow, especially in dry periods. In order to conduct the comparison between simulated and measured base flow the estimation of both is needed. For the base flow calculation a Baseflow program was used that includes a method of Automated Base Flow Separation from total flow (Arnold et al., 1995). Separation for the measured flow showed that base flow contributes 28 to 45% or on average about 37% of the river Reka flow and 37 to 55% or on average about 46% of the river Dragonja flow (Table 5, Table 6; Fig. 2).

	Base flow separation pass Fr 1		Base flow separation pass Fr2	
Catchment	Measured	Simulated	Measured	Simulated
Reka	0.45	0.43	0.28	0.27
Dragonja	0.55	0.58	0.37	0.43

Table 5. Calculation of the percentages of the base flow from total flow with Baseflow Program for the calibration period 2001–2005 for the rivers Reka and Dragonja

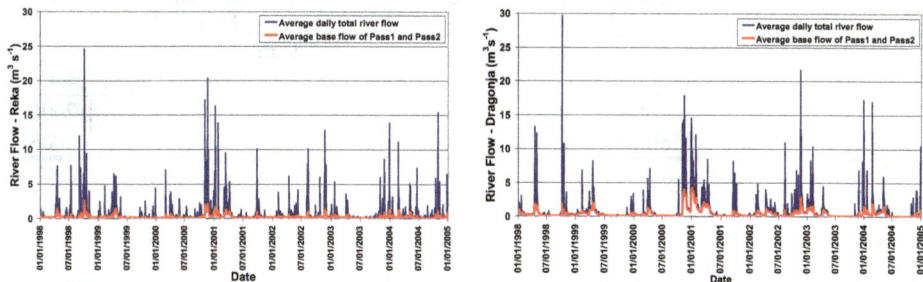

Fig. 2. Separation of base flow from the measured average daily flow (m³ s⁻¹) of the rivers Reka and Dragonja (1998–2005)

Catchment	Year	Measured Flow (mm)		Simulated (mm)		Ratio Simulated/Measured	
		Total	Base	Total	Base	Total	Base
Reka	2001	700	289	761	317	1.09	1.10
	2002	587	215	512	177	0.87	0.82
	2003	455	162	415	122	0.91	0.76
	2004	717	267	614	202	0.86	0.76
	2005	513	187	461	175	0.90	0.93
	Average	595	224	553	199	0.93	0.89
Dragonja	2001	380	222	367	228	0.97	1.03
	2002	327	148	273	124	0.83	0.84
	2003	217	93	231	109	1.07	1.18
	2004	246	114	264	138	1.07	1.21
	2005	205	107	174	93	0.85	0.87
	Average	275	137	262	139	0.95	1.02

Table 6. Comparison of measured and simulated average annual sums of total and base flow (mm) for the calibration period (2001–2005) for the rivers Reka and Dragonja

Daily calibration objective functions show that the simulated total flows are within the acceptable range (Table 7, Fig. 3). Correlation coefficient (R^2) for a daily flow is influenced by low flows. Official measurements of a flow showed that on certain days the flow was not present or it was negligible. Model does not neglect extremely low flows, as is evident from the cumulative distribution of the flow (Fig. 3). Errors in flow measurements, in the worst case may be up to 42 % and in best case up to 3 % of the total flow (Harmel et al., 2006).

The E_{NS} values of flow fall into the category of satisfactory results (Moriasi et al., 2007, Henriksen et al., 2003), R^2 values fall into the category of good results, RMSE into the category of very good results (Henriksen et al., 2003) and PBIAS into the category of very good, good and satisfactory results (Moriasi et al., 2007). The reasons for lower results of the objective functions in the validation lie in the representation of the soil, rainfall and in the river flow data uncertainty.

	Reka				Dragonja			
	Calibration		Validation (Total Flow)		Calibration		Validation (Total Flow)	
Objective function	Base Flow	Total Flow	1993 - 1997	2006 - 2008	Base Flow	Total Flow	1994 - 1996	2006 - 2008
E_{NS}	0.61	0.61	0.39	0.69	0.55	0.57	0.45	0.42
R^2	0.72	0.64	0.57	0.70	0.66	0.59	0.49	0.49
RMSE	0.13	0.82	1.21	0.74	0.35	1.06	1.98	1.50
PBIAS	-12.79	7.04	-14.19	19.40	1.49	4.69	23.15	-3.31

Table 7. Daily time step river flow performance statistics for the rivers Dragonja and Reka for the calibration (2001-2005) and validation periods

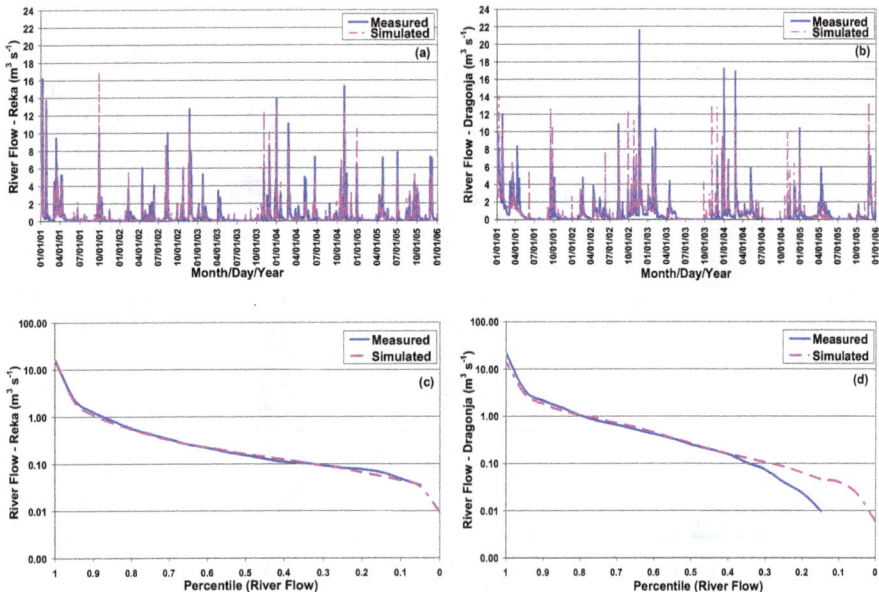

Fig. 3. Comparison between simulated (SWAT) and daily measured flows (m³ s⁻¹) (a, b) and cumulative distribution by percentiles (c, d) of daily river flows for the calibration period (2001-2005)

4.2 Sediment

Parameters used for the sediment calibration were Spcon, Spexp, Ch_Erod, Ch_Cov and Usle_P (USLE equation support practice (P) factor). Simulation results for the river Reka show lower E_{NS} = 0.23 and a good result in predicting the variability of $E_{NSpercentile}$ = 0.83 (Table 8, Fig. 4). In the case of Dragonja, model achieved good results for E_{NS} = 0.70 and $E_{NSpercentile}$ = 0.73. Simulated PBIAS values fall for both catchments within the category of very good results as deviation is less than 15% (Moriasi et al., 2007) (Table 8, Fig. 4). Errors in typically measured sediment data can be on average around 18% and in worst and best case up to 117% or down to 3%, respectively (Harmel et al., 2006).

Strengths, Weaknesses, Opportunities and Threats of Catchment Modelling with Soil and Water Assessment Tool (SWAT) Model

15

Parameter		Default	Range	Calibrated values	
Sediment				Reka – Kožbanjšček	Dragonja
1	Spcon	0.0001	0.0001–0.01	0.002	0.002
2	Spexp	1	1–1.5	1.3	1
3	Ch_Erod	0	0–1	0.092	0,06
4	Ch_Cov	0	0.05–0.6	0.1	0,1
5	Usle_P	1	0–1	slope dependent	slope dependent
E_{NS}				0.23	0.70
E_{NS} percentile				0.83	0.73
R^2				0.24	0.80
RMSE				10.35	19.81
PBIAS				–0.15	–6.33

Table 8. Calibrated sediment parameters, their ranges and the final values that are chosen for the model calibration periods (Reka Jul. 2008 – Jun. 2009; Dragonja 1994 - 2008)

Majority of the sediment in the Reka catchment eroded in the winter and early spring, when soils are in combination with tillage and weather conditions highly exposed to erosion (Fig. 4). In the Dragonja catchment it is apparent that monitoring scheme until the year 2004 was not adequately optimised in a way to capture the full range of possible daily situations, causing problems during calibration and disabling validation. Daily eroded curve shows the maximum erosion in the autumn, winter and early spring, when the area receives the

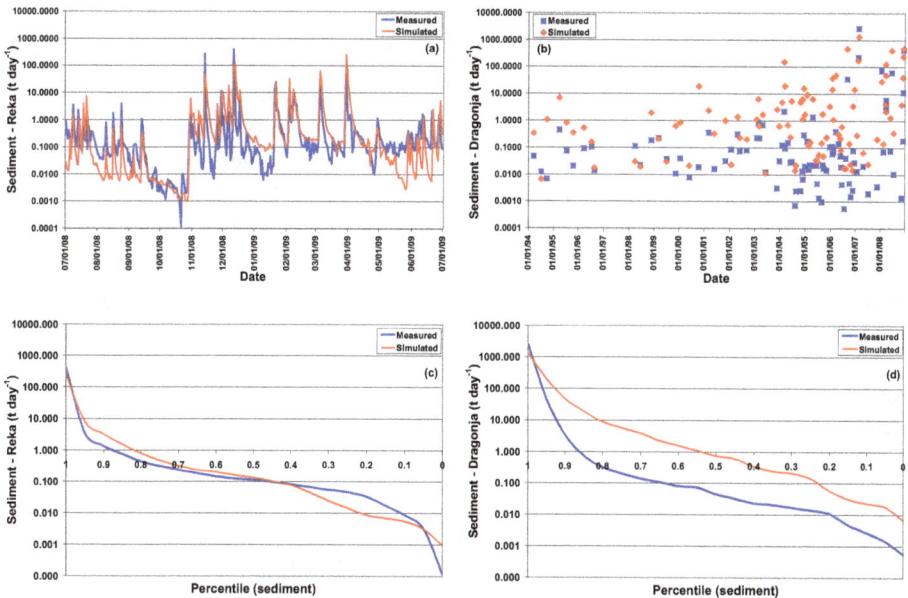

Fig. 4. Comparison between simulated and measured daily loads (t day^{-1}) of sediment (a, b) and cumulative distribution by percentiles (c, d) for the river Reka - Kožbanjšček (Jul. 2008-Jun. 2009) and river Dragonja (1994–2008) on logarithmic scale

majority of precipitation. During this period soils get quickly saturated with low evapotranspiration. Low hydraulic conductivity of the soils (surface flow dominates) accompanied with higher rainfall intensities can result in flash floods. However, it is necessary to draw attention to the lack of soil data that would better describe the processes in the catchments. It is important to note that the measured sediment loads (day t^{-1}) are calculated from a certain part of the day sampled concentration (mg l^{-1}) using the mean daily flow (m^3 day^{-1}), creating uncertainty in the application of this type of data. Terracing of steep slopes where agricultural production takes place is common practice in both areas. As terraces significantly reduce surface runoff and consequently erosion Usle_P parameter values are adapted (Table 9).

Slope (%)	Usle_P (arable, olive grove, vineyard, orchard)	Usle_P (grassland)
1–10	0.55	0.55
11–20	0.70	0.70
21–35	0.06 - terraces	0.75
36–50	0.07 - terraces	0.80
> 51	0.08 - terraces	0.85

Table 9. Usle_P factor for agricultural land use at different slopes and for terraces

4.3 Model performance indicators

Important model performance indicators of the water balance for correct representation of flow, sediment and chemicals transport and losses are evapotranspiration (ET) and Soil Water Content (SWC) through infiltration (lateral, groundwater flow) and surface run-off. As evapotranspiration is a function of crop growth only a proper simulation of crop growth and management can ensure realistic modelling of evapotranspiration and nutrients within a river catchment. Sediment erosion and infiltration of chemicals through the soil profile depends on the share of water between surface runoff and water entering and moving through the soil profile in terms of percolation. Before sediment and chemicals modelling is attempted a correct partitioning of water in these three phases is required, apart from the requirement for a match of predicted and observed stream flow.

Evapotranspiration is a primary mechanism by which water is removed from the catchment. It depends on air temperature and soil water content. The higher the temperature, the higher is potential evapotranspiration (PET) and consequently ET, if there is enough of water in the soil. A simple monthly water balance between monthly precipitation and PET showed that average monthly water balance in the Reka catchment (station Bilje) is negative between May and August (Fig. 5). In the Dragonja catchment (station Portorož) water balance is negative from April to August (growing season) (Fig. 5).

Water that enters the soil may move along one of the several different pathways. It may be removed by plant uptake or evaporation; it may percolate past the bottom of the soil profile or may move laterally in the profile. However, plant uptake removes the majority of water that enters the soil profile (Neitsch et al., 2005). The SWC will be represented correctly if crops are growing at the expected rate and soils have been correctly parameterized. Figure 6 shows the SWC and precipitation of selected HRUs No. 38 (Reka) and No. 182 (Dragonja), with a silt clay soils, with the prevailing surface runoff and slow lateral subsurface flow. Water in soil exits the field capacity in the spring and return to that state in the autumn (Fig. 6). Soils in the summer are often completely dry with occasional increasing induced by storms.

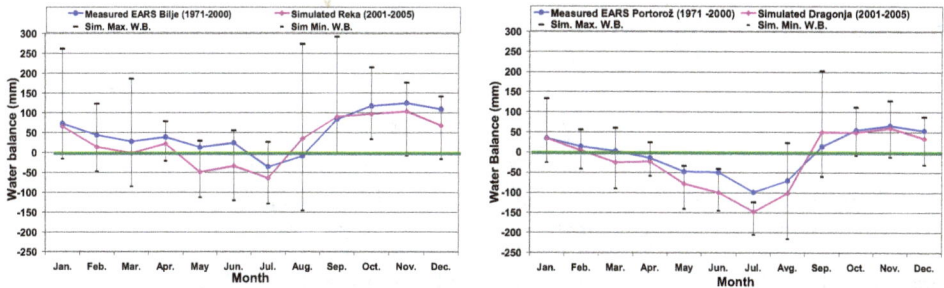

Fig. 5. Comparison of simulated and measured (Environment Agency of Republic of Slovenia - EARS) water balance (mm) for the Reka subcatchments 8 and Dragonja subcatchment 14.

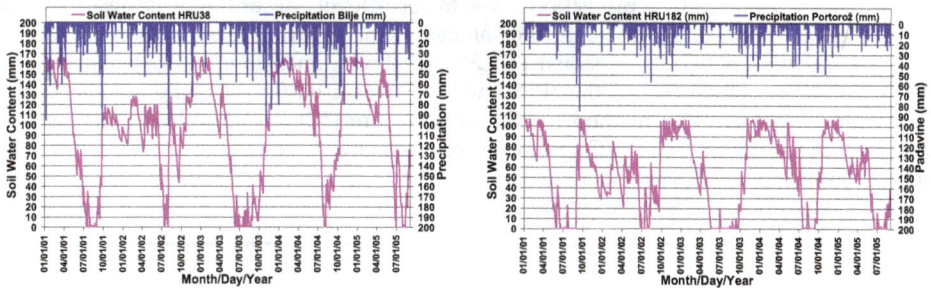

Fig. 6. Comparison of simulated soil water content (mm) for the HRU No. 38 (Reka) and HRU No. 182 (Dragonja) and observed precipitation (mm) in the calibration period (2001–2005)

The plant growth component of SWAT is a simplified version of the plant growth model. Phenological plant development is based on daily accumulated heat units, leaf area development, potential biomass that is based on a method developed by Monteith, a harvest index is used to calculate yield, and plant growth that can be inhibited by temperature, water, N or P stress. (Neitsch et al., 2005). In the crop database a range of parameters can be changed to meet the requirements for optimal plant growth. We used default SWAT database parameters that were additionally modified (Frame, 1992). An example crop growth profile for development of leaf area index (LAI) and plant biomass (BIOM) for vineyard is presented on Fig. 7.

5. Climate change scenarios

The base of climate scenarios was taken from the research on the transfer of global climate simulations on the local level (Bergant, 2003). This is currently the only one climate change study for both research areas where climate changes are described with specific numeric

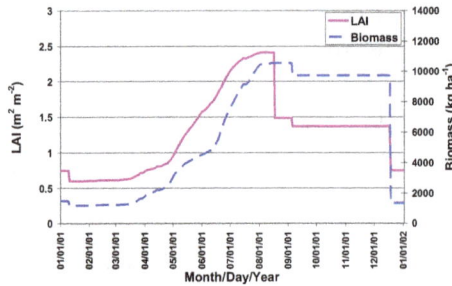

Fig. 7. Simulated vineyard biomass growth (kg ha^{-1}) and leaf area index (m^2 m^{-2}) for the HRU No. 38 in the river Reka catchment

data. Majority of SWAT model studiesthat predict the effects of climate changes on the environment are focused on the proper projection of global, general and regional circulation climate models on the local level (Gassman et al., 2007).

Bergant (2003) research is concentrated on the empirical reduction of global regional scale climate models from REG 3 prediction level to local level. Accordingly he used several methods, from which one was for further analysis where partial least squares regression (PLS) method with a two-dimensional (2PLS) and three-dimensional (3PLS) matrix were selected. Average value between the 3PLS and 2PLS was used for calculation of changes in precipitation (%) and 3PLS method for the temperature (°C) change calculation. Further, average change between two IPCC socio-economic scenarios (A2 - global-economic and B2 - regional environmental) was calculated for the base scenario (Bergant, 2003). The final results (Table 10) are absolute change in temperature (°C) and percentage change in

Period of the year / IPCC Scenario	Scenario					
	2030		2060		2090	
	Reka	Dragonja	Reka	Dragonja	Reka	Dragonja
Temperature (change in °C)						
Warm period / A2	1.70	1.00	2.65	1.55	5.90	2.70
Warm period / B2	1.60	0.90	2.65	1.60	4.30	2.10
Average / A2B2	1.65	0.95	2.65	1.58	5.10	2.40
Cold period / A2	1.25	1.35	2.85	2.60	3.85	4.15
Cold period / B2	1.40	1.40	2.45	2.25	3.30	3.05
Average /A2B2	1.33	1.38	2.65	2.43	3.58	3.60
Precipitation (change in %)						
Warm period / A2	–12.00	–6.00	–22.50	–13.50	–35.00	–21.50
Warm period / B2	–15.50	–7.50	–23.00	–14.00	–31.50	–17.50
Average / A2B2	–13.75	–6.75	–22.75	–13.75	–33.25	–19.50
Cold period / A2	–16.00	+3.50	–16.00	+8.50	–24.00	+16.00
Cold period / B2	–13.50	+5.00	–21.00	+10.00	–18.00	+14.00
Average /A2B2	–14.75	+4.25	–18.50	+9.25	–21.00	+15.00

Table 10. The used climate change data on temperature (°C) and precipitation (%) – adapted from Bergant (2003)

precipitation (%) for warm (April - September) and cold (October - March) half of the year for the three time periods (2001-2030, 2031-2060, 2061-2090). These changes were used for the modification of existing databases of daily weather data. When transferring the regional models on the local level there are a number of uncertainties (Bergant, 2003) related to downscaling and representation of local features. Forecasts of temperature are more reliable than precipitation since research areas are located in the rough hilly terrain and at the interface between Mediterranean and Alpine climate. This strongly influences the local weather, especially in the warm half of the year characterized by storms. One of the uncertainties is associated with the input data, since we only had information on changes in precipitation and temperatures, but none on wind, solar radiation and relative humidity. For the research areas they are predicted to be, in terms of agriculture and water balance, the most problematic years, when temperatures are higher and rainfall are lower than long-term average (Bergant & Kajfež Bogataj, 2004).

For each catchment were designed three scenarios (2030, 2060 and 2090). In the process of modelling climate change we have not changed any of the calibrated SWAT model parameters. The purpose of the scenarios is to investigate the impacts of changing climatic conditions in catchments with Flysch soils on the river flow and sediment load in the rivers.

6. Results and discussion

Base and climate scenarios simulation were carried out for the Reka catchment for the period of 18 years (1991-2008) with three years warm up period (1991-1993) and for the Dragonja catchment for the period of 17 years (1992-2008) with two years warm up period (1992-1993). Long time period is important for smoothing the effects caused by extreme meteorological events (storms) or to exclude effects of dry or wet periods. Warm up period is essential for stabilisation of parameters as the initial results can vary significantly from the observed values. In this period the model deposits sediment in the river network and fills the soil profile with water before simulation results can be considered realistic. All outputs from this period of time are excluded from statistical analysis.

The base scenario indicates a high average annual variability in the transport of the sediment in the river flow (Table 11). The standard deviations for the Reka subcatchment 8 reveal that 2/3 of the transported sediment quantities are expected in the interval 1 844 ± 1 075 t sediment year^{-1}, and for the Dragonja subcatchment 14 in the interval 4 804 ± 1 576 t sediment year^{-1}.

Catchment/subcatchment	Average	Median	Standard deviation	Min.	Max.
Flow (m³ s⁻¹)					
Reka/8	0.57	0.56	0.21	0.27	1.00
Dragonja/14	0.80	0.78	0.21	0.42	1.11
Sediment (t year⁻¹)					
Reka/8	1 844	1 576	1 075	571	4 185
Dragonja/14	4 804	4 934	1 576	1 917	7 734

Table 11. Average annual flow (m³ s⁻¹) and river load of sediment (t year⁻¹) for the Reka subcatchment 8 and Dragonja subcatchment 14 (1994–2008)

6.1 Climate change and river flow

Climate scenarios results relative to the base scenario show a reduction of average annual flows in the river in all periods (2030, 2060, 2090) for 29%, 41% and 55%, respectively (Table 12). Reduction of average flow was detected in the Dragonja catchment however they are less pronounced. Figure 8 shows the distinct change towards lower average flow in the summer and towards higher average flow in the winter months. Student t-statistics for comparison between base and climate scenarios, the average annual total flow shows differences between scenarios for the river Reka are highly statistically significant (Table 13). Differences for the river Dragonja are not statistically significant.

Catchment/ subcatchment	Average annual percentage change (%)		
	2030	2060	2090
River Flow			
Reka/8	−29.08	−40.69	−55.07
Dragonja/14	−3.16	−5.46	−6.53
Sediment			
Reka/8	−36.70	−51.60	−69.58
Dragonja/14	−29.93	−27.32	−28.12

Table 12. Impacts (change in %) of climate change scenarios on the river flow and sediment load in the watercourse; compared to the baseline scenario

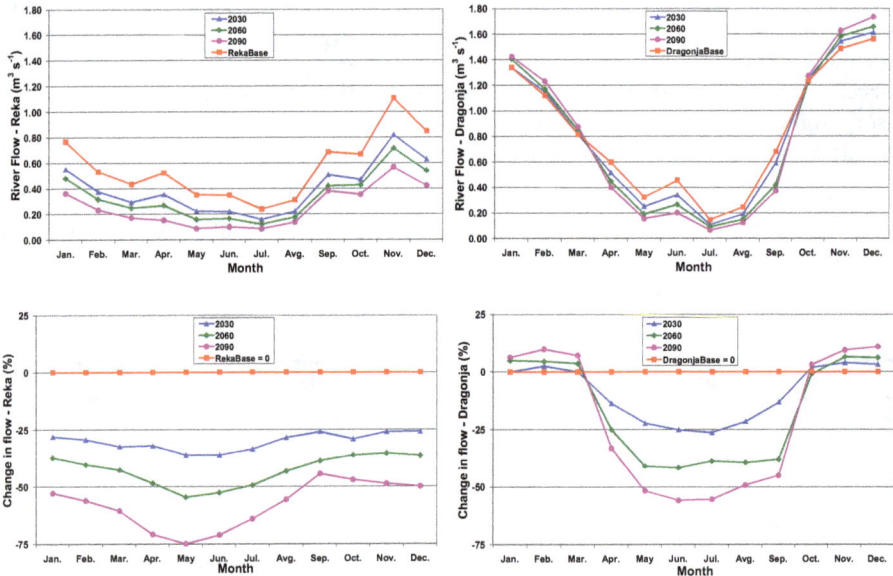

Fig. 8. Change in average monthly flow (m^3 s^{-1}, %) between the base (Base = 0) and climate change scenarios for the Reka subcatchment 8 and Dragonja subcatchment 14 (1994–2008)

	Student t-test (Significance level 0.05) Student distribution of the sample with n-1 degrees of freedom $\alpha=0.025$, SP=14, $t_\alpha=2.145$			
	Reka – subcatchment 8		Dragonja – subcatchment 14	
Scenario	Flow	Sediment	Flow	Sediment
2030	–4.046	–3.545	–0.464	–4.656
2060	–6.788	–6.888	–0.783	–3.427
2090	–12.545	–13.450	–0.842	–3.313

Note: The results of the scenarios are statistically significantly different from the base scenario, if the value of Student t-test exceeds $t_\alpha = 2.145$. If the value is negative, scenario is reducing the quantities in the river flow, and vice versa.

Table 13. Review of statistically significant results of Student t-statistics for average annual flow and average annual load of sediment

Climate changes in the river Reka catchment are forecasted to be more drastic, because of the temperatures rise and precipitation decrease by 500 mm by the end of the 21st century. Furthermore, the average annual potential ET is forecasted to be increased from 1044 mm (Base) to 1219 mm (2090) and actual ET is forecasted to be reduced by more than 100 mm. Consequently, average annual flow would be reduced by 340 mm. Average annual flow for the river Reka in 2090 (268 mm) would get closer to the present flow of the river Dragonja (265 mm). In the Dragonja catchment, there are lower summer precipitations aligned with higher winter precipitations, leading to annual differences between scenarios of only a few millimetres.

6.2 Climate change and sediment

Climate scenarios percentage change in average annual sediment load in the river flow is lower as compared to the base scenario. For Reka catchment the sediment load for the scenarios 2030, 2060 and 2090 in the river flow are lower by 37%, 52% and 70%, respectively (Table 12). Statistical method showed the climate scenario differences for sediment load in the river Reka which is significantly different from baseline scenario (Table 13; Fig. 9)

For the Dragonja catchment the average annual sediment loads for the scenarios 2030, 2060 and 2090 in the river flow are 30%, 27% and 28% lower, respectively (Table 12). Statistical method showed that the differences in sediment loads transferred in the Dragonja river flow significantly differ from the base scenario (Table 13; Fig. 9).

6.3 Discussion and scenario evaluation

Statistically significant differences between the base and climate scenarios which are particularly pronounced in the Reka catchment are expected, as the decline in precipitation contributes to the reductions in surface flow and thereby to the reductions in erosion of soil particles. Climate models forecast less summer and more winter precipitation for the Dragonja catchment (Bergant, 2003); however the differences between seasons are only a few percent (2-5%) (Table 14). Lower summer precipitation, when majority of agricultural activities are taking place, and high proportion of modelled forest (74%) and grassland (19%),

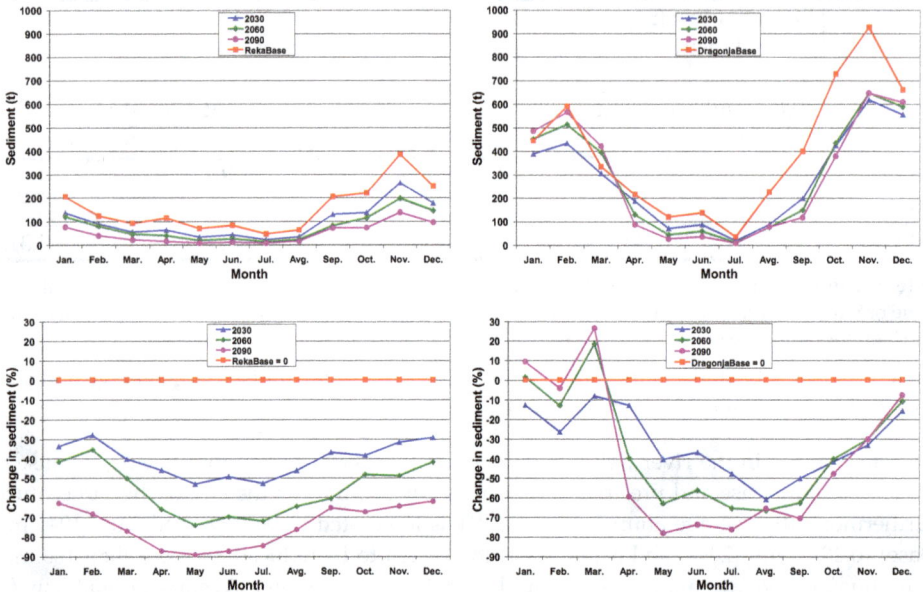

Fig. 9. Change in average monthly river loads of sediment (t, %) between the base (Base = 0) and climate change scenarios for the Reka subcatchment 8 and Dragonja subcatchment 14 (1994–2008)

both less susceptible for erosion, have resulted in a decrease of the average annual sediment loads. But positive consequences reflected in the sediment load reduction, may also be of concern due to reduction of the water quantities (Xu et al., 2008).

Simulated sediment loads in the climate scenarios are also lower because of less precipitation in the summer when agricultural activities are taking place (fertilizer, mulching, tillage). With increasing temperatures the shift in the growing season is expected like earlier times in the year (Ficklin et al., 2009) providing better conditions for growth of vegetation and preventing soil erosion. This means cultivation of the land during the major precipitation events, which can result in an increased release of soil particles and nutrients (Bouraoui et al., 2002). It is expected that there will be strong influence of the climate changes on higher frequency and quantity of extreme precipitation events and drought lengths, which will greatly affect the water cycle (Wilby et al., 2006). One or two major events may substantially contribute to total average annual quantity of sediment (Ramos & Martinez-Casanovas, 2008).

When interpreting the concentrations, we need to have in mind the geological and soil characteristics of the catchment. There is also a question, whether to consider set guide levels for the rivers that does not represent an economic interest (Lohse, 2008); however rivers are not only economic asset. Careful evaluation of each scenario has to be performed according to its positive and negative issues on the environment, agriculture, social life and economy (Glavan et al., 2011, Volk et al., 2009). Climate change scenarios have a significant impact on the concentration of sediment, since both catchments would decrease by the legislation recommended concentration of 25 mg l[-1] (Table 14).

| | Average annual sediment concentration (mg l⁻¹) | |
	REKA – subcatchment 5	DRAGONJA – subcatchment 14
Measured	32.6	29.3
Scenarios 2030	20.5	20.5
Scenarios 2060	15.5	21.3
Scenarios 2090	9.9	21.1

A guide concentration for the sediment in the rivers set legislation is **25 mg l⁻¹** (Regulation of surface water quality for freshwater fish species - Official Gazette of Republic of Slovenia 46/02).

Table 14. Impacts of the climate change scenarios on the average annual sediment concentration (mg l⁻¹)

Based on the obtained data and model calculations it can be concluded that climate changes, if used values are realized, would significantly affect quantity of water in the river Reka flow and significantly reduce sediment loads in the rivers Reka and Dragonja. The data shows that by the end of the century the average annual precipitation in the Reka catchment would approach to the precipitation in the Dragonja catchment. Forecasted reductions would result in a significant decline of flow and extension of the dry riverbed of the River Reka, which is now rarely longer than a week or two, and change ecological conditions for the organisms. Climate change in the river Dragonja would lead to extension of the dry riverbed period in the warm half of the year, which would negatively affect the ecosystem.

In the future, the changing status of ecosystem conditions of water bodies as a result of climate change will require special attention by the relevant public agencies and governmental departments (Purkey et al., 2007), by enforcement of appropriate legislation and regulations (Wilby et al., 2006). The results of these scenarios should be considered only as one possible scenario, but only if climate change would really change in the direction of used temperature and precipitation data.

7. Conclusions

The application of the SWAT model in the Reka and Dragonja catchment has demonstrated that SWAT is able to represent the hydrological behaviour of the heterogeneous catchments. Within the constraints of the available data the model was able to represent the sediment loads, concentrations and cumulative distributions. However, there are a number of issues that these model results can demonstrate as important in the water pollution control.

7.1 Strengths

The SWAT model is easily available on-line and enables water managers to model the quantity of surface water and quality of catchments worldwide. It is a comprehensive model integrating surface land and channel environmental processes. It combines studies of water quantity (river discharge, surface flow, subsurface flow, lateral flow, base flow, drains, irrigation, reservoirs, lakes), water quality (weather, erosion, plant growth, nutrients cycles, pesticides, soil temperatures, agricultural land management, crop production, urban land management, agri-environmental measures) and climate change. It is capable of yearly, monthly, daily or sub-hourly simulation over long periods.

Over 20 years of work on model development has resulted in several tools, interfaces and support software for SWAT. SWATeditor is a standalone program which reads the project database generated by ArcSWAT interface to edit SWAT input files, execute SWAT run, perform sensitivity, autocalibration and uncertainty analysis. VIZSWAT is a visualization and analysis tool which analyzes model results generated from all versions of the model. MWSWAT is an interface to the SWAT 2005 model and a plug-in for MapWindow, an Open Source GIS system which runs under the Windows operating system. The SWAT Check program helps to identify potential model input parameters issues. Sensitivity Analysis and Manual and Auto-calibration tool incorporated in to the model greatly helps modellers with automated procedures to define and change the model parameters and perform optimal calibration. SWAT-CUP is a standalone, public domain computer program for sensitivity analysis, calibration, validation, and uncertainty analysis of SWAT models. The program links GLUE, ParaSol, SUFI2, MCMC, and PSO procedures to SWAT.

Flexible framework that allows simulation of agri-environmental measures and best management practices is essential strength of the SWAT model. Simulations of measures and practices can be in the majority of cases be directly linked with changing model parameters. Management files allow the modeller to model crops in rotations, fertilizer and manure application rates and timing, tillage, sowing and harvesting time, type of farming tools and machinery, irrigation management, buffer strips, terraces etc.

7.2 Weaknesses

The main weakness of the model is a non-spatial representation of the HRU inside each subcatchment. This kept the model simple and supported application of the model to almost every catchment. Land use, soil and slope heterogeneity of the model is accounted through subcatchments. This approach ignores flow and pollutants routing between HRUs.

Wide range of different data needs to be obtained to run the model and numerous parameters needed to be modified during the calibration which discourage modellers to use SWAT. However, environment is a complex system and disregarding or underestimation of importance of parameters could lead to inaccurate model results and evaluations.

More extensive use of the model would be expected with adding more groundwater routines and algorithms or with permanent coupling of the model with groundwater model.

The model does not allow simulations of multicultural plant communities which are common in organic farming, grasslands and forests as they were originally developed for monocultures.

Sensitivity analysis, manual and auto-calibration tools in the SWAT model is time demanding when modelling complex catchments with numerous HRUs. This tool should be upgraded at least with visual and objective functional representation of the results. The SWAT-CUP tool is a significant improvement for the calibration procedures, however coupling of SWAT and SWAT-CUP is needed to increase efficiency of modelling.

7.3 Opportunities

There are many opportunities outside the SWAT model that provide unique possibility for growth and change of the model in future. Numerous environmental problems due to industrial, mining and land use policy resulted in stricter government legislations around the world (USA – Clean Water Act, EU - WFD) encouraging the use of the models like SWAT.

Non-spatial representation of the HRUs requires new studies on water, sediment and nutrients, routing across landscapes by surface, subsurface and base flow, to allow the model to carry out realistic simulation of the HRU pollutants inter-transfer, source areas, buffer zones, etc.

7.4 Threats

Threats may adversely impact the model performance and use, if they are not addressed. In the process of building a model several adjustments of the parameters need to be made in order to improve simulations. Adjustments are usually not measurable and are made using the modellers' experiences, best knowledge and subjective assessment of the study area. This can have important implications on overall performance and outcome of the model and its suitability for certain case studies which are difficult to quantify.

Measured monitoring data is usually expressed and presented in concentrations (mg l^{-1}) and has to be recalculated with the measured flow data to make loads comparable with model outputs. This can be a source of errors especially if the monitored data, e.g. flow, sediment, nutrients, is not measured properly or is sampled at low frequency rate.

In highly managed catchments the unknown and unaccountable activities like transport construction sites, water abstraction, reservoirs, dams, waste treatment plants, waste and chemicals dumping in the rivers can add substantial error to the model outputs.

Spatial resolution of land use, soil, slope and weather data are usually not set on the same scale. This leads us to the conclusion that spatial results are as good as the data with lowest resolution. Actual on-site distribution of crops, crop rotations and actual management practices (sowing and harvest dates and fertiliser application dates and rates) is in large and medium catchments a challenge to represent spatially. Combined with uncertainties in the soil spatial and attribute data may significantly affect proper modelling of water balance and sediment transport in the catchments.

Model results and their interpretation by the modeller must lead to constructive discussion, which aims to achieve and maintain good water quality in research catchments, which is the objective of the Water Framework Directive and other legislations related to water.

8. Acknowledgments

Financial support for this study was provided by the Slovenian Research Agency founded by the Government of the Republic of Slovenia. Contract number: 1000-06-310163.

9. References

Abbaspour, K.C.; Yang, J.; Maximov, I.; Siber, R.; Bogner, K.; Mieleitner, J.; Zobrist, J. & R. Srinivasan (2007). Modelling hydrology and water quality in the pre-alpine/alpine Thur watershed using SWAT. *Journal of Hydrology*, Vol.333, pp. 413-430

Arnold, J.G.; Allen, P.M.; Muttiah, R. & Bernhardt, G. (1995). Automated Base Flow Separation and Recession Analysis Techniques. *Ground Water*, Vol.33, No.6, pp. 1010–1018

Arnold, J.G.; Srinivasan, R.S.; Muttiah, R.S. & Williams, J.R. (1998). Large area hydrological modelling and assessment Part I: Model development. *Journal of the American Water Resources Association*, Vol.34, No.1, pp. 73–89

Barlund, I.; Kirkkala, T.; Malve, O. & Kamari, J. (2007). Assessing SWAT model performance in the evaluation of management actions for the implementation of the Water Framework Directive in a Finnish catchment. *Environmental Modeling Software*, Vol.22, pp. 719–724

Bouraoui, F.; Galbiati, L. & Bidoglio, G. (2002). Climate change impacts on nutrient loads in the Yorkshire Ouse catchment (UK). *Hydrology and Earth System Sciences*, Vol.6, No.2, pp. 197–209

Bergant, K. (2003). Projections of global climate simulations to local level and their use in agrometeorology. Doctoral thesis. University of Ljubljana, Ljubljana

Bergant, K.; Kajfež Bogataj, L. (2004). Nekatere metode za pripravo regionalnih scenarijev podnebnih sprememb = Empirical downscaling method as a tool for development of regional climate change scenarios. *Acta agriculturae Slovenica*, Vol.83, No.2, pp. 273–287

Bergant, K.; Kajfež Bogataj, L. (2005). N–PLS regression as empirical downscaling tool in climate change studies. *Theoretical and Applied Climatology*, Vol.81, pp. 11–23

Di Luzio, M.; Arnold, J. G. & Srinivasan, R. (2005). Effect of GIS data quality on small watershed streamflow and sediment simulations. *Hydrolgical Processes*, Vol.19, No.3, pp. 629–650

Ficklin, D.L.; Luo, Y.; Luedeling, E. & Zhang, M. (2009). Climate change sensitivity assessment of a highly agricultural watershed using SWAT. *Journal of Hydrology*, Vol.374, pp. 16–29

Frame, J. (1992). *Improved grassland management*. Farming Press Books, Wharfedale

Garen, D.C. & Moore, D.S. (2005). Curve number hydrology in water quality modeling: uses, abuses, and future directions. *Journal of the American Water Resources Association*, Vol.41, No.6, pp. 1491–1492

Gassman, P.W.; Reyes, M.R.; Green, C.H. & Arnold, J.G. (2007). The soil and water assessment tool: Historical development, applications, and future research direction. *Transactions of the ASABE*, Vol.50, No.4, pp. 1211–1250

Glavan, M.; White, S. & Holman, I. (2011). Evaluation of river water quality simulations at a daily time step – Experience with SWAT in the Axe Catchment, UK. *CLEAN – Soil, Air, Water*, Vol.39, No.1, pp. 43–54

Harmel, R.D.; Potter, S.; Ellis, P.; Reckhow, K.; Green, C.H. & Haney, R.L. (2006). Compilation of measured nutrient load data for agricultural land uses in the US. *Journal of American Water Resources Association*, Vol.42, pp. 1163–1178

Hejzlar, J.; Anthony, S.; Arheimer, B.; Behrendt, H.; Bouraoui, F.; Grizzetti, B.; Groenendijk, P.; Jeuken, M.; Johnsson, H.; Lo Porto, A.; Kronvang, B.; Panagopoulos, Y.; Siderius, C.; Silgram, M.; Venohrd, M. & Žaloudíka, J. (2009). Nitrogen and phosphorus retention in surface waters: an inter-comparison of predictions by catchment models of different complexity, *Journal of Environmental Monitoring*, Vol.11, pp. 584–593

Henriksen, H.J.; Troldborg, L.; Nyegaard, P.; Sonnenborg, O. T.; Refsgaard, J. C. & Madsen, B. (2003). Methodology for construction, calibration and validation of a national hydrological model for Denmark. *Journal of Hydrology*, Vol.280, pp. 52–71

Knutti, R.; Allen, M.R.; Friedlingstein, P.; Gregory, J.M.; Hegerl, G.C.; Meehl, G.A.; Meinshausen, M.; Murphy, J.M.; Plattner, G.-K.; Raper, S.C.B.; Stocker, T.F.; Stott, P.A.; Teng, H. & Wigley, T.M.L. (2008). A review of uncertainties in global temperature projections over the twenty-first century. *Journal of Climate*, Vol.21, No.11, pp. 2651–2663

Krause, P.; Boyle, D.P. & Bäse, F. (2005). Comparison of different efficiency criteria for hydrological model Assessment. *Advances in Geosciences*, Vol.5, pp. 89–97

Kronvang, B.; Borgvang, S. A. & Barkved, L. J. (2009a). Towards European harmonised procedures for quantification of nutrient losses from diffuse sources – the EUROHARP project. *Journal of Environmental Monitoring*, Vol.11, No.3, pp. 503–505

Kronvang, B.; Behrendt, H.; Andersen, H.; Arheimer, B.; Barr, A.; Borgvang, S.; Bouraoui, F.; Granlund, K.; Grizzetti, B.; Groenendijk, P.; Schwaiger, E.; Hejzlar, J.; Hoffman, L.; Johnsson, H.; Panagopoulos, Y.; Lo Porto, A.; Reisser, H; Schoumans, O.; Anthony, S.; Silgram, M.; Venohr, M. & Larsen, S. (2009b). Ensemble modelling of nutrient loads and nutrient load partitioning in 17 European catchments. *Journal of Environmental Monitoring*, Vol.11, pp. 572–583

Krysanova, V. & Arnold, J.G. (2008). Advances in ecohydrological modelling with SWAT – a review. *Hydrological Sciences–Journal des Sciences Hydrologiques*, Vol.53, No.5, pp. 939–947

Lohse, K.A.; Newburn, D.A.; Opperman, J.J. & Merenlender, A.M. (2008). Forecasting relative impacts of land use on anadromous fish habitat to guide conservation planning. *Ecological Applications*, Vol.18, No.2, pp. 467–482

Meehl, G.A.; Washington, W.M.; Collins, W.D.; Arblaster, J.M.; Hu, A.; Buja, L.E.; Strand, W.G. & Teng, H. (2005). How much more global warming and sea level rise? *Science*, Vol.307, pp. 1769–1772

Mihelič, R.; Čop, J.; Jakše, M.; Štampar, F.; Majer, D.; Tojnko S. & Vršič S. (2010). *Smernice za strokovno utemeljeno gnojenje = Guidelines for professionally justified fertilisation.* Ministry of Agriculture, Forestry and Food of Republic of Slovenia, Ljubljana

Montana Department of Environmental Quality (2005). *Flathead Basin Program, quality assurance project plane (QAPP).* Land and Water Quality Consulting/PBS&J, Montana

Moriasi, D.N.; Arnold, J.G.; Van Liew, M.W.; Bingner, R.L.; Harmel, R.D. & Veith, T.L. (2007). Model evaluation guidelines for systematic quantification of accuracy in watershed simulations. *Transactions of the ASABE*, Vol.50, No.3, pp. 885–900

Nash, J. & Sutcliffe, J. (1970). River flow forecasting through conceptual models: I. A discussion of principles. *Journal of Hydrology*, Vol.10, pp. 374–387

Neitsch, S.L.; Arnold, J.G.; Kiniry, J.R. & Williams, J.R. (2005). *Soil and water assessment tool theoretical documentation – Version 2005.* Texas Agricultural Experiment Station, Blackland Research Center, Agricultural Research Service, Grassland, Soil and Water Research Laboratory, Texas, Temple

Pedosphere (2009). Soil Texture Triangle Hydraulic Properties Calculator, 21. July 2010, Available from http://www.pedosphere.com/

Purkey, D.R.; Huber-Lee, A.; Yates, D.N.; Hanemann, M. & Herrod-Julius, S. (2007). Integrating a climate change assessment tool into stakeholder-driven water management decision-making processes in California. *Water Resources Management*, Vol.21, pp. 315–329

Ramos, M.C. & Martinez-Casasnovas, J.A. (2006). Nutrient losses by runoff in vineyards of the Mediterranean Alt Penede`s region (NE Spain). *Agriculture, Ecosystems and Environment*, Vol.113, pp. 356–363

Sardans, J. & Penuelas, J. (2004). Increasing drought decreases phosphorus availability in an evergreen Mediterranean forest. *Plant and Soil*, Vol.267, pp. 367–377

Saxton, K.E.; Rawls, W.J.; Romberger, J.S. & Papendick, R.I. (1986). Estimating generalized soil-water characteristics from texture. *Soil Science Society of America Journal*, Vol.50, No.4, pp. 1031–1036

van Griensven, A.; Meixner, T.; Grunwald, S.; Bishop, T.; Di Luzio, M. & Srinivasan, R. (2006). A global sensitivity analysis tool for the parameters of multi-variable catchment models. *Journal of Hydrology*, Vol.324, pp. 10–23

Volk, M.; Liersch, S. & Schmidt, G. (2009). Towards the implementation of the European Water Framework Directive? Lessons learned from water quality simulations in an agricultural watershed. *Land Use Policy*, Vol.26, pp. 580–588

Wilby, R.L.; Orr, H.G.; Hedger, M.; Forrow, D. & Blackmore, M. (2006). Risks posed by climate change to the delivery of Water Framework Directive objectives in the UK. *Environment International*, Vol.32, pp. 1043–1055

Xu, Z.X.; Zhao, F.F. & Li, J.Y. (2009). Response of streamflow to climate change in the headwater catchment of the Yellow River basin. *Quaternary International*, Vol.208, pp. 62–75

2

Tools for Watershed Planning – Development of a Statewide Source Water Protection System (SWPS)

Michael P. Strager

Division of Resource Management, West Virginia University,
USA

1. Introduction

A Surface Water Protection System (SWPS) was developed to bring spatial data and surface water modeling to the desktop of West Virginia Bureau of Public Health (WVBPH), Office of Environmental Health Services (OEHS), Environmental Engineering Division (EED). The SWPS integrates spatial data and associated information with the overall goal of helping to protect public drinking water supply systems.

The SWPS is a specialized GIS project interface, incorporating relevant data layers with customized Geographic Information Systems (GIS) functions. Data layers have been assembled for the entire state of West Virginia. Capabilities of the system include map display and query, zone of critical concern delineation, stream flow modeling, coordinate conversion, water quality modeling, and susceptibility ranking. The system was designed to help meet the goals of the Surface Water Assessment and Protection (SWAP) Program.

The goal of the SWAP program is to assess, preserve, and protect West Virginia's source waters that supply water for the state's public drinking water supply systems. Additionally, the program seeks to provide for long term availability of abundant, safe water in sufficient quality for present and future citizens of West Virginia. The SWPS was designed to help meet this goal by addressing the three major components of the SWAP program: delineating the source water protection area for surface and groundwater intakes, cataloging all potential contamination sources, and determining the public drinking water supply system's susceptibility to contamination.

This chapter outlines the functions and capabilities of the SWPS and discusses how it addresses the needs of the SWAP program. The following sections discuss the application components. The components consist of:

1. A customized interface for study area selection
2. Integration of the EPA WHAEM and MODFLOW models
3. Delineation of groundwater public supply systems
4. Watershed delineation and zone of critical concern delineation for surface water sites
5. Stream flow model from multivariate regression
6. The environmental database

7. UTM latitude/longitude conversion utility
8. Statewide map/GIS data layers
9. Water quality modeling capability
10. Groundwater and surface water susceptibility model

Component 1. A customized interface for study area selection

Using customized programming we were able to create a GIS interface to allow users to quickly find locations or define study areas for further analysis in the state. The locations may be selected in three ways: by geographical extent (e.g. county, watershed, 1:24,000 quad map, major river basin), by area name or code (e.g. abandoned mine land problem area description number, stream or river name, WV Division of Natural Resource (WVDNR) stream code, public water identification number or name), or by typing in the latitude and longitude coordinates. Once the study area is defined, the system zooms automatically to the extent of the selected feature and all available spatial data layers are then displayed. A discussion of the spatial data layers included is discussed in Component 8 of this document.

Component 2. Integrating EPA WHAEM and MODFLOW models

The SWPS application has the ability to read output from either EPA WHAEM or MODFLOW models. It does this by importing dxf file formats directly into SWPS from a pull down menu choice. Data can also be converted to shapefile format from SWPS to be read directly into WHAEM and MODFLOW. The data being read into SWPS needs to be in the UTM zone 17, NAD27 projection (with map units meters) for the new data to overlay on the current data existing within SWPS. Consequently, any data exported from SWPS will automatically be in the UTM zone 17 NAD27 coordinate system.

Component 3. Delineation of groundwater public supply systems

A fixed radius buffer zone was created around each groundwater supply site based on the pumping rate. If the pumping rate was less than or equal to 2,500 gpd, a radius of 500 feet was used. If the pumping rate was greater than 2,500 gpd but less than or equal to 5,000 gpd, a radius of 750 feet was used. If the pumping rate was greater than 5,000 gpd and less than or equal to 10,000 gpd, a radius of 1,000 feet was used. If the pumping rate was greater than 10,000 gpd and less than or equal to 25,000 gpd, then a radius of 1,500 feet was used.

There were two exceptions to this fixed radius buffer procedure. The first was for any groundwater site less than or equal to 25,000 gpd that was in a Karst or mine area. These locations regardless of their pumping rate less than 25,000 gpd were buffered 2,000 feet. The second exception was for sites over 25,000 gpd. For these sites, hydro geologic and/or analytical mapping delineations will be done by personnel at the Bureau of Public Health. These were only identified in SWPS as being a well location and are left to more sophisticated groundwater modeling software.

To perform buffers automatically, the user can use the GIS to create buffers dialog within SWPS susceptibility ranking menu option. The automatic fixed radius buffering requires knowledge about the pumping rate and fixed radius distance. This information is provided in a pulldown text information box within the susceptibility ranking menu option.

Component 4. Watershed delineation and zone of critical concern delineation for surface water sites

The ability to interactively delineate watersheds and zones of critical concern is built into SWPS. In this section, the watershed delineation tool is discussed, followed by the zone of critical concern delineation tool.

Watershed Delineation

SWPS allows the user to delineate a watershed for any mapped stream location in the state. The watershed is delineated based on the user-clicked point and it is added to the current view's table of contents as a new theme or map layer labeled "Subwatershed." The drainage area is reported back to the user as well. If only drainage area is requested, a separate tool allows for quick query of stream drainage area in acres and square miles, without waiting for the watershed boundary to be calculated.

The watershed delineation is driven by a hydrologically correct digital elevation model (DEM). The DEM is corrected using stream centerlines for all 1:24,000 streams. The stream centerlines are converted to raster cells and DEM values are calculated for each cell. All off-stream DEM cells are raised by a value of 20 meters to assure the DEM stream locations are the lowest cells in the DEM. This step is necessary to assure of more accurate watershed delineations especially at the mouth of the watersheds. After the DEM is filled of all spurious sinks, flow direction and flow accumulation grids are calculated. These grids help determine the direction of flow and the accumulated area for each cell in the landscape. These grids were necessary for watershed delineation to occur and are important inputs for finding the zones of critical concern for surface water intakes.

Surface Water Zones of Critical Concern

Stream velocity is the driving factor for determining a five-hour upstream delineation for each surface water intake in WV. Only with stream velocity calculated was it possible to include factors such as high bank-full flow, average flow, stream slope, and drainage area all at once. The velocity equation used in this study came from a report titled "Prediction of Travel Time and Longitudinal Dispersion in Rivers and Streams" (US Geological Survey, Water-Resources Investigations Report 96-4013, 1996). In this report, data were analyzed for over 980 subreaches or about 90 different rivers in the United States representing a wide range of river sizes, slopes, and geomorphic types. The authors found that four variables were available in sufficient quantities for a regression analysis. The variables included the drainage area (D_a), the reach slope (S), the mean annual river discharge (Q_a), and the discharge at the section at time of the measurement (Q). The report defines peak velocity as:

$$V'_p = V_p D_a / Q$$

The dimensionless drainage area as:

$$D'_a = D_a^{1.25} * sqrt(g) / Q_a$$

Where g is the acceleration of gravity. The dimensionless relative discharge is defined as:

$$Q'_a = Q/Q_a$$

The equations are homogeneous, so any consistent system of units can be used in the dimensionless groups. The regression equation that follows has a constant term that has specific units, meters per second. The most convenient set of units for use with the equation are: velocity in meters per second, discharge in cubic meters per second, drainage area in square meters, acceleration of gravity in m/s2, and slope in meters per meter.

The equation derived in the report and the equation used in this study for peak velocity in meters per second was the following:

$$Vp = 0.094 + 0.0143 * (D'_a)^{0.919} * (Q'_a)^{-0.469} * S^{0.159} * Q/D_a$$

The standard error estimates of the constant and slope are 0.026 m/s and 0.0003, respectively. This prediction equation had an R^2 of 0.70 and a RMS error of 0.157 m/s.

Once a velocity grid was calculated as described above, it was used as an inverse weight grid in the flowlength ArcGIS (ESRI, 2010) command. The flowlength command calculates a stream length in meters. If velocity is in meters per second, the inverse velocity as a weight grid will return seconds in our output grid. This calculation of seconds would track how long water takes to move from every cell in the state where a stream is located to where it leaves the state. The higher values will exist in the headwater sections of a watershed. By querying the grid, it is possible to add the appropriate travel time to the cell value and this will the time of travel for an intake. All cells above an intake by 18,000 seconds (5 hours) will be the locations in which water would take to reach the intake.

To use this methodology, GIS data layers had to be calculated for drainage area, stream slope, annual average flow, and bank-full flow for all of WV. The sections below describe how each of these grids was created.

Drainage area

To obtain a drainage area calculation for every stream cell in the state required a hydrogically correct DEM. The process of creating a hydrologically correct DEM was covered in the watershed delineation component described earlier. Essentially, from the DEM the flow direction and flow accumulation values for each stream cell are derived. The output of the flow direction request is an integer grid whose values range from 1 to 255. The values for each direction from the center are:

32	64	128
16	X	1
8	4	2

For example, if the direction of steepest drop were to the left of the current processing cell, its flow direction would be coded as 16. If a cell is lower than its 8 neighbors, that cell is given the value of its lowest neighbor and flow is defined towards this cell (ESRI, 2010).

The accumulated flow is based upon the number of cells flowing into each cell in the output grid. The current processing cell is not considered in this accumulation. Output cells with a high flow accumulation are areas of concentrated flow and may be used to identify stream channels. Output cells with a flow accumulation of zero are local topographic highs and may be used to identify ridges. The equation to calculate drainage area from a 20-meter cell sized flow accumulation grid was:

(cell value of flow accumulation grid + 1) * 400 = drainage area in meters squared

Stream slope

Stream slope was calculated for each stream reach in the state. A stream reach is not necessarily an entire stream but only the section of a stream between junctions. The GIS command streamlink was first used to find all unique streams between stream intersections or junctions. For each of these reaches, the length was calculated from the flowlength GIS command. Having the original DEM allowed us to find the maximum and minimum values for each of the stream reaches. The difference in the maximum and minimum elevations for the stream reach divided by the total reach length gave us our stream reach slope in meters per meter.

Annual average flow

Annual average flow for each stream cell location was found based on a relationship between drainage area and gauged stream flow. For 88 gauging stations in WV, covering many different rainfall, geological, and elevation regions, we assembled a table of drainage area for the gauges versus the historic annual stream flow for the gauge. After fitting a linear regression line for this data set, we found the following equation for annual stream flow setting the y intercept to zero.

Annual stream flow in cfs = 2.05 * drainage area in square miles

This equation had a corrected R^2 of .9729. The XY plot and equation are shown in Figure 1.

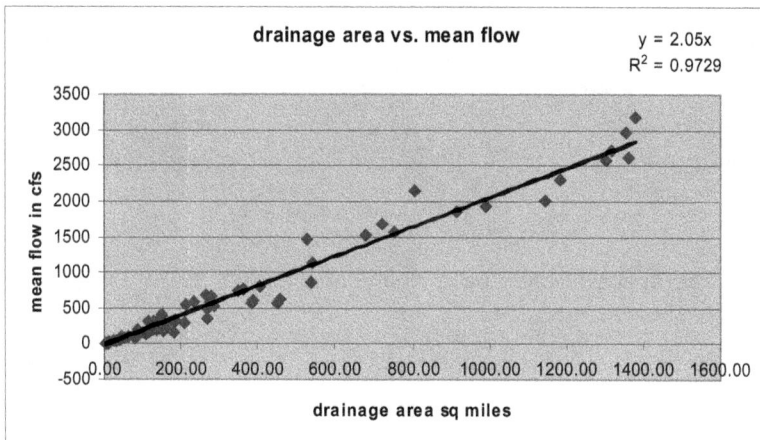

Fig. 1. Annual stream flow from gauged stations and drainage area at the gauges

Since drainage area is already calculated for each stream cell location, this equation incorporated the drainage area grid to compute a separate grid layer of annual stream flow. This would be another input for the velocity calculation.

Bank-full flow

The last input for the velocity equation was the bank-full flow measure. Just as with annual average flow, this required a modeled value for every raster stream cell in WV. Using the same approach to regressing drainage area to gauged stream flow as performed to find an annual average flow equation, this equation was used to find bank-full flow. Bank-full flow as defined by the Bureau of Public Health, is 90% of the annual high flow. To find the 90% of high flow for each gauging station, all historic daily stream flow data was downloaded for each of the 88 gauging stations. This data was then sorted lowest to highest and then numbered lowest to highest after removing repeating values. The value of flow at the 90% of the data became the bank-full flow value for that gauge. These values were then regressed against drainage area at the gauge. The linear regression equation for bank-full stream flow setting the y intercept to zero is listed below.

Bank-full stream flow in cfs = 4.357 * drainage area in square miles

This equation had a corrected R^2 of .9265. The XY plot and equation are shown in Figure 2.

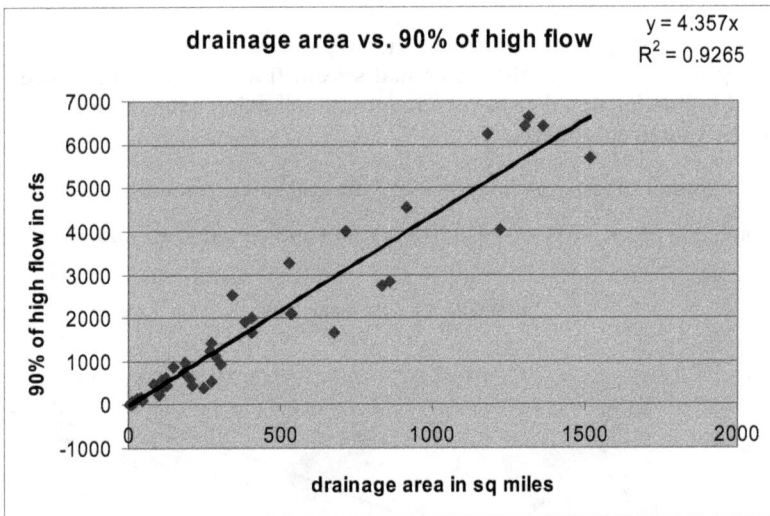

Fig. 2. Bank-full stream flow from gauged stations and drainage area at the gauges

This equation could be applied to the drainage area grid to calculate the bank-full flow for any stream cell in the state. It was the final input needed in the velocity calculation.

The interactive zone of critical concern ability of SWPS delineates the upstream contributing area for a surface water intake in the following way. First, the user locates the surface water intake and makes sure the intake is on the raster stream cell. A button on the interface then initiates the model. The model will query the time of travel value for the intake and then add 18,000 seconds (5 hours) to the queried value upper range. All cells which fit this range are identified and the stream order attribute retrieved for those cells. All cells that are on the main stem stream where the intake existed are buffered 1000 feet on each side of the stream. All tributaries to the main stem are buffered 500 feet on each side of the stream. Next, a watershed boundary for the location of the intake is delineated and used to clip any areas of the buffer that may extend beyond ridgelines. And lastly, the surface water intake is buffered 1000 feet and combined with the clipped buffer to include areas 1000 feet downstream of the intake.

This interactive ability allows zones of critical concern to be delineated for any river or stream in WV. Only large rivers which border WV, such as the Ohio, Tug, and Potomac can not be interactively delineated using this method. This is due to unknown drainage areas for these bordering rivers and unknown tributaries to these major rivers coming from the bordering states. This is the major limitation of this modeling approach for WV at this time and in the next version of this watershed tool will account for all outer drainage influences.

The Ohio River Sanitation Commission (ORSANCO) is responsible for delineating zones of critical concern for the intakes along the Ohio River. ORSANCO uses uniform 25-mile upstream distances for zones of critical concern for intakes along the Ohio River. This same approach could be applied to other rivers such as the Tug and Potomac in WV.

For reservoirs and lakes within the watershed delineation area, a set of standards was set by the Bureau of Public Health and was used in this study. For a reservoir, a buffer of 1000 feet on each bank and 500 feet on each bank of the tributaries that drain into the lake or reservoir was used. When a lake or reservoir is encountered within the five-hour time of travel, a specific delineation was used. If the length of the lake/reservoir was less than or equal to the five hour calculated time of travel distance from the intake, then the entire water body was included. If the length of the lake/reservoir was greater than the calculated five hour time of travel distance from the intake, then the section of water body within the five hour time of travel distance was used to establish the zone of critical concern.

Component 5. Stream flow model from multivariate regression

Overview

This project component for SWPS used multivariate techniques to evaluate stream flow estimation variables in West Virginia. The techniques included correlation analysis, multiple regression, cluster analysis, discriminant analysis and factor analysis. The major goal was to define watershed scale factors to estimate the stream flow at recorded USGS gauges. To do this, the contributing area upstream of each gauge was first delineated. Next, annual averages of precipitation and temperature and landscape based variables for the contributing upstream area were calculated and regressed against 30-year average annual flow at the USGS gauge. Results from the statistical analysis techniques found the most important variables to be upstream drainage area, 30-year annual maximum temperature, and stream slope. While this analysis was limited by the availability of data and assumptions to predict stream flow, the results indicate that stream flow can be modeled with reasonably good results.

The following sections include a review of the literature on stream flow estimation techniques, a description of the variables used in this study to predict stream flow, the multivariate statistical methods, and a discussion of results and limitations of the study.

Literature Review

The intent of this literature review was to determine variables that were used to estimate stream flow in other studies, identify different statistical procedures, and to find limitations in this study based on other papers.

The impact of land-use, climate change and groundwater abstraction on stream flow was examined by Qerner et al. (1997). They analyzed the effects of these factors using physical models BILAN, HBVOR, MODFLOW and MODGROW. The models were used to simulate the impact of afforstation, climate warming by 2 and 4 degrees Celsius in combination with an adoption of the precipitation changes in groundwater recharge and groundwater abstractions on stream flow droughts. The authors found that all the physical models can be used to assess the impacts of human activities on stream flow. They also concluded that based on some climate change scenarios they followed out, that the deficit volume of water is very sensitive to both an increase in temperature and a change in precipitation. Even in basins with abundant precipitation, the warming of 2 degrees Celsius would result in a rise in the deficit volume of water by 20 percent. Their findings also acknowledge the importance of using precipitation, temperature, groundwater recharge and groundwater abstractions along with water storage holding capacity of watersheds.

Timofeyeva and Craig (1998) used Monte Carlo techniques to estimate month by month variability of temperature and precipitation for drainage basins delineated by a digital elevation model. They also used a runoff grid from the digital elevation model to estimate discharge at selected points and compared this to known gauge station data. The variance of temperature was modeled as the standard error of the regression from the canonical regression equation. For precipitation, they modeled the variance as the standard error of the prediction. This was done to achieve unbiased estimators. When comparing the climate and resulting runoff and stream flow estimators calculated by Monte Carlo estimation, to the observed flow, the simulated results were within the natural variability of the record (Timofeyeva and Craig, 1998).

Long-range stream flow forecasting using nonparametric regression procedures was developed by Smith, (1991). The forecasting procedures, which were based solely on daily stream flow data, utilized nonparametric regression to relate a forecast variable to a covariate variable. The techniques were adopted to develop long-term forecasts of minimum daily flow of the Potomac River at Washington, D.C. Smith's key finding was that to implement nonparametric regression requires the successful specification of "bandwidth parameters." The bandwidth parameters are chosen to minimize the integrated mean square error of forecasts. Basically, his stream flow technique focussed on examining past history of stream flow and making nonparametric regression forecasts based on what is likely to occur in the future. No additional variables besides historic flow were used to model future conditions.

Another nonparameteric approach to stream flow simulation was done by Sharma et al. (1997). They used kernal estimates of the joint and conditional probability density functions to generate synthetic stream flow sequences. Kernal density estimation includes a weighted moving average of the empirical frequency distribution of the data (Sharma, et al. 1997). The reason for this method is to estimate a multivariate density function. This is a nonparametric method for the synthesis of stream flow that is data driven and avoids prior assumptions as to the form of dependence (linear or non linear) and the form of the probability density function. The authors main finding was that the nonparametric method was more flexible for their study than the conventional models used in stochastic hydrology and is capable of reproducing both linear and nonlinear dependence. In addition, their results when applied to a river basin indicated that the nonparametric approach was a feasible alternative to parametric approaches used to model stream flow.

Garren (1992) noted that although multiple regression has been used to predict seasonal stream flow volumes, typical practice has not realized the maximum accuracy obtainable from regression. The forecasting methods he mentions which can help provide superior forecasting include: (1) Using only data known at forecast time; (2) principal components regression; (3) cross validation; and (4) systematically searching for optimal or near-optimal combinations of variables. Some of the variables he used included snow water equivalent, monthly precipitation, and stream flow. The testing of selection sites for a stream flow forecasting study, he feels should be based on data quality, correlation analyses, conceptual appropriateness, professional judgement, and trial and error. The use of principal components regression provides the most satisfactory and statistically rigorous way to deal with intercorrelation of variables. He concluded that the maximum forecast accuracy gain is obtained by proper selection of variables followed by the use of principal components regression and using only known data (no future variables).

The results of a multiple-input transfer function modeling for daily stream flow using nonlinear inputs was studied by Astatkie and Watt (1998). They argue that since the relationship between stream flow and its major inputs, precipitation and temperature, are nonlinear, the next best alternative is to use a multiple input transfer function model identification procedure. The transfer function model they use includes variables such as type of terrain, drainage area, watercourse, the rate of areal distribution of rainfall input, catchment retention, loss through evapotranspiration and infiltration into the groundwater, catchment storage, and melting snow. When comparing their modeling technique for stream flow to that of a nonlinear time series model, they found their transfer function model to be direct and relatively easy for modeling multiple inputs. They also found it more accurate in head to head tests against the nonlinear time series model.

Since stream flow modeling is an outcome of many runoff estimation models, the literature for deriving runoff grids is applicable to stream flow studies. Anderson and Lepisto (1998) examined the links between runoff generation, climate, and nitrate leaching from forested catchments. One of the things they sought out to prove in their study was that climate will influence the amount of nitrate that can be leached from the soil and the water flow that will transport it to the streams. They found that a negative correlation existed between stream flow and temperature. Significant positive correlation between modeled surface runoff and concentrations of nitrate was found when they considered periods of flow increases during cold periods. Their study identified the importance for identifying and calculating the surface runoff fraction, daily dynamics of soil moisture, groundwater levels, and extensions of saturated areas when doing a contaminant transport or flow estimation study.

In another study, Moore (1997) sought to provide an alternative to the matching strip, correlation, and parameter-averaging methods for deriving master recession characteristics from a set of recession segments. The author then choose to apply the method to stream flow recession segments for a small forested catchment in which baseflow is provided by drainage of the saturated zone in the shallow permeable soil. The plots indicated the recessions were non-linear and that the recessions did not follow a common single valued storage outflow relation. The final decision was a model with two linear reservoirs that provided substantially better fit than three single reservoir models, indicating that the form of the recession curve probably depends on not just the volume of subsurface storage, but also on its initial distribution among reservoirs.

Gabriele et al. (1997) developed a watershed specific model to quantify stream flow, suspended sediment, and metal transport. The model, which estimated stream flow, included the sum of three major components: quick storm flow, slow storm flow, and long-term base flow. Channel components were included to account for timing effects associated with waters, sediments, and metals coming from different areas. Because of relatively good results from the modeling process, the conceptualizations supported that the study area river was strongly influenced by three major components of flow: quick storm flow, slow storm flow, and long-term base flow. Therefore, sediment inputs can be associated with each of those stream flow components and assign metal pollution concentrations to each flow and sediment input.

From this review of other studies, variables were determined that have been used successfully in stream flow estimation. Examining the limitations of other studies has also provided insight into data layers that may not be able to include. Of the statistical techniques used, the multivariate approach, in which components are added or subtracted to achieve the best fit possible, is a sound statistical procedure. In addition to this approach, testing the correlations between variables is another way of finding a model for estimating stream flow in WV.

Methodology

The first step in assembling data for this study was to delineate the total upstream contributing area for each of thirteen USGS gauge stations in West Virginia. Figure 3 displays the location of each gauge and the defined upstream drainage area for that gauge.

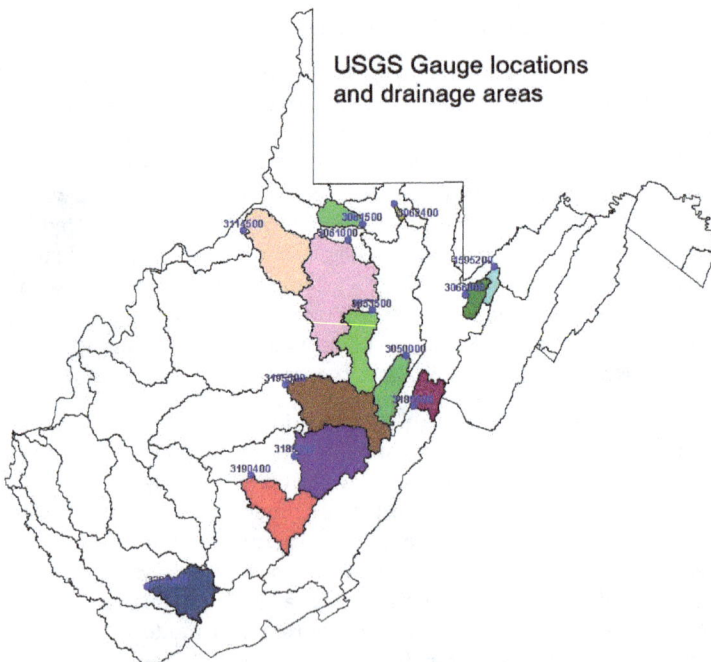

Fig. 3.

For every drainage area, the following criteria were calculated; total area, 30 year average annual precipitation, 30 year average annual maximum temperature, 30 year average annual minimum temperature, average drainage area slope, and stream slope. These variables were explanatory variables, which would be regressed against the dependent variable, the 30year average annual flow recorded at the gauge stations. The figures 4 to 7 show the distribution of 30-year precipitation, maximum temperature, minimum temperature, and elevation across the different areas. By using GIS techniques, it was possible to find the average value in the drainage areas along with drainage area slope and stream slope for each of the variables. The data for each gauge area and assembled variables is summarized in table 1.

id#	USGS Gauge name	Upstream drainage area (acres)	30yr annual precip ave (inches)	30yr annual ave temp max(F)	30yr annual ave temp min(F)	30yr annual Stream flow (cfs)	stream elevation drop max-min in (meters)	Watershed Slope average (degree)
g1	1595200	31296	52	54	35	99.68	418	5
g5	3050000	120352	50	57	37	379.37	607	15
g7	3053500	176708	46	60	38	613.56	643	11
g10	3061000	484507	43	62	39	1158.14	26	13
g11	3061500	74501	42	61	39	168.99	130	13
g12	3062400	7146	43	60	37	16.54	189	9
g13	3066000	55068	53	54	36	210.40	289	6
g17	3114500	289609	42	62	40	665.40	57	15
g19	3180500	85166	53	55	36	273.49	435	14
g21	3189100	338131	53	57	37	1445.61	743	13
g22	3190400	232990	50	59	38	750.36	830	11
g24	3195500	346231	48	59	38	1176.66	933	17
g26	3202400	196645	47	63	39	421.78	583	18

Table 1. Data used in study

The first step in analyzing the data in table 1 was to perform some basic statistics. The values across the different gauging station locations were investigated. The summarized statistical data is shown in table 2.

Variable	N	Mean	Median	TrMean	StDev	SE Mean	Minimum	Maximum	Q1	Q3
area	13	187565	176708	176972	144629	40113	7146	484507	64784	313870
precip	13	47.85	48	47.91	4.34	1.2	42	53	43	52.5
maxtemp	13	58.692	59	58.727	3.066	0.85	54	63	56	61.5
mintemp	13	37.615	38	37.636	1.446	0.401	35	40	36.5	39
flow	13	568	422	538	455	126	17	1446	190	954
strmslop	13	452.5	435	447.6	299.3	83	26	933	159.5	693
wsslope	13	12.31	13	12.45	3.88	1.08	5	18	10	15

Table 2. Basic statistics

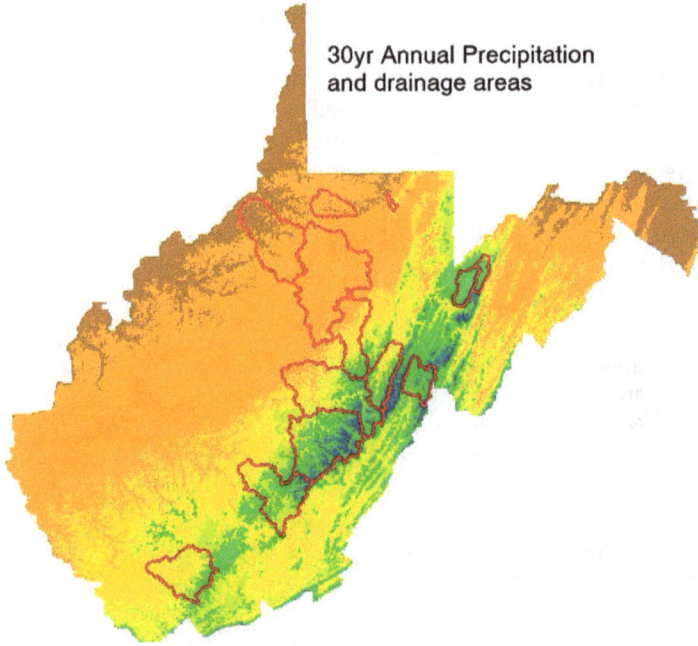

Fig. 4.

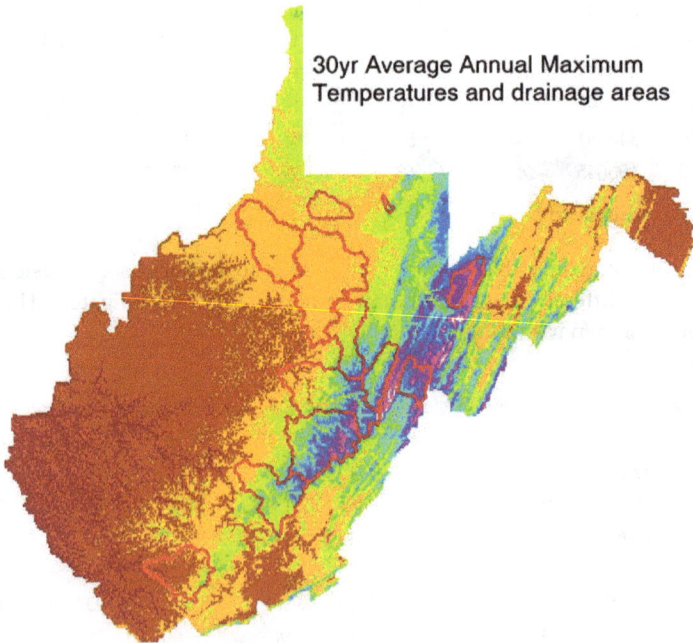

Fig. 5.

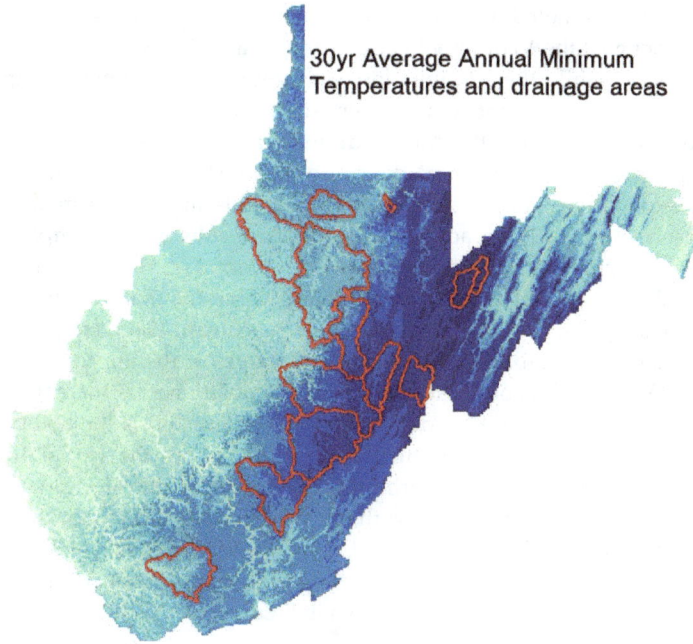

Fig. 6.

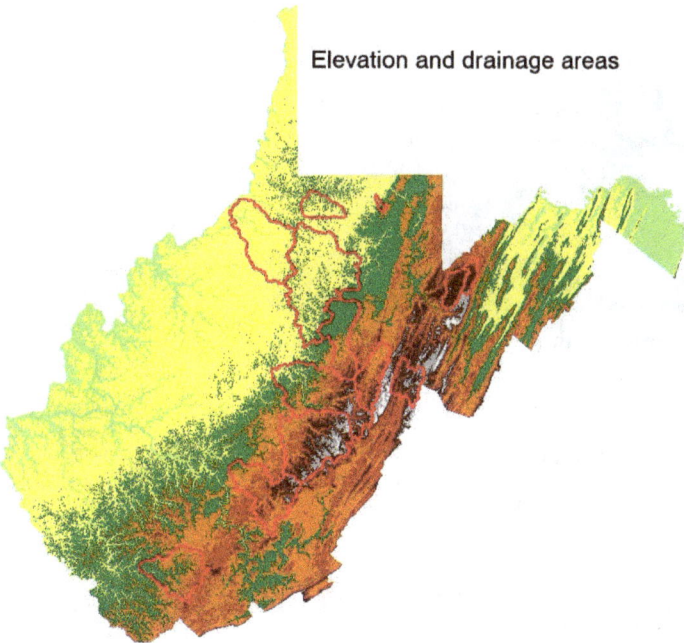

Fig. 7.

From table 2, it was noted which variables were closely grouped and which varied significantly among all the 13 different gauges. The area and flow variables have the highest standard deviation while the precipitation, maximum and minimum temperatures, and watershed slope have the lowest standard deviation. Other simple statistical graphs, which were used to gain insights into the data distribution and spreads, are shown in figures 8 to 14. The figures provided a graphical display of the distribution of values across the 13 gauges. Data exploration is important to determine trends and outliers in data that may bias results (Johnson, et al 2001). In addition, regression results may be impacted from large variations in data values. A common technique is to normalize data with a simple equation such as the value of interest minus the minimum value for that variable divided by the maximum minus minimum within the data range (Kachigan, 1986). However, in this study the values were not normalized due to the spatial nature of the information source. It was necessary to identify and incorporate the spatial variability across the entire study area at the statewide level. The end use of our regression relationship is the ability to query any raster stream cell and report all the unique information from the spatial analysis. Stream flow and water quality decisions for permitting may occur in high elevation cold headwater segments as well as large river systems with much accumulated drainage. Because the study area had unique topographic features that were to be regressed against representative stream flow information, the gauge driven delineated watersheds were chosen to represent this differentiation as best as possible as shown in Figure 3.

Descriptive Statistics

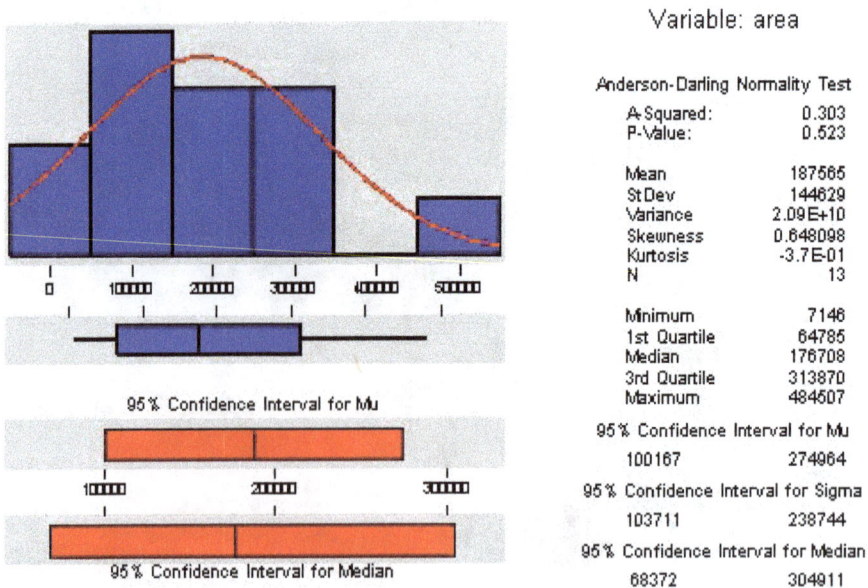

Variable: area

Anderson-Darling Normality Test

A-Squared:	0.303
P-Value:	0.523
Mean	187565
StDev	144629
Variance	2.09E+10
Skewness	0.648098
Kurtosis	-3.7E-01
N	13
Minimum	7146
1st Quartile	64785
Median	176708
3rd Quartile	313870
Maximum	484607

95% Confidence Interval for Mu

100167	274964

95% Confidence Interval for Sigma

103711	238744

95% Confidence Interval for Median

68372	304911

Fig. 8.

Descriptive Statistics

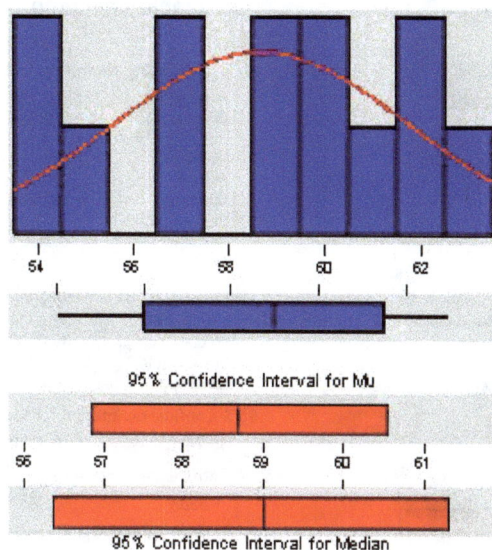

Variable: maxtemp

Anderson-Darling Normality Test

A-Squared:	0.340
P-Value:	0.438
Mean	58.6923
StDev	3.0655
Variance	9.39744
Skewness	-3.3E-01
Kurtosis	-1.13292
N	13
Minimum	54.0000
1st Quartile	56.0000
Median	59.0000
3rd Quartile	61.5000
Maximum	63.0000

95% Confidence Interval for Mu

56.8398   60.5448

95% Confidence Interval for Sigma

2.1982   5.0604

95% Confidence Interval for Median

56.3693   61.3154

Fig. 9.

Descriptive Statistics

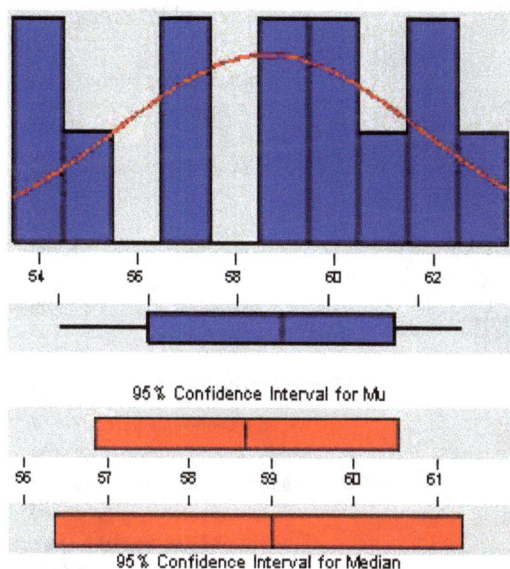

Variable: maxtemp

Anderson-Darling Normality Test

A-Squared:	0.340
P-Value:	0.438
Mean	58.6923
StDev	3.0655
Variance	9.39744
Skewness	-3.3E-01
Kurtosis	-1.13292
N	13
Minimum	54.0000
1st Quartile	56.0000
Median	59.0000
3rd Quartile	61.5000
Maximum	63.0000

95% Confidence Interval for Mu

56.8398   60.5448

95% Confidence Interval for Sigma

2.1982   5.0604

95% Confidence Interval for Median

56.3693   61.3154

Fig. 10.

Descriptive Statistics

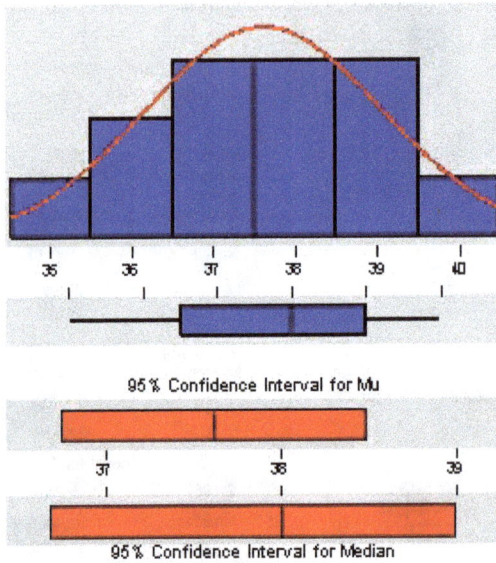

Variable: mintemp

Anderson-Darling Normality Test

A-Squared:	0.298
P-Value:	0.534
Mean	37.6154
StDev	1.4456
Variance	2.08974
Skewness	-1.7E-01
Kurtosis	-6.4E-01
N	13
Minimum	35.0000
1st Quartile	36.5000
Median	38.0000
3rd Quartile	39.0000
Maximum	40.0000

95% Confidence Interval for Mu

36.7418   38.4889

95% Confidence Interval for Sigma

1.0366   2.3863

95% Confidence Interval for Median

36.6846   39.0000

Fig. 11.

Descriptive Statistics

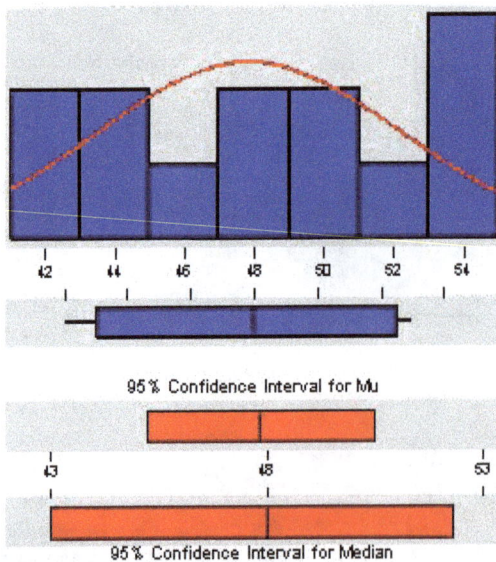

Variable: precip

Anderson-Darling Normality Test

A-Squared:	0.523
P-Value:	0.148
Mean	47.8462
StDev	4.3368
Variance	18.8077
Skewness	-1.6E-01
Kurtosis	-1.62409
N	13
Minimum	42.0000
1st Quartile	43.0000
Median	48.0000
3rd Quartile	52.5000
Maximum	53.0000

95% Confidence Interval for Mu

45.2255   50.4668

95% Confidence Interval for Sigma

3.1098   7.1589

95% Confidence Interval for Median

43.0000   52.3154

Fig. 12.

Descriptive Statistics

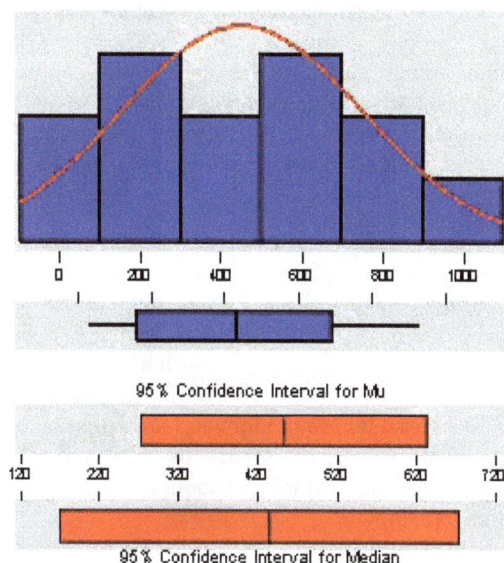

Variable: strmslope

Anderson-Darling Normality Test

A-Squared:	0.228
P-Value:	0.764
Mean	452.538
StDev	299.299
Variance	89579.8
Skewness	2.30E-02
Kurtosis	-1.23275
N	13
Minimum	26.000
1st Quartile	159.500
Median	435.000
3rd Quartile	693.000
Maximum	933.000

95% Confidence Interval for Mu

271.674   633.403

95% Confidence Interval for Sigma

214.623   494.063

95% Confidence Interval for Median

170.393   674.537

Fig. 13.

Descriptive Statistics

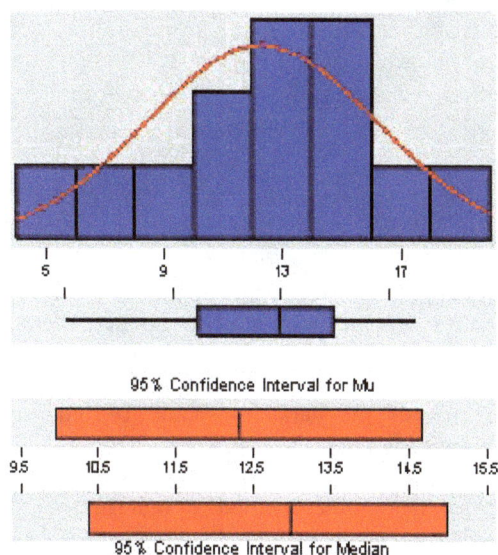

Variable: wsslope

Anderson-Darling Normality Test

A-Squared:	0.319
P-Value:	0.493
Mean	12.3077
StDev	3.8813
Variance	15.0641
Skewness	-5.9E-01
Kurtosis	-1.5E-01
N	13
Minimum	5.0000
1st Quartile	10.0000
Median	13.0000
3rd Quartile	15.0000
Maximum	18.0000

95% Confidence Interval for Mu

9.9623   14.6531

95% Confidence Interval for Sigma

2.7832   6.4069

95% Confidence Interval for Median

10.3693   15.0000

Fig. 14.

The next step in analyzing the data was to use generate a best-fit line plot for each of the independent variables in table 1 regressed against the dependent variable stream flow. These plots are shown in figures 15 to 20. From these best-fit line plots, the area, stream slope, and watershed slope variables had the best R squared values and positive linear relationship. The maximum and minimum temperature variables along with precipitation had the worst linear fit with stream flow. Their R squared values were very low with the precipitation variable looking very random in describing stream flow. At this point in the analysis it appeared that the area, stream slope and watershed slope will be the better variables to predict stream flow.

While the linear regression plots provided some idea of the extent of the relationship between two variables, the correlation coefficient gives a summary measure that communicates the extent of correlation between two variables in a single number (Kachigan, 1986). The higher the correlation coefficient, the more closely grouped are the data points representing each objects score on the respective variables. Some important assumptions of the correlation coefficient are that the data line in groupings that are linear in form. The other important assumptions include that the variables are random and measured on either an interval or a ratio scale. In addition, the last assumption for the use of the correlation coefficient is that the two variables have a bivariate normal distribution. The correlation matrix for the data used in this study is shown in table 3.

	area	precip	maxtemp	mintemp	flow	strslope	wsslope
area	1	-0.212	0.470	0.571	0.922	0.138	0.516
precip	-0.212	1	-0.850	-0.781	0.039	0.560	-0.279
maxtemp	0.470	-0.850	1	0.930	0.245	-0.226	0.590
mintemp	0.571	-0.781	0.930	1	0.356	-0.217	0.647
flow	0.922	0.039	0.245	0.356	1	0.392	0.435
strslope	0.138	0.560	-0.226	-0.217	0.392	1	0.245
wsslope	0.516	-0.279	0.590	0.647	0.435	0.245	1

Table 3. Correlation matrix

The variables with significant correlations (R > .7) are shaded in table 3. The variables listed in order of highest correlation to lowest significance are mintemp and maxtemp, flow and area, precip and maxtemp, and precip and mintemp. The correlations between the weather data were expected. In areas of higher precipitation, the temperature will be cooler (the annual averages for maximum temperature will be lower and the annual average for minimum temperatures will be lower) hence the high negative correlation. The other high positive correlated variables indicate that the variation in one variable will lead to variation in the other variable. For regression analysis the variables should be independent. Collinearity refers to linear relationships within the variables. The amount of multicollinearity across variables can be examined with principal component analysis of a sample correlation matrix (Sundberg, 2002) among other methods to remove dependence. This study examined the smallest eigenvalue and eliminated variables with values less than 0.05 as an indication of substantial collinearity (Hocking, 1996). As expected the precipitation variables were not independent to the elevation data and therefore removed.

Regression Plot

Y = 23.4281 + 2.90E-03X
R-Sq = 84.9 %

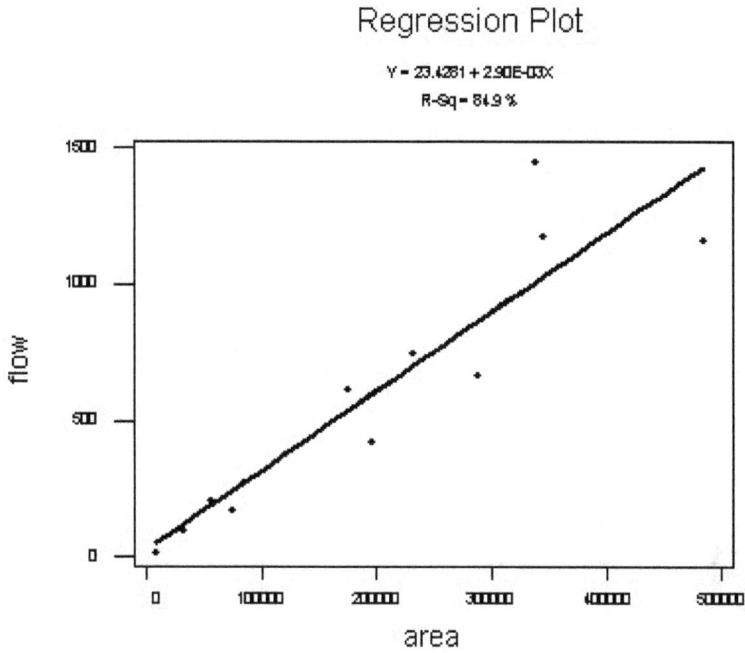

Fig. 15.

Regression Plot

Y = -1567.40 + 36.3777X
R-Sq = 6.0 %

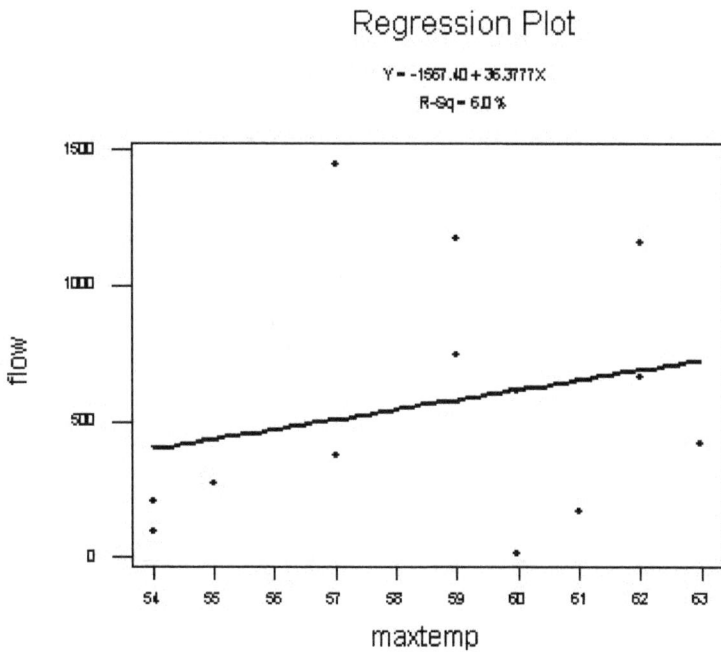

Fig. 16.

Regression Plot

Y = -3547.07 + 112.049X

R-Sq = 12.7 %

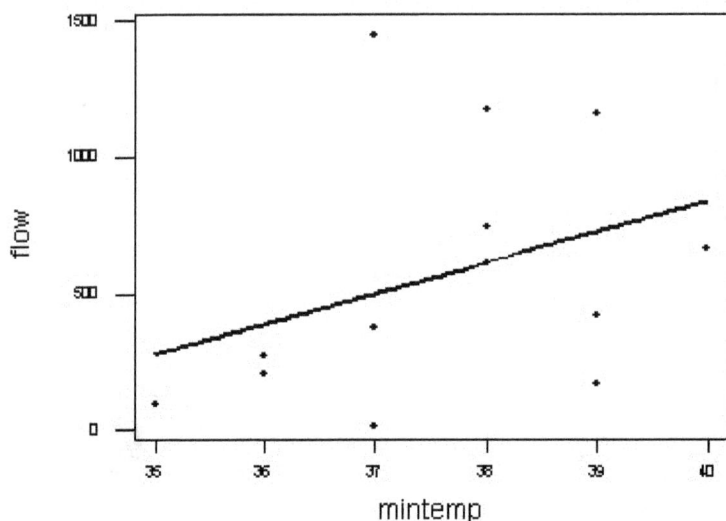

Fig. 17.

Regression Plot

Y = 374.266 + 4.04263X

R-Sq = 0.1 %

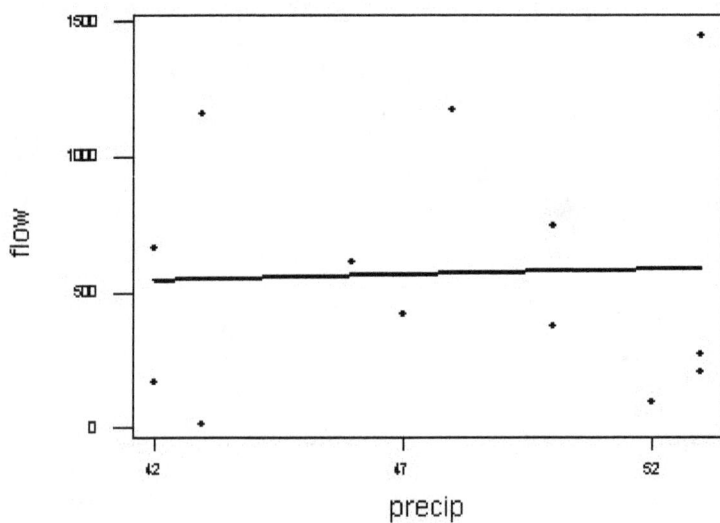

Fig. 18.

Regression Plot

Y = 298.154 + 0.595610X

R-Sq = 15.3 %

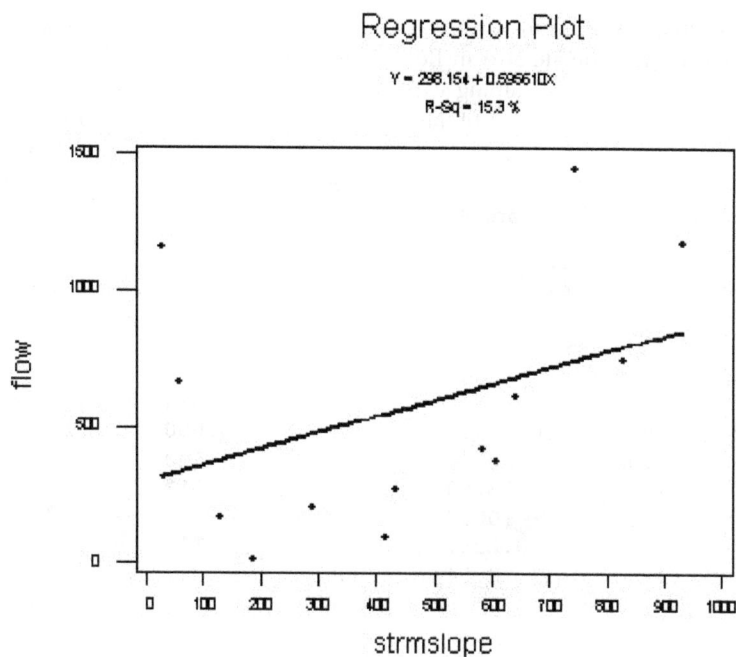

Fig. 19.

Regression Plot

Y = -60.2974 + 51.0240X

R-Sq = 18.9 %

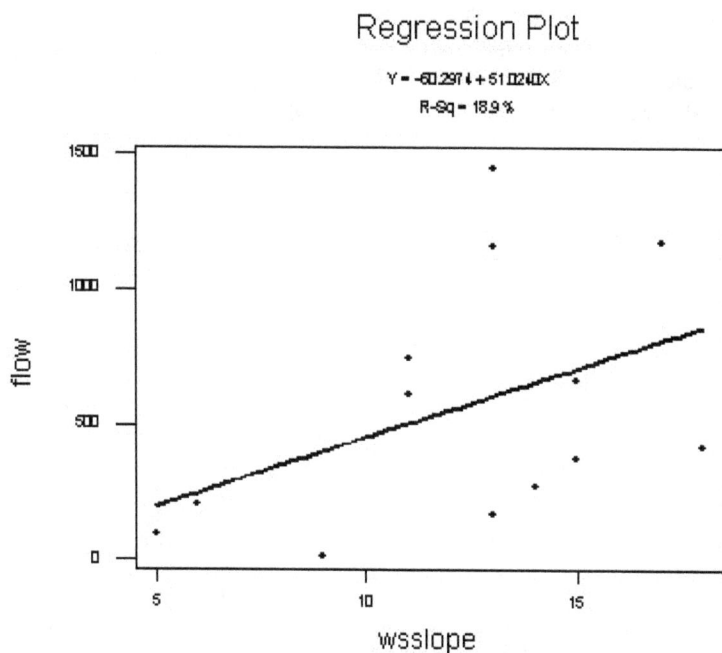

Fig. 20.

Performing regression analysis on the data was the next step in formulating a relationship and model to predict and estimate stream flow. Using the technique by Garren (1992) a regression equation with all the remaining variables was created, evaluate the P values of each variable, and eliminate variables until the highest adjusted R square is found. The first run with the regression analysis indicates that the variables area, strmslop and maxtemp will have the most influence on flow because of their low P values. Table 4 shows the regression analysis including all the variables.

```
The regression equation is
flow = 2325 + 0.00310 area - 12.0 precip - 37.3 maxtemp + 8 mintemp
           + 0.423 strmslope - 4.8 wsslope
```

Predictor	Coef	StDev	T	P
Constant	2325	4370	0.53	0.614
area	0.0030987	0.0004281	7.24	**0.000**
precip	-12.02	30.84	-0.39	0.710
maxtemp	-37.31	53.50	-0.70	**0.512**
mintemp	7.8	100.7	0.08	0.941
strmslop	0.4235	0.2281	1.86	**0.113**
wsslope	-4.83	18.54	-0.26	0.803

```
S = 156.3       R-Sq = 94.1%      R-Sq(adj) = 88.2%
```

Table 4. Regression analysis including all variables

By systematically removing the variables with a high P value and noting the R squared adjusted value, it was possible to arrive at a final set of variables to use in a regression equation to estimate stream flow. Table 5 shows the regression analysis results after removing the variable with the highest P value (mintemp).

```
The regression equation is
flow  =  2492  +  0.00311  area  -  12.3  precip  -  35.0  maxtemp  +  0.421
strmslope
           - 4.3 wsslope
```

Predictor	Coef	StDev	T	P
Constant	2492	3522	0.71	0.502
area	0.0031121	0.0003628	8.58	0.000
precip	-12.35	28.30	-0.44	0.676
maxtemp	-35.00	41.23	-0.85	0.424
strmslop	0.4208	0.2089	2.01	0.084
wsslope	-4.33	16.11	-0.27	0.796

```
S = 144.7       R-Sq = 94.1%      R-Sq(adj) = 89.9%
```

Table 5. Regression analysis with mintemp removed

The R squared adjusted improved slightly to 89.9% with mintemp removed. This process of removing the current highest P value variable and re-running of the model was repeated six times. The associated R squared values were noted and table 6 was created from the results.

As table 6 indicates, the combination of variables that provided the highest R squared adjusted value were area, maxtemp, and strslope. The associated regression equation with the optimal set of variables is:

$$flow = 1232 + 0.00304 \text{ area} - 23.6 \text{ maxtemp} + 0.338 \text{ strmslope}$$

Variables included in the regression	R squared adjusted
Area, mintemp, maxtemp, strslope, wsslope	88.2
Area, maxtemp, strslope, wsslope	89.9
Area, maxtemp, strslope	91.1
Area, maxtemp, strslope	91.8
Area, strslope	90.5
Area	83.6

Table 6. Multiple regression results

The next procedure used in the analysis was discriminant analysis. This technique was used to identify relationships between qualitative criterion variables and the quantitative predictor variables in the dataset. The objective was to identify boundaries between the groups of watersheds that the gauges were associated. The boundaries between the groups are the characteristics that distinguish or discriminate the objects in the respective groups. Discriminant analysis allows the user to classify the given objects into groups – or equivalently, to assign them a qualitative label – based on information on various predictor or classification variables (Kachigan, 1992).

The gauge station dataset was assigned a qualitative variable based on which major drainage basin in West Virginia the area was located. The major basins used were the Monongahela (m), Gauley (g) and Other (x). The class "other" was assigned to gauges that did not fall in the Monongahela or Gauley drainage basins. Running the discriminant analysis in Minitab produced the results shown in table 7.

Only gauge one and gauge five were reclassified from the discriminant analysis results. It should be noted however that the discriminant function should be validated by testing its efficacy with a fresh sample of analytical objects. Kachigan (1986) notes that the observed accuracy of prediction on the sample upon which the function was developed will always be spuriously high, because we will have capitalized on chance relationships. The true discriminatory power of the function will be found when tested with a completely separate sample.

By using discriminant analysis, it enabled the investigation of how the given groups differ. In the next analysis step, cluster analysis, the goal is to find whether a given group can be partitioned into subgroups that differ. The advantage of the approach is in providing a better feel of how the clusters are formed and which particular objects are most similar to one another.

The cluster analysis was performed with distance measures of Pearson and Average and link methods of single and Euclidean. The Average and Euclidean choices worked the best in identifying clusters. Figure 21 shows the dendrogram results and table 8 lists the computation results.

```
Linear Method for Response:    class

Predictors:   area  precip  maxtemp  mintemp  flow  strslope  wsslope

Group            g          m          x
Count            2          5          6

Summary of Classification
Put into       ....True Group....
Group            g          m          x
g                2          0          0
m                0          4          1
x                0          1          5
Total N          2          5          6
N Correct        2          4          5
Proportion     1.000      0.800      0.833

N =    13      N Correct =    11      Proportion Correct = 0.846
Squared Distance Between Groups
                 g          m          x
g           0.0000    14.5434    17.0393
m          14.5434     0.0000     4.5539
x          17.0393     4.5539     0.0000

Linear Discriminant Function for Group
                 g          m          x
Constant  -7379.7    -7053.9    -7003.2
area          -0.0       -0.0       -0.0
precip        85.6       83.5       83.6
maxtemp       86.5       84.2       84.3
mintemp      157.5      155.0      153.0
flow           0.8        0.7        0.7
strslope      -0.3       -0.3       -0.3
wsslope      -38.0      -37.0      -36.6

Summary of Misclassified Observations
Observation    True      Pred     Group    Squared     Probability
               Group     Group              Distance
    1 **          x         m        g       20.956        0.000
                                     m        5.163        0.578
                                     x        5.796        0.421
    2 **          m         x        g       23.906        0.000
                                     m        5.223        0.229
                                     x        2.790        0.771

gauge id   majshed              class     FITS1
g1         NorthBranch          x         m
g5         Tygart               m         x
g7         Tygart               m         m
g10        WestFork             x         x
g11        MonRiver             m         m
g12        MonRiver             m         m
g13        Cheat                m         m
g17        MiddleOhio           x         x
g19        Greenbrier           x         x
g21        Gauley               g         g
g22        Gauley               g         g
g24        Elk                  x         x
g26        UpGuyandotte         x         x
```

Table 7. Discriminant analysis

```
Standardized Variables, Euclidean Distance, Average Linkage
Amalgamation Steps
Step Number of Similarity  Distance  Clusters   New    Number of obs.
     clusters    level      level    joined  cluster in new cluster
  1    12       85.24      0.933    1    7     1           2
  2    11       80.06      1.261    3   11     3           2
  3    10       78.24      1.376    2    9     2           2
  4     9       70.39      1.872    5    6     5           2
  5     8       69.18      1.948    4    8     4           2
  6     7       68.16      2.013   10   12    10           2
  7     6       61.42      2.439    3   10     3           4
  8     5       56.79      2.732    1    2     1           4
  9     4       54.13      2.900    3   13     3           5
 10     3       45.06      3.473    4    5     4           4
 11     2       41.04      3.727    3    4     3           9
 12     1       35.49      4.078    1    3     1          13
```

```
Final Partition
Number of clusters:   2
                Number of     Within cluster  Average distance Maximum distance
              observations  sum of squares   from centroid    from centroid
Cluster1          4              8.340          1.414            1.816
Cluster2          9             46.098          2.188            2.934
```

```
Cluster Centroids
Variable        Cluster1      Cluster2    Grand centrd
area            -0.7923        0.3522       -0.0000
precip           0.9578       -0.4257       -0.0000
maxtemp         -1.2045        0.5353       -0.0000
mintemp         -1.1175        0.4966        0.0000
flow            -0.7181        0.3191       -0.0000
strslope        -0.0511        0.0227       -0.0000
wsslope         -0.5946        0.2643       -0.0000
```

```
Distances Between Cluster Centroids

               Cluster1      Cluster2
Cluster1        0.0000        3.2672
Cluster2        3.2672        0.0000
```

Table 8. Hierarchical cluster analysis of observations

From the clustered results, gauges 1 and 7 (g1 and g13) are the most alike and merge into a cluster at around 85 on the similarity scale. Gauges 3 and 11 (g7 and g22) are the next most similar at the 78 level. However, these objects do not form the same cluster until a lower level of similarity around the 35 level. By clustering the objects, we were able to identify groups that are alike and because of the small dataset, it was easy to examine the data table and discover values that make the objects similar.

After cluster analysis, the choice was made to perform a factor analysis as an aid in data reduction. Although there were only seven variables, the possibility existed to gain insight into removing the duplicated information from among the set of variables. The results were assembled as a loading plot – figure 22, a score plot – figure 23, and a scree plot – figure 24. The output session data is listed in table 9.

Similarity

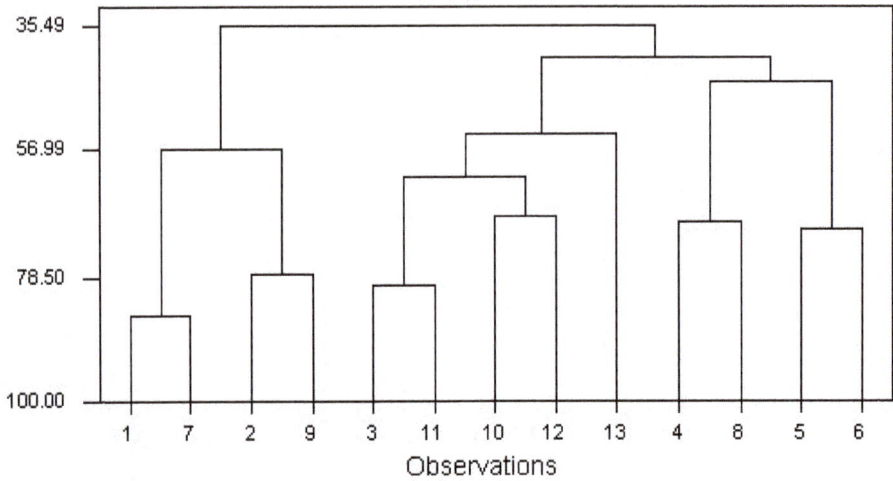

Fig. 21.

Loading Plot of area-wsslope

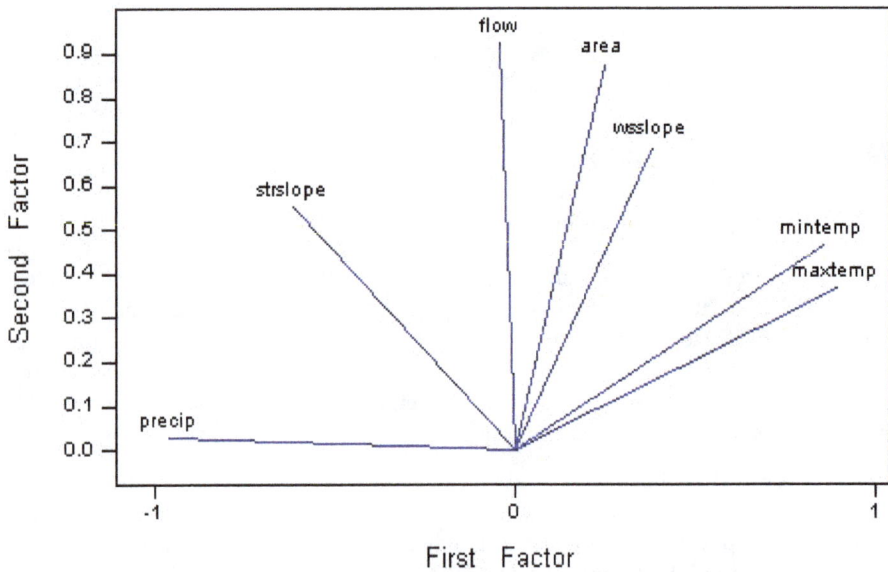

Fig. 22.

Score Plot of area-wsslope

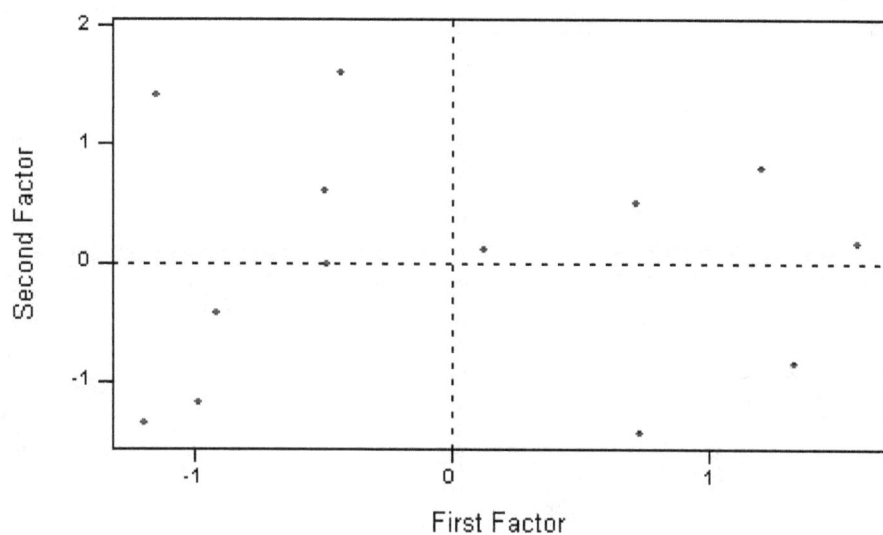

Fig. 23.

Scree Plot of area-wsslope

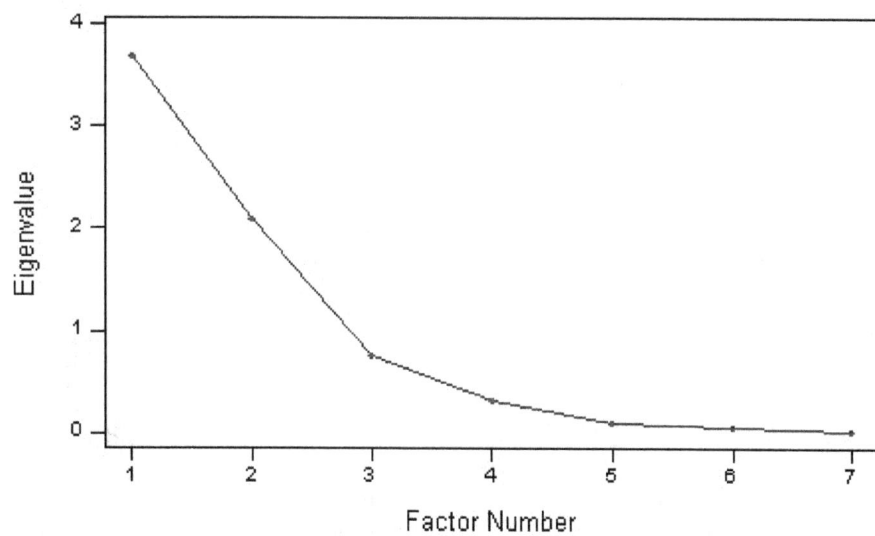

Fig. 24.

Principal Component Factor Analysis of the Correlation Matrix

Unrotated Factor Loadings and Communalities

Variable	Factor1	Factor2	Communality
area	-0.748	-0.512	0.822
precip	0.720	-0.639	0.927
maxtemp	-0.913	0.291	0.919
mintemp	-0.949	0.192	0.937
flow	-0.559	-0.737	0.855
strslope	0.117	-0.821	0.687
wsslope	-0.731	-0.286	0.616
Variance	3.6732	2.0898	5.7630
% Var	0.525	0.299	0.823

Rotated Factor Loadings and Communalities
Varimax Rotation

Variable	Factor1	Factor2	Communality
area	0.243	0.874	0.822
precip	-0.962	0.025	0.927
maxtemp	0.886	0.365	0.919
mintemp	0.849	0.465	0.937
flow	-0.047	0.924	0.855
strslope	-0.618	0.552	0.687
wsslope	0.375	0.689	0.616
Variance	3.0164	2.7466	5.7630
% Var	0.431	0.392	0.823

Factor Score Coefficients

Variable	Factor1	Factor2
area	-0.002	0.319
precip	-0.347	0.108
maxtemp	0.280	0.054
mintemp	0.257	0.096
flow	-0.111	0.368
strslope	-0.277	0.280
wsslope	0.064	0.233

Table 9. Factor analysis

From these results, the variables high in loadings on a particular factor would be those which are highly correlated with one another, but which have little or no correlation with the variables loading highly on the other factors. The negative loading variable has a meaning opposite to that of the factor. The size of the loading is an indication of the extent to which the variable correlates with the factor.

Limitations, and Discussion of Results

The limitations with this study can be attributed to the number of gauges used and the variables used to predict stream flow. With more complete data over the state, it would have been able to assemble more gauges for this component of the project. Also, if possible it would have been good to include variables used to describe interception, evapotranspiration, infiltration, interflow, saturated overland flow, and baseflow from groundwater. The rate and areal distribution of rainfall input would have been helpful in establishing the catchment retention.

Other issues with the data collection make the estimation of stream flow difficult. First, there is very high variability in recording stream flow data. The stream flow variable exhibited the highest standard deviation and variation across the year. Second, taking yearly annual averages was a crude method in which to characterize the varying conditions that occur across seasons, months, weeks, and even days. Third, the precipitation and temperature data used in the study needed to be better allocated to the gauge drainage areas (as compared to using the drainage area average for the variable) because of the amount of variability that is present in the entire watershed for the precipitation and temperature data. Overall, the choice of variables to analyze were appropriate based on the success other studies found. In the study the results of the multivariate regression indicated that stream flow could best be estimated using area, stream slope and 30 year annual average maximum temperature. Other data analysis techniques revealed the correlation present between the two temperature variables, flow and area, and precipitation to the two temperature variables.

The last important summary from the tests came from the cluster analysis that grouped the gauge station objects based on similarity. The grouped gauges shared the same ecoregions. Ecoregions are defined as "regions of relative homogeneity in ecological systems or in relationships between organisms and their environments" (Omernik 1987). Omernik (1987) mapped the ecoregions of the conterminous United States, based on regional patterns in individual maps of land use, land surface form, potential natural vegetation, and soils. A discriminant analysis using the ecoregion of each gauge station catchment area would have been a better choice than the using the major river basins used in this study. The similar gauge station catchment areas identified by the cluster analysis and the associated ecoregion borders in West Virginia are displayed in figure 25.

Component 6. The environmental database

An environmental database of point data was included within SWPS. These points are found in the shapefiles directory of SWPS and are loaded for viewing when a user defines a study area location in the state. A brief listing of some of the files in the environmental database follows:

- National pollution discharge elimination system sites
- Landfills
- Superfund sites (CERCLIS)
- Hazardous and solid waste sites (RCRIS)
- Toxic release inventory sites
- Coal dams

- Abandoned mine land locations
- Animal feed lots
- Major highways
- Railroads

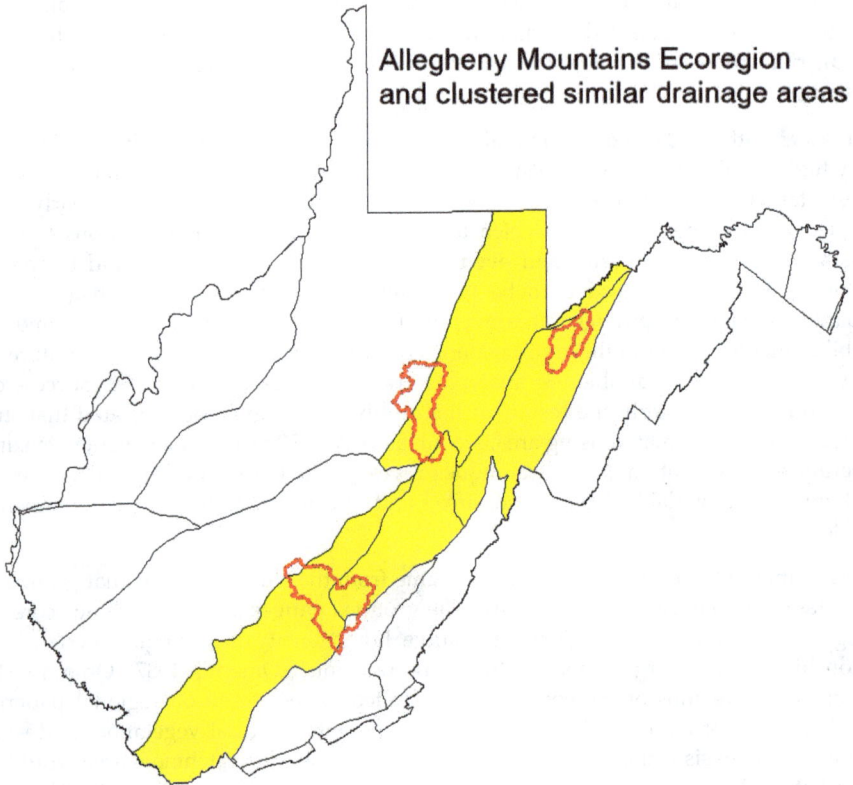

Allegheny Mountains Ecoregion and clustered similar drainage areas

Fig. 25.

Component 7. UTM latitude/longitude conversion utility

This capability of SWPS allows the user to map coordinates in degrees, minutes and seconds by using an input dialog screen. The user's points are then mapped in the UTM zone 17 projection. Points may be added to an existing point feature theme or a new point theme can be created. The ability to type coordinates and have the points reprojected saves the user many extra steps. In addition to mapping points from user input, a point can be queried for its x and y locations in UTM, stateplane, or latitude and longitude coordinates. The user can identify locations quickly by clicking anywhere in the display to report this information.

Component 8. Statewide map/GIS data layers

All GIS data is organized in the shapefiles and grids directories of SWPS. This data is listed below. These datasets are provided in addition to the data listed in the environmental database discussed in component 6.

- Coded hydrology
- Watersheds
- Lakes/impoundments
- Counties
- 1:24K quads
- Major river basins
- Abandoned mine lands
- Watersheds with fish collection data
- Stream orders
- Major Rivers
- WV GAP land use/ cover
- Bond forfeiture sites
- Expected mean concentration grid
- Public wells
- Groundwater wells
- Surface water wells

- Roads (1:100K scale)
- Cities
- USGS gauging stations
- Public wells
- Runoff grid
- Flow direction
- Flow accumulation
- Cumulative runoff
- Coal Geology
- Override 7Q10 streams
- Landfills
- NW Wetlands
- Springs over 500gpm
- 1950 land use/cover
- Shreve stream orders
- Stream length from mouth

- EPA MRLC land use/cover
- Digital Raster Graphics
- SPOT imagery
- Digital Elevation Model
- Hillshaded relief
- Elevation TIN
- 303-D listed streams
- DRAFT 14 digit HUCS
- Wet weather streams
- Surface mine inventory sites
- Public wells
- Strahler stream orders
- Stream slope
- Max and min stream elevations
- Surface water zones of critical concern
- Streamflow

Component 9. Water quality modeling capability

The water quality modeling capabilities of SWPS are built using a landscape driven approach that uses a predefined runoff and cumulative runoff grid to drive the analysis. It is essentially a weighted mass balance approach that will show changing concentrations and loadings based on changing flow conditions only. The runoff grid is based on a relationship between rainfall and stream flow. It is the main factor that directs flow directions to the stream or steepest path direction and estimates the stream flow.

The assumptions/limitations of this water quality modeling approach are the following:

1. Streams have the same **hydrogeometric properties** (stream slope, roughness, width, and depth).
2. Also assumed are that the streams have the same **ecological rate constants** (reareation rates, pollution decay rates and sediment oxygen demand rate).
3. Transport of pollutants is considered to be conservative (values get averaged over changing flow conditions only) -> no loss or decay of pollutants is considered Does not consider infiltration, or ground water flow additions
5. Does not include atmospheric conditions such as evapotranspiration

The water quality model in SWPS can be used in two different ways. The first is when the user has collected point locations of water quality data and wants to associate the sampled data to instream concentrations and loadings downstream of the sampling points. This is essentially a weighted mass balance approach using the stream flow and sampled locations to associate the point location information to stream condition. The input data using this method needs to be in Mg/L. The resultant modeled levels are reported back as stream values in Mg/L for concentration and Kg/Yr. for loading. The advantage of this first method of using sampled data is that it allows the user to see how the data location information can be used to estimate downstream conditions away from the sampling site.

The second way the water quality model in SWPS can be used is in estimating total nitrogen, phosphorous and total suspended solids as concentrations and loadings in the stream based on expected mean concentrations from land use/cover classes. This method does not require any sampled water quality but uses the cover classes from a land use/cover grid (30meter-cell size). The thirteen classes for West Virginia from this data set were aggregated to six general classes because loading values for nitrogen, phosphorous and total suspended solids were only available for those six classes. The aggregated classes and the corresponding classes included:

-    Urban (low intensity developed, high intensity developed, residential)
-    Open/Brush (hay, pasture grass, mixed pasture, other grasses)
-    Agriculture (row crops)
-    Woodland (conifer forest, mixed forest, deciduous forest)
-    Barren (quarry areas, barren transitional areas)
-    Wetland (emergent and woody wetlands)

The classes are associated with expected loadings based on the acreage size of the class. The loadings are annual averages and when used with the modeled stream flow can give concentration and loading results for the stream. The cover classes and associated expected mean concentrations levels used in the model are shown below.

	Total Nitrogen	Total Phosphorous	Total Suspended Solids
Urban	1.89	0.009	166
Open/Brush	2.19	0.13	70
Agriculture	3.41	0.24	201
Woodland	0.79	0.006	39
Barren	3.90	0.10	2200
Wetland	0.79	0.006	39

The nutrient export coefficients above are multiplied by the amount (area) of a given land cover type. It is used as a simulation to estimate the probability of increased nutrient loads from land cover composition. It should be noted that there are factors other than land cover that contribute to nutrient export and these are rarely known with certainty. Some of the factors that may vary across watersheds and may change the expected mean concentration results include:

•   year to year changes in precipitation
•   soil type
•   slope and slope morphology (convex, concave)
•   geology
•   cropping practices
•   timing of fertilizer application relative to precipitation events
•   density of impervious surface

The loading and concentration results in consideration of these assumptions however can still give insight in comparing expected pollutant values for watersheds. The results should be thought of in most cases as the worst case scenarios for stream water quality levels.

Component 10. Groundwater and surface water susceptibility model

The susceptibility ranking model within SWPS was constructed using the steps defined within the West Virginia Surface Water Assessment Program (SWAP) document. The susceptibility ranking for ground water systems was based on the physical integrity of the well and spring infrastructure; hydrologic setting; inventory of potential contaminant sources and land uses, and water quality. The susceptibility ranking for surface water systems was based on water quality and the inventory of potential contaminant sources. A more detailed explanation of the ground and surface water susceptibility models follows.

Groundwater susceptibility model

To determine the groundwater susceptibility for a site, the physical barrier effectiveness is first calculated. Physical barrier effectiveness is the Tier 1 assessment. It is used to note if there is a known impact on water quality, evaluate the source integrity as low or high, and to find the aquifer vulnerability. Based on these results, the physical barrier effectiveness can be determined as having high, moderate, or low potential susceptibility. If there is a known impact on water quality, then the model automatically goes to the Tier Two assessment and sets the groundwater susceptibility as being high. If there is no known impact on water quality then the source integrity and aquifer vulnerability set the physical barrier effectiveness for the Tier Two assessment. The aquifer vulnerability is determined from the different scenarios listed below;

If	Then
All springs	High Aquifer Sensitivity
Alluvial Valleys	
Unconfined	High Aquifer Sensitivity
Confined	Moderate Aquifer Sensitivity
Appalachian Plateau Province (fracture)	Moderate Aquifer Sensitivity
Folded Plateau Area (fracture)	Moderate Aquifer Sensitivity
Karst Areas	High Aquifer Sensitivity
Valley and Ridge Province (fracture)	Moderate Aquifer Sensitivity
Coal Mine Areas	High Sensitivity

From the above scenarios, an aquifer vulnerability was determined. Using this with the source integrity rating can provide physical barrier effectiveness. Physical barrier effectiveness is

- High if there is low source integrity and high aquifer sensitivity.
- Low if there is high source integrity and moderate aquifer sensitivity
- Moderate if there is high source integrity and high aquifer sensitivity or low source integrity and moderate aquifer sensitivity

Again, if there is no known impact on water quality, this method will determine the physical barrier effectiveness as being high, moderate or low susceptibility. If there is a known water quality impact then the final groundwater susceptibility is high.

Using the physical barrier effectiveness with the land use concern level determines the final groundwater susceptibility. The land use concern level is determined from the percentage of land use in the buffered groundwater site. The percentage of land use was found for every buffered location. In cases where the groundwater site had a pumping rate over 25,000 gpd, no buffer was created. For these groundwater sites no land use percentage was calculated.

The final groundwater susceptibility is rated as:

- High if the physical barrier effectiveness is high
- High if the land use concern is high and the physical barrier effectiveness is moderate
- Moderate if the land use concern is medium or low and the physical barrier effectiveness is moderate
- Moderate if the land use concern is high or medium and the physical barrier effectiveness is low
- Low if the land use concern is low and the physical barrier effectiveness is low

The percentage of land use is reported to the user before the tier one assessment appears. He or she needs to know the associated concern levels with the percentage of land uses that are reported for each groundwater site. By not hard coding in the land use concern levels for each buffer, the user has the ability to perform "what if" type scenarios if existing land use changes or is different than what currently exists.

Surface water susceptibility model

The surface water susceptibility model is slightly less complicated than the groundwater model. For the surface water susceptibility determination, the percent of land use was calculated for each of the zone of critical concern. The land use concern level, and if there is a known water quality impact, are the two factors which are used to determine the surface water susceptibility. As in the groundwater model, the percent land use is presented to the user before the model is run so the user can make a determination and perform "what if" type scenarios with differing land use within the zone of critical concern. If there is a known water quality impact, then the surface water susceptibility is automatically high. If there is no known water quality impact, then the final surface water susceptibility is:

- High if land use concern level is high
- High if land use concern level is medium
- Low if land use concern level is moderate

Summary and Conclusion

Drinking water is a critical resource that continues to need protection and management to assure safe supplies for the public. Since agencies to protect water resources operate at mostly state jurisdictions, it is important to implement a system at a statewide level. This chapter discussed watershed tools that integrate spatially explicit data and decision support to assist managers with both surface and ground water resources. It has three major components which include; an ability to delineate source water protection areas upstream of supply water, an inventory of potential contamination sources within various zones of critical concern, and the determination of the public drinking water supply systems susceptibility to contamination. The current system provides the ability to assess, preserve, and protect the states source waters for public drinking.

2. References

Anderson, L. and A. Lepisto. 1998. "Links Between Runoff Generation, Climate and Nitrate-N Leaching From Forested Catchments." Water, Air, and Soil Pollution. 105: pp227-237.

Astatkie, T. and W. E. Watt. 1998. "Multiple-Input Transfer Function Modeling of Daily Stream flow Series Using Nonlinear Inputs." Water Resources Research. Vol. 34, No.10, pp2217-2725.

Environmental Systems Research Institute (ESRI). 2010. ArcInfo ArcMap. Redlands, CA.

Gabriele, H. M., and F. E. Perkins. 1997. "Watershed-Specific Model for Stream flow, Sediment, and Metal Transport." Journal of Environmental Engineering. pp61-69.

Garren, D. C. 1992. "Improved Techniques in Regression-Based Stream flow Volume Forecasting." Journal of Water Resources Planning and Management. Vol. 118, No. 6. pp654-669.

Hocking, R. R. 1996. Methods and Applications of Linear Models. Wiley and Sons, New York, NY.

Johnston, K., J. M. Ver Hoef, K. Krivoruchko, and N. Lucas. 2001. Using ArcGis Geostatistical Analyst. Environmental System Research Institute, Redlands, CA.

Kachigan, S. K. 1986. Statistical Analysis. Radius Press, New York, NY.

Moore, R. D. 1997. "Storage-Outflow Modeling of Stream flow Recessions, with Application to Shallow-Soil Forested Catchments." Journal of Hydrology. 198(1997) pp260-270.

Omernik, J. M. 1987. "Ecoregions of the conterminous United States." Annals of the Association of American Geographers Vol. 77, pp118-125.

Querner, E. P., Tallaksen, L. M., Kasparek, L., and H. A. J. Van Lanen. 1997. "Impact of Land-Use, Climate Change and Groundwater Abstraction on Stream flow Droughts Using Physically-Based Models." Regional Hydrology: Concepts and Models for Sustainable Water Resource Management. IAHS Publ. No. 246, pp171-179.

Sharma, A., D. G. Tarboton, and U. Lall. 1997. "Stream flow Simulation: A Nonparametric Approach." Water Resources Research. Vol. 33, No. 2, pp291-308.

Smith, J. A. 1991. "Long-Range Stream flow Forecasting Using Nonparametric Regression." Water Resources Bulletin. Vol. 27, No. 1, pp39-46.

Sundberg, R. 2002. Collinearity. Encyclopedia of Environmetrics, Vol. 1. John Wiley and Sons, Ltd, Chichester, West Sussex UK.

Timofeyeva, M. and R. Craig. 1998. "Using Monte Carlo Technique for Modelling the Natural Variability of Stream flow in Headwaters of the Sierra Nevada, USA." Hydrology, Water Resources and Ecology in Headwaters. 1998. No. 248, pp59-65.

U. S. Geological Survey. 1996. "Prediction of Travel Time and Longitudinal Dispersion in Rivers and Streams." Water-Resources Investigations Report 96-4013.

West Virginia Department of Health and Human Resources. August 1, 1999. "State of West Virginia Source Water Assessment and Protection Program." Office of Environmental Health Services, Charleston, WV.

3

Modelling in the Semi Arid Volta Basin of West Africa

Raymond Abudu Kasei
University for Development Studies,
Ghana

1. Introduction

A region is said to be semi-arid when it receive precipitation below potential evapotranspiration thus insufficient water in the soil (excess to the wilting point) which to the extent support less growth and development of plant and animal life as compared to arid region which is characterized by a severe lack of available water, to the extent of hindering or even preventing the growth of primary life. The arid and semi-arid regions are fairly distributed within the earth's surface (Figure 1)

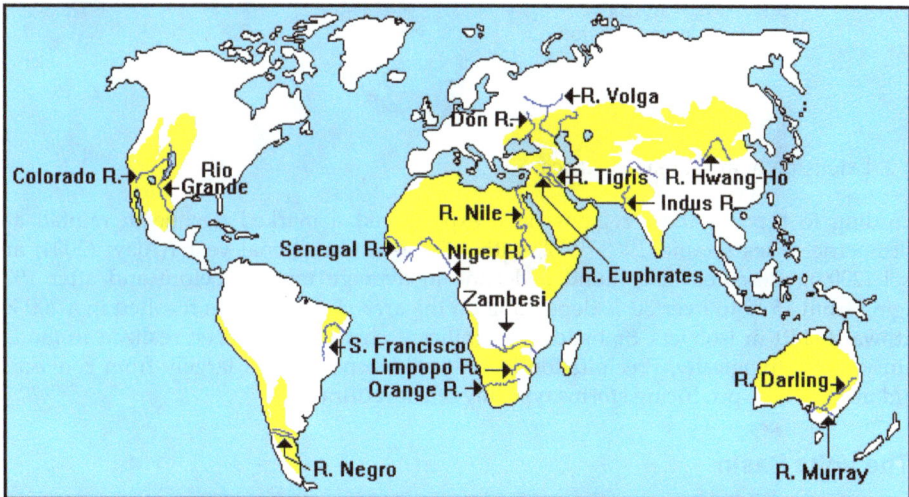

Fig. 1. Spatial distribution of arid and semi-arid regions of the earth (Shaded)

Semi-arid climates tend to support short or scrubby vegetation, with areas usually being dominated by either grasses or shrubs which can be called xeric. The climate of West Africa, particularly in the Sahelian zone has been undergoing recurrent variations of significant magnitude, particularly since the early 1970's.

Countries in Africa boarded with arid and semi-arid zones extend to north and south of the equator including, Senegal, Upper Volta and Chad Morocco, Algeria, Libya, and Egypt,

Lesotho, parts of the Cape, Botswana; Namibia The zones extend southeast through Somalia and Northern Kenya and parts of Zimbabwe. Semi-arid zones also extend to the northern part of Ghana and Burkina Faso of the Volta Basin in West Africa (Figure 2).

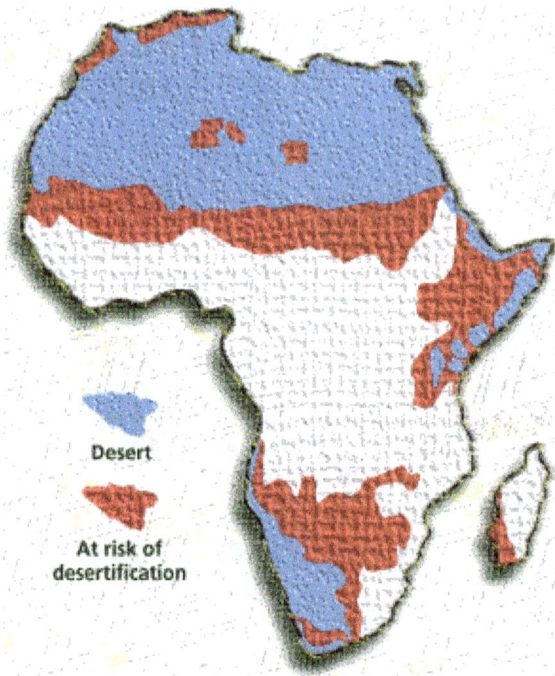

Fig. 2. Extension of arid areas in Africa

According to reports, these regions have experienced a marked decline in rainfall and hydrometric series around 1968–1972, with 1970 as a transitional year ((Ajayi (2004) and Kasei (2009)). Research also found a decline in average rainfall, before and after 1970, ranging from 15 % to over 30 % depending on the area. This situation resulted in a 200 km southward shift in isohyets. Statistically, significant changes have been realized in the last century. Climatologically, Precipitation in semi-arid zones results largely from convective cloud mechanisms producing storms typically of short duration.

2. The Volta Basin

The Volta River Basin is located between latitudes 5oN and 14oN and longitudes 2oE and 5°W (Figure 3). It has a surface area of about 414 000 km2 covering areas in six riparian West African countries (Benin to the east, Burkina Faso to the north, Côte d'Ivoire to the west, Mali, Togo and Ghana to the south). The total basin population is estimated at a little over 19 million inhabitants, with an annual growth rate estimated at 2.9% (Green Cross International, 2001). The hydrographical network of the basin is delineated into three main subcatchments: the Mouhoun (Black Volta), the Nakambé (White Volta) and the Oti River. Apart from the huge network of rivers, the basin is dotted with a number of reservoirs, ponds and dugouts. In areas where surface water is inadequate, groundwater

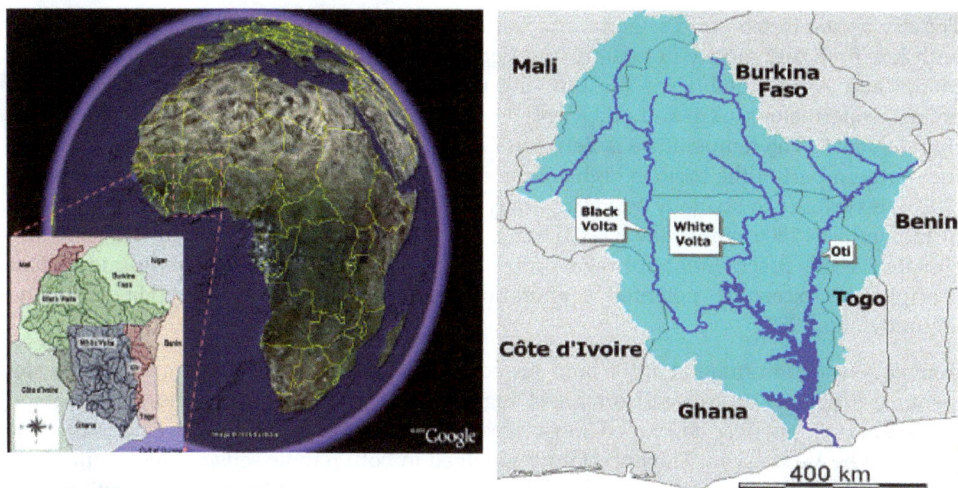

Fig. 3. Volta River Basin of West Africa (Source: GLOWA Volta project and Google map)

resources are used by those small communities for domestic and irrigation purposes; and since the communities depend on rainfed agriculture which have recently failed due to poor rains, irrigation has become the best alternative to food production. According to the World Bank report (1992), groundwater resources are relatively of good quality and usually need minimum treatment. Many communities, both rural and urban, within the basin depend largely on groundwater for their water needs. Data is scarce on groundwater level fluctuation and recharge, but in some areas a high recharge is observed. Runoff is essential for hydropower generation which is a major source of energy for the countries within the Basin. Reduction in flows has rendered the hydropower systems vulnerable and shows no sign of ending any time soon. Since the Akosombo hydroelectric dam was constructed in 1964, discharge has barely reached 1000 m3 s-1 and recent records show a further decline.

The semi-arid regions have variable rainfall pattern with extreme cases of drought and sporadic flood and has an annual average rainfall between 1,150 mm in the north and 1,380 mm in the south, According to Ajayi (2004), the semi-arid regions have one rainy season which begins from April and peaks up in July, August, and September and gradually end with some showers in October and part rainfall distribution in the northern part of the basin spatially variable with the intensity ranging between 2 mm/hr to 240 mm/hr and a median intensity of about 70 mm/hr. Rainfall duration is generally short with an average of about 30 to 50 minutes, but some events are longer especially the monsoon rainfall which is also common in the study area. In contrasting, other regions in the basin such as the humid forest have bimodal rainy seasons with a mean annual rainfall between 1500 mm and 2000 mm. Generally, a good amount of rainfall normally occurs and measured during the wet season which is influenced by the South-West Monsoons. The wet season is characterized by two main rainfall regimes with two distict modes, The first occuring from March to July with a peak in June, and the second from September to November with its peak in September/October.

The dry season in semi-arid region starts from November to April with dry hamattan winds with low humidity mostly in January and February and high sunshine in part of February, March and April suitable for the growing of horticultural crops like tomatoes, pepper, onions, watermelons, okro and other leafy vegetables. Most of the rains in the region fall as thunderstorm. The temperature is between 27°C and 36°C. Potential evapotranspiration in the northern part of the basin is high compared to the southern part. According to Amisigo (2005), it varies both spatially and temporally with an annual mean varying from 2500 mm in the north to 1800 mm in the coastal zone. In other words, evaporation in this semi-arid climate exceeds precipitation for 6-9 months, in the sharply contrasting, sub-humid climate precipitation exceeding potential evaporation in 6-9 months of a year (Hayward and Oguntoyinbo, 1987).

Remote sensing information of the land surface is important in solving and managing water resource problems in arid and semi-arid regions. It is important that regional water and energy balance models developed most especially in arid and semi-arid regions of West Africa in the Volta Basin. The uncertainty involved in both remote sensing information and regional water and energy balance models, it is crucial that data assimilation methods are applied to improve the accuracy of these management tools

A thorough literature review on models used on large watersheds that incorporated land-use changes, runoff and soil characteristics of watersheds and basins explains that hydrological models are inherently imperfect in many ways because they abstract and simplify "real" hydrological patterns and processes. This imperfection is partly due to the scale of the catchment against the backdrop that most hydrological modeling is based on regionalization of hydrologic variables, with constituent process and theories essentially derived at the laboratory or other small scales (Blöschel, 1996, Brown and Heuvelink 2005). The underpinned assumption of catchment homogeneity and uniformity and time invariance of various flow paths over watersheds and through soils and vegetation underscores the embedded processes such as channel hydraulics, soil physics and chemistry, groundwater flow, crop micrometeorology, plant physiology, boundary layer meteorology, etc.(Brown and Heuvelink, 2005). All the flaws of hydrological modeling notwithstanding, process-based distributed models have proven to simulate fairly well the spatial variability of the water balance among other processes when the main hydrological parameters and processes are known (Schellekens, 2000). Until recently, there has not been an alternative to hydrological simulation of watersheds that incorporate spatial scenarios such as land use changes. A typical hydrological model follows the protocols of figure 4

3. Regional climate modelling

Given the discernable evidences of climate change due to natural or/and human activity, there is a growing demand for the reliable climate change scenario in response to future carbon dioxide emission forcing climate viriables. One of the most significant impacts of climate change can be that on the hydrological process. Changes in the seasonality and the low and high rainfall extremes can influence the water balance of river basins, with several consequences for cities and ecosystems. In fact, recent studies have reported that West Africa is regarded to be a highly vulnerable region under global warming, especially for water resources; coupled with population rises in communities. Given the discernible evidences of climate change due to natural or/and human activity, there is a growing

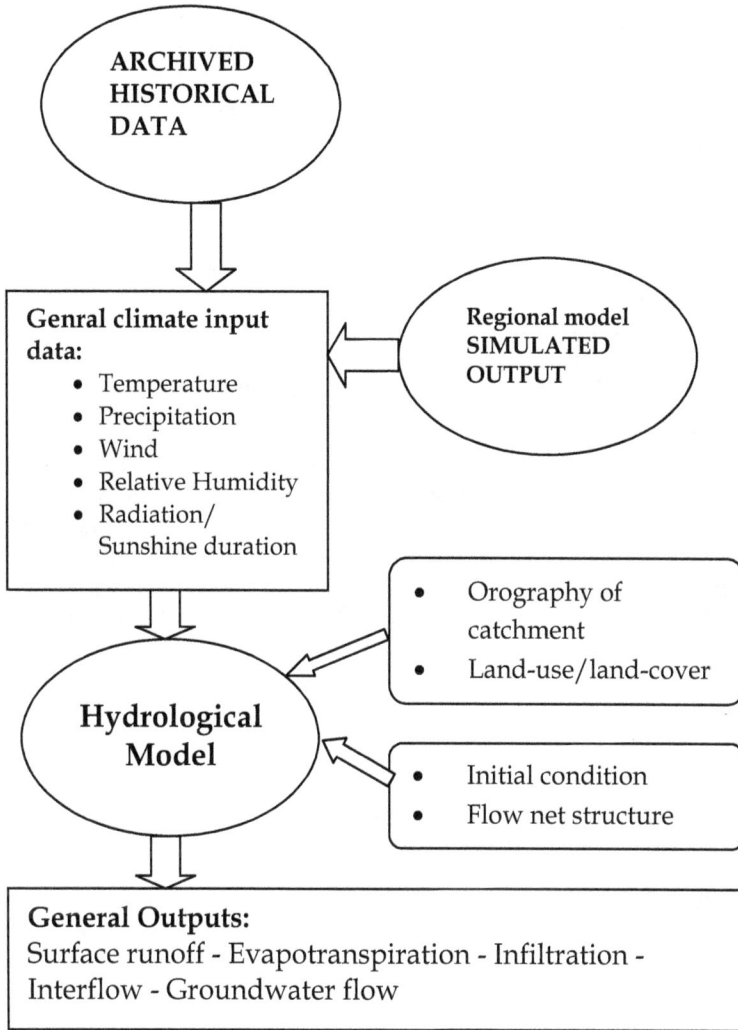

Fig. 4. Common protocol of hydrological modelling

demand for the reliable climate change scenario in response to future carbon dioxide emission forcing climate variables downscaling system using the Regional climate models with mosaic-type parameterization of subgrid-scale topography and land use are preferred.

Downscaling climate data is a strategy for generating locally relevant data from Global Circulation Models (GCMs). The overarching strategy is to connect global scale predictions and regional dynamics to generate regionally specific forecasts. Downscaling can be done in several ways. The two most prefered methods are: i) Nesting a regional climate model into an existing GCM is one way to downscale data. To do this, a specific location is defined and certain driving factors from the GCM are applied to the regional climate model. A regional climate model is a dynamic model, like a GCM, but it can be thought of as being composed of three layers. One layer is largely driven by the GCM, another layer builds on some locally specific data, and the third layer uses its own physics based equations to resolve the model based on data from the other two. The results are comparatively local predictions that are informed by both local specifics and global models. This process requires significant computational resources because it is dependent on the use of complex models. Currently Canada has just one Regional Climate Model (CRCM). ii) the use of statistical regressions which has a variety of such methods ranging from multiple regressions that link local variables to particular drivers in GCMs, to more complex methods using statistics designed for neural networks. The general strategy of these methods is to establish the relationship between large scale variables, such as the driving factors derived from GCMs, to local level climate conditions. Once these relationships have been developed for existing conditions, they can be used to predict what might happen under the different conditions indicated by GCMs.

Derived from the atmospheric global circulation models (AGCMs) and the atmosphere-ocean coupled GCMs (AOGCMs) are the HADCM, GFDL, CM2.X, ECHAM, among others, used for the study and simulation of the present climate and for the projections of the future climate. In the simulation of the hydrology of a watershed, credible input parameters of climate are essential for good results. Outputs of GCMs, however, have a spatial resolution of 250 km, offering only very coarse data for the study of small watersheds. According to Sintondji (2005) and Busche et al. (2005), GCMs have flaws for events of heavy rainfall in respect to their exceeding thresholds and frequencies. It is also evident that local or regional climates are influenced not only by the atmospheric processes, but are also greatly influenced by land-sea interaction, land use and the topography, which are poorly presented in GCMs due to their coarse spatial resolutions (Storch et al. 1993)

Regional Climate Models (RCMs) have been derived from the coarse GCMs to much higher resolutions. The process of downscaling of GCMs to meso-scales or regional scales enables the downscaled regional climate model to adequately simulate the physical processes consistently with GCMs on a large scale (Mearns et al. 2004). Since parameters of land use and topography are crucial in the efficiency of the RCMs, the higher the resolution of the RCMs, the better the simulation of the climate and ultimately, a better hydrological simulation is achieved. Some of the RCMs that have been popular in West Africa are REMO, MM5 and PRECIS.

4. Regional climate scenarios – MM5

The regional climate model MM5 is a meso-scale model derived from the GCM-ECHAM4 recently developed for the assessments of the impacts of environmental and climate change on water resources on the Volta Basin of West Africa. The MM5 is a brain child of the cooperation of the Pennsylvania State University (PSU) and the National Center for Atmospheric Research (NCAR) of the USA. According to Grell et al. (1995), the MM5 non-hydrostatic or hydrostatic (available only in version 2) is designed with the initial and lateral boundary conditions of a region to simulate or predict meso-scale and regional-scale atmospheric circulation.

The GLOWA-Volta project (GVP) executed by the Center for Development Research (ZEF), Germany, ran MM5 with the initial and lateral boundary conditions derived from the

ECHAM4 runs of the time slice 1860-2100, and based on IPCC's IS92a (assuming an annual increase in CO_2 of 1 %, and doubling of CO2 in 90 years (May and Rockner, 2001; cited in Jung, 2006). Using future climate scenario and girded monthly observational dataset from the East Anglia Climate Research Unit (CRU), UK, the model was calibrated to $0.5°$ x $0.5°$ resolution. GVP further down-scaled the MM5 model for the Volta Basin to finer resolutions of 9 km x9 km, and for some watersheds within the basin to 3 km x 3 km. Details of the setup, coupling and simulation are available in Kunstmann and Jung (2005) and Jung (2006).

A good agreement was reported between the ECHAM4-MM5 simulated climate and the CRU data sets for 1961-1990. According to Jung (2006), simulated temperature was slightly higher in the Sahara during the wet seasons and for the humid south during the dry season, while rainfall was generally comparable except for higher rainfall events that were underestimated. Between ECHAM4 and MM5, 1990-2000 simulations revealed that temperature was generally over estimated and rainfall under estimated by the latter even though the spatial representation was relatively good. However, the future simulations of the models were almost the same. Generally, MM5-Volta estimated an increase in rainfall in the Sahel zone (10-30 %), an increase mean annual rainfall of 45 mm (5 %) between 1990-2000 and 2030-2039, and a 1.2°C mean temperature rise (Jung, 2006).

GVP produced two 10-year simulated time slices of MM5 (1991-2000 and 2030-2039). For this section of the study, the 1991-2000 outputs were considered as the present climate and the time slice 2030-2039 as the future climate.

Changes in the hydrological cycle of a basin hinge largely on, among others, changes in climate. 10-year time slices were used to represent three windows of the past, present and future conditions. Climatic inputs for the past are data obtained from the meteorological agencies of Ghana and Burkina Faso through the GLOWA Volta project for the period 1961-1970. The present and future climate conditions are outputs of the MM5-Volta after Jung (2006) and Kunstmann and Jung (2005); these are 1991-2000 and 2030-2039, respectively. For these analyses, the Pwalugu watershed (Savannah) represents the north of the basin and the Bui watershed (transition zone) represents the south.

5. Highlights of MM5 on the Volta Basin Rainfall

Results of the GLOWA-Volta climate studies showed some significant changes, most especially in rainfall over the entire basin across the periods (Figure 5. In general, average monthly rainfall decreased in the period 1991-2000 for the whole of the rainy season (May-September) compared to the previous climate years.

The Frequency Distribution Curves (FDCs) of 1991-2000 and 2030-2039 do not differ significantly except for the extremely high and low rainfall events. However, both simulated time slices differ considerably from those of the 1961-1970 data records. This accounts in particular for the high percentiles in the high rainfall range. For example, the extreme daily precipitation amounts of two locations of the basin, Pwalugu in the north and Bui in the south will nearly double from the past to the future (80 mm-160 mm) and (52 mm-78 mm), respectively. The frequency of the daily mean of 10 mm of rainfall, for example, is expected to reduce from 17 % to 8 % in the north, and from 14 % to 7 % in the south.

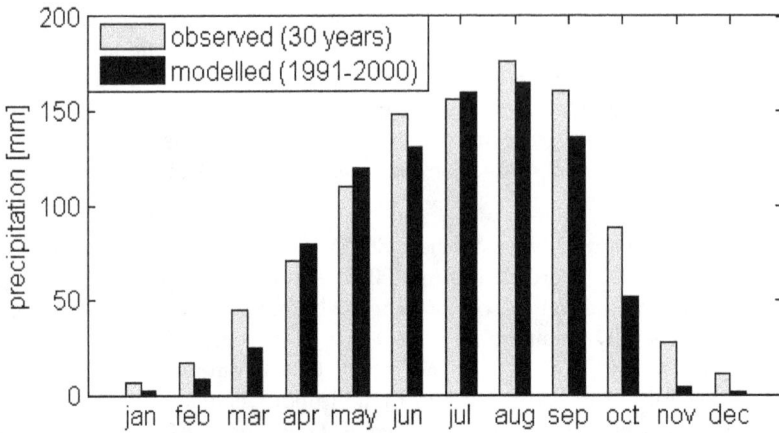

Fig. 5. Spatially averaged, simulated mean monthly precipitation (1991-2000) versus observed long-term means [mm] for the Volta Basin area (Jung, 2006)

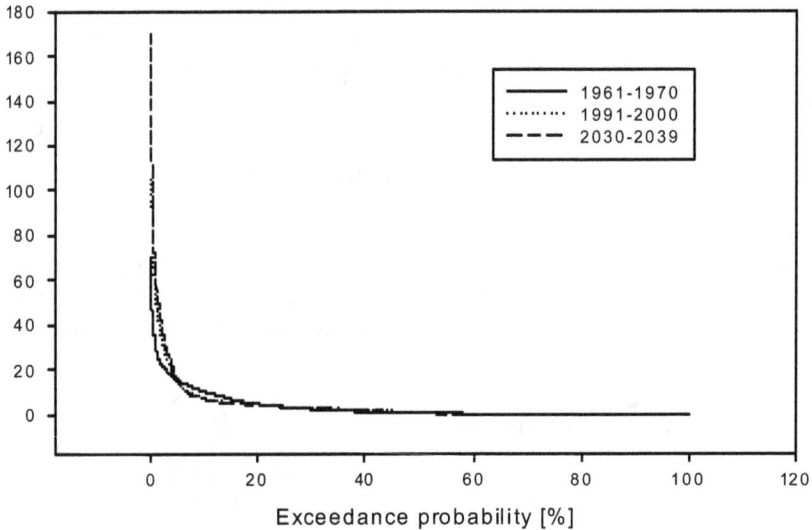

Fig. 6. Frequency distribution curves of rainfall for the north of the Volta Basin at the Pwalugu (56,760 km²) catchment for the time slices 1961-1970 (observed), 1991-2000 and 2030-2039 (both MM5 simulated)

6. Evapotranspiration

Actual evapotranspiration and temperature are closely related in the basin. As predicted by almost all climatic models and the IPCC, increases in the global and the basin mean temperature are eminent. Jung (2006) reported that monthly mean temperatures will increase in the Volta Basin by an average of 1.3°C by 2030-2039 compared to the present

1991-2000 data sets (Table 1) conforming to the assumptions of IPCC's IS92a of an annual increase in CO_2 of 1 %, and doubling of CO_2 in 90 years (May and Rockner, 2001).

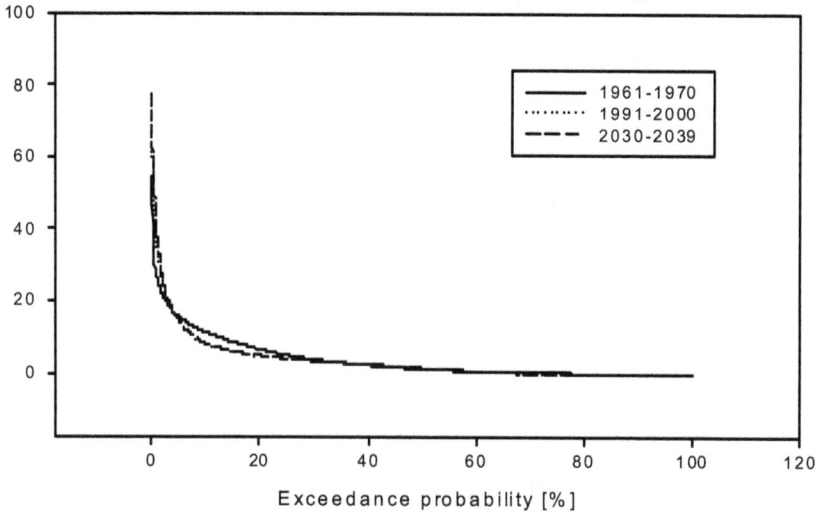

Fig. 7. Frequency distribution curves of rainfall for the south of the Volta Basin at the Bui (99,360 km²) catchment for the time slices 1961-1970 (observed), 1991-2000 and 2030-2039 (both MM5 simulated).

	Average(1991-2000)°C	Average(2030-2039)°C	Temperature Change °C
January	28.10	28.90	0.80
February	30.10	31.30	1.20
March	32.40	33.60	1.20
April	32.20	34.30	2.10
May	30.10	32.20	2.00
June	28.80	30.10	1.30
July	28.40	29.60	1.20
August	28.20	29.30	1.10
September	29.10	29.80	0.70
October	30.90	31.90	1.00
November	30.40	32.00	1.60
December	27.90	29.60	1.60

Table 1. Mean monthly and annual temperatures (1991-2000 and 2030-2039) for the Volta Basin (after Jung, 2006)

Many reports on the Volta Basin estimated evapotranspiration between 70 % - 90 % of total rainfall (e.g., Andreini et al. 2000, Martin, 2005). According to Jung (2006), significant increases are expected based on the use of the MM5-Volta model for the 2030-2039 periods for nearly all the months of the rainy seasons except May. Increases range from an average 64 mm to 79 mm per month (Figure 8).

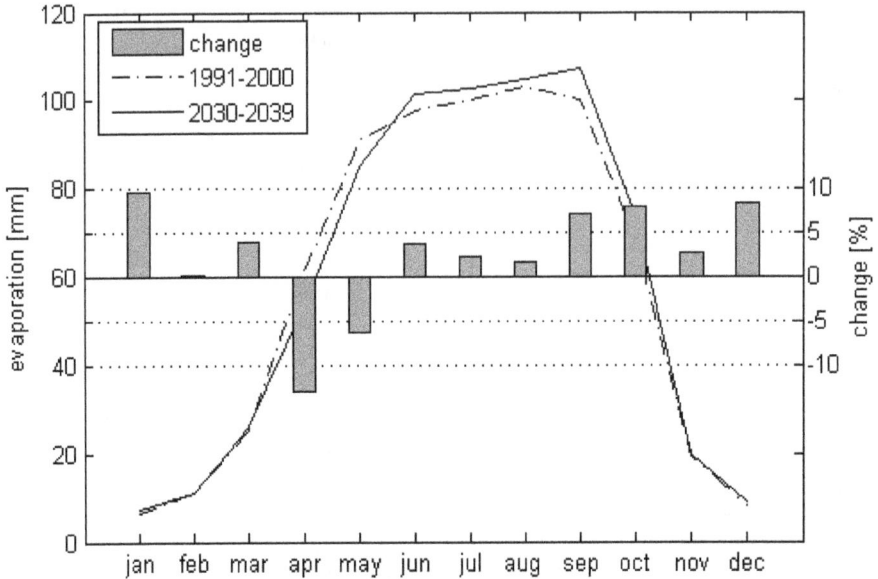

Fig. 8. Spatially averaged monthly mean real evapotranspiration [mm], (1991-2000 and 2030-2039) and evapotranspiration change [%] for the Volta Basin (Jung, 2006))

Calculating evapotranspiration using the Penman-Montheith method requires temperature, wind run, radiation and humidity data. An analysis carried out on the Basin based on archived data from the metrological agencies of Ghana and Burkina Faso, indicated that this was not adequately a representative of the entire basin. Nevertheless, the trend and amounts of increase in evapotranspiration of the time slices is phenomenal. The dryer north is expected to experience much higher increases in evaporation and transpiration e.g., the probability of 3 mm of water lost to evapotranspiration per day is increased from 0.07 in 1961-1970 to 0.21 in 2030-2039. The transition zone (south) is expected to lose less water compared to the north but with similar trends, while the frequency of higher evapotranspiration days will increase. The spatial distribution of evapotranspiration shows a general increase over nearly the whole basin, with only the south showing little or no increase in evapotranspiration for the period from the 1990s to the 2030s.

7. Runoff

Runoff in the Volta Basin is closely associated with the pattern of local precipitation, and changes in runoff frequency can reflect changes in climate, vegetation, or land use. An analysis of a stream in the Pwalugu catchment north of the basin showed an increase in the frequency of extreme high flows, but more profound are the expected increase in low flow events that will ultimately have severe impacts on the water resources of the basin in terms of drought frequencies. Jung (2006) attributed the higher values in the future climate scenario run (Figure 11) to the increase in direct runoff especially in the rainfall-intense month of September. Discharge in the south of the basin shows a similar pattern (Figure 13), but differences are more pronounced between the past and future time slices. While the

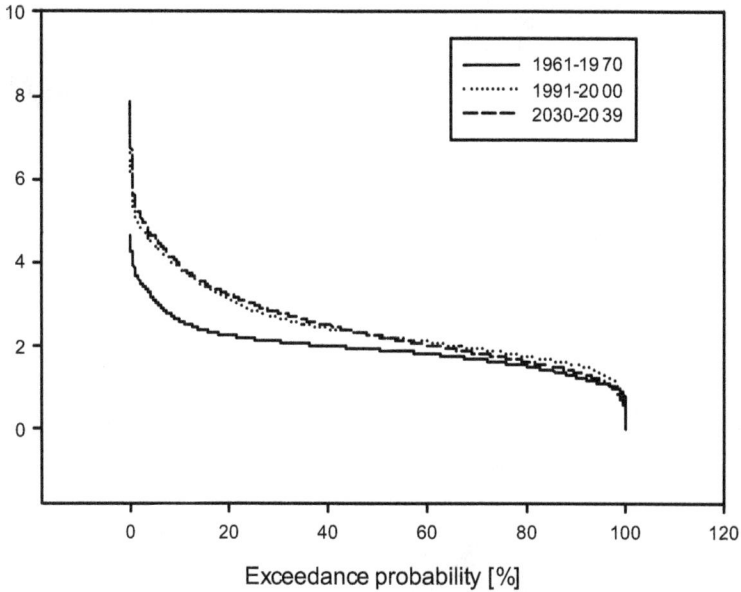

Fig. 9. Frequency distribution curves of evapotranspiration for the north of the Volta Basin at the Pwalugu (56,760 km²) catchment for the time slices 1961-1970 (gauged), 1991-2000 and 2030-2039 (both MM5 simulations).

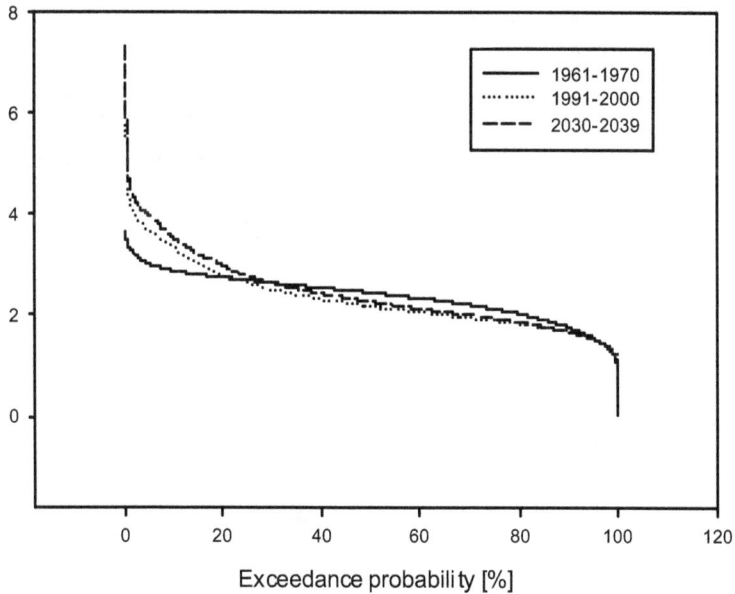

Fig. 10. Frequency distribution curves of evapotranspiration for the south of the Volta Basin at the Bui (99,360 km²) catchment for the time slices 1961-1970 (gauged), 1991-2000 and 2030-2039 (both MM5 simulations)

probability of daily discharge exceeding 15 mm increases in the future, there is also an equally significant increase in low flow events. For example, the probability of daily discharge exceeding 1 mm in south for the past time slice was 0.4, but is expected to rise sharply to 0.8 for the future time slice of 2030-2039.

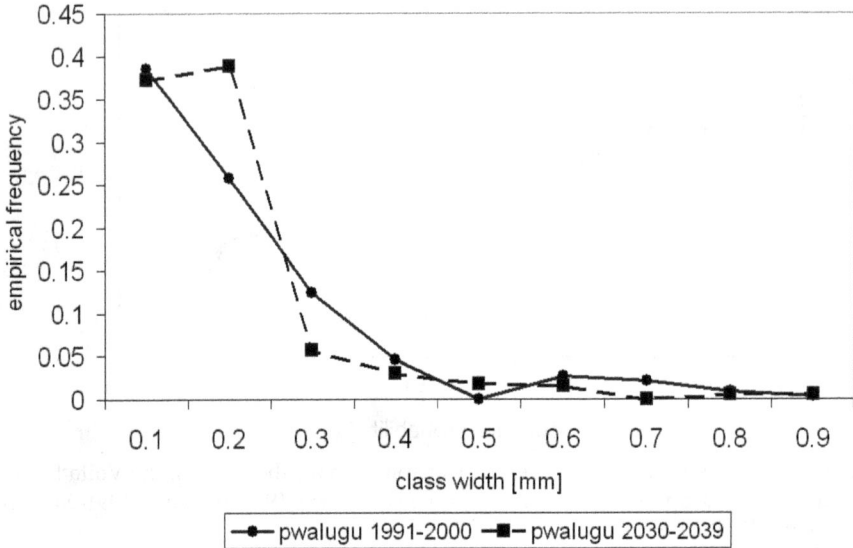

Fig. 11. Normalized frequency distribution of daily runoff values [mm] (1991-2000 and (2030-2039), Pwalugu, Volta Basin (Jung, 2006)

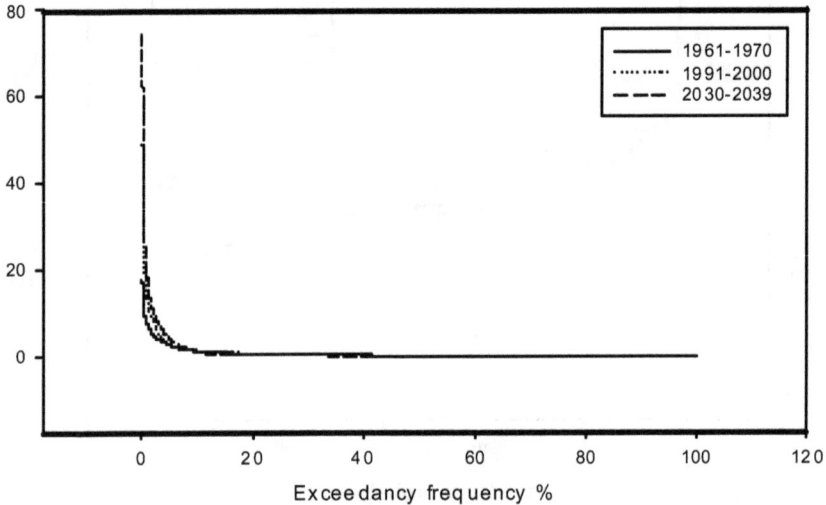

Fig. 12. Frequency distribution curves of discharge for the north of the Volta Basin at the Pwalugu (56,760 km²) catchment for the time slices 1961-1970 (observed), 1991-2000 and 2030-2039 (both MM5 simulations).

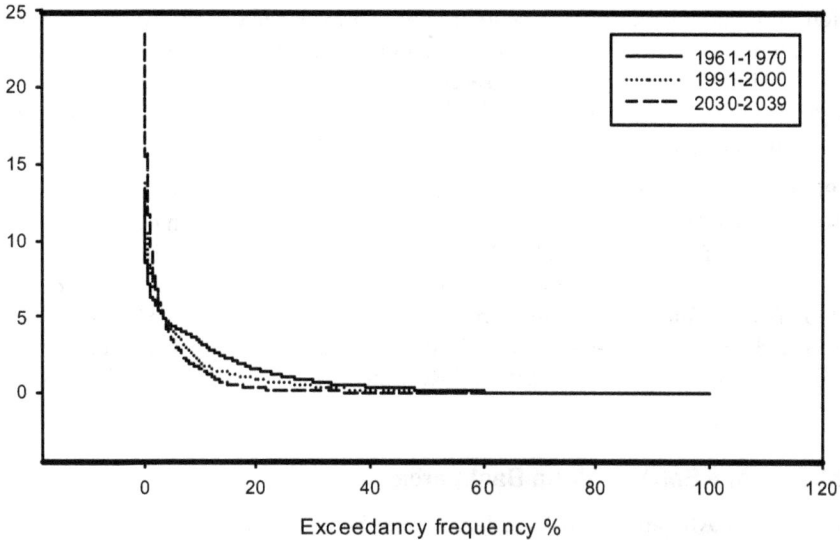

Fig. 13. Frequency distribution curves of discharge for the south of the Volta Basin at the Bui (99,360 km²) catchment for the time slices 1961-1970 (observed), 1991-2000 and 2030-2039 (both MM5 simulated).

8. Regional climate scenarios – REMO

REMO is a hydrostatic regional climate model, initially developed at the Max-Planck-Institute for Meteorology (MPI) in Hamburg, Germany, on the foundation of the operational weather forecast model Europa-Modell of the German Weather Service (DWD) (Majewski 1991). According to Jacob et al. (2001) cited in Paeth (2005), the dynamical kernel is based on primitive equations with temperature, horizontal wind components, surface pressure, water vapor content and cloud water content as prognostic variables.

REMO simulations are driven according to Roeckner et al. (2003) by recent global coupled climate model simulations of ECHAM5/MPI-OM, which are known to be forced by enhanced greenhouse and sulphate aerosol conditions and are synonymous with the modeling approaches of the fourth assessment report of the Intergovernmental Panel on Climate Change (IPCC).

The simulation outputs used for this study are those produced and used in the GLOWA IMPETUS project whose focus was on western and northern Africa. The horizontal resolution is 0.5º, equivalent to about 55-km grid spacing at the equator, and 20 hybrid vertical levels are resolved. These levels follow the orography near the surface and correspond to pressure levels in the upper troposphere (Paeth, 2005).

The global climate model ECHAM4 (Roeckner et al. 1996) was adjusted to the 0.5º model grid scale of REMO to account for atmospheric processes like deep convection, cloud formation, convective rainfall, radiation and microphysics from the sub grid scale of

ECHAM4. Similarly, land surface parameters such as soil characteristics, orography, vegetation, roughness length, and albedo are derived from the GTOPO30 and NOAA data sets (Hagemann et al. 1999) and partly modified according to the scenarios of land degradation. Underlying an idealized seasonal cycle over West Africa, the same daily interpolated surface parameters are prescribed each year using a model output statistics (MOS) system (Paeth, 2005).

The IMPETUS project made some changes to the default parameter of REMO to adapt to the tropical-subtropical West African region. The focus is on the key region of the West African monsoon system of which the Volta Basin is part. Some results of the adopted REMO correlated well with observed extreme climate year For example, the driest years derived from simulated rainfall in the basin were 1981, 1983, 1990, 1992 and 1998, whereas 1979, 1984, 1988, 1989, and 1991 were characterized by abundant monsoon rainfall. Parker and Alexander (2002) basically confirm that the CRU time series data set reveals almost the same composite years.

9. Highlights of REMO on Volta Basin area

Data on the West African region have large gaps, hence the CRU data set is deficient in regions with low data coverage. This problem applies to all available gridded rainfall data sets. REMO, in an attempt to resolve the handicap of data gaps, uses two statistical post processing steps:

1. Monthly rainfall is adjusted to the observations by constructing a model output statistics (MOS) system.
2. Daily rainfall intensity is corrected from grids by fitting Γ distributions to the simulated and observed time series, and then taking the ratios between both distributions as weighting factors in the combined correction algorithm. The MOS system is a stepwise multiple linear regression analysis.

The MOS-corrected precipitation by the REMO model reveals a good performance according to Paeth et al. (2005) of the model in terms of the basic features of African climate, including the complex mid-tropospheric jet and wave dynamics and the climate seasonality of the Volta Basin area.

Available scenario runs of REMO (Jacob et al. 2001, 2007) in 0.5º resolution over tropical Africa (Paeth et al. 2005) consider three scenarios of GHG (Figure 14) and emissions and land-cover changes in order to evaluate the range of options given by different achievements in mitigation policy and to quantify the relative contribution of land degradation in line with IPCC projections.

The 2007 report of the IPCC lacks new information on the African climate change but highlights model uncertainties, particularly over tropical Africa. The inconsistency of different model projections reflects in the low values of the regional climate change index in Giorgi (2006), which relies on regional temperature and precipitation changes in the IPCC multi-model ensemble framework. This model discrepancy clouds the interpretation of the results of uncertain model parameters, which may impact specifically on the simulation of African climate.

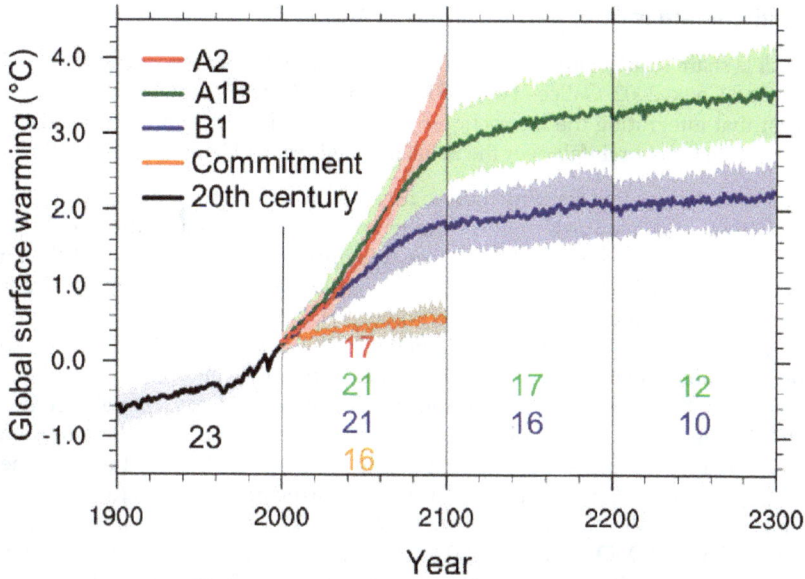

Fig. 14. Multi-model means of surface warming for the scenarios A2, A1B and B1, shown as continuations of the 20th century simulation. Values beyond 2100 are for the stabilization scenarios. Linear trends from the corresponding control runs have been removed from these time series. Lines show the multi model means, shading denotes the plus minus one standard deviation range. (Source: http://ipcc-wg1.ucar.edu/wg1/wg1-report.html)

Within the adopted REMO, the spatial mean of the change of forest to agricultural land under the A2, B1 and A1B scenarios have been considered. For example, under the A1B (all)[1] scenario the estimate of the FAO (2006), assumes a decrease in forest coverage of about 30 % until 2050 for the entire Africa region. Associated albedo changes between 2000 and 2050 are in the order of 5-10 %. Forests transformation into grasslands and agricultural areas in the order of 10-15 % were incorporated. The B1 storyline and scenario family describes a convergent world with the same global population, that peaks in mid-century and declines thereafter, as in the A1 storyline, but with rapid change in economic structures toward a service and information economy, with reductions in material intensity and the introduction of clean and resource-efficient technologies. The emphasis is on global solutions to economic, social and environmental sustainability, including improved equity, but without additional climate initiatives, whereas the A2 storyline and scenario family describes a very heterogeneous world. The underlying theme is self-reliance and preservation of local identities. Fertility patterns across regions converge very slowly, which results in continuously increasing population. Economic development is primarily regionally oriented and per capita economic growth and technological change more fragmented and slower than other storylines. The available REMO outputs from the IMPETUS project have three ensemble runs for the various scenarios and time slices. These are for the periods 1961-2000; the A1B scenarios with land-use changes are for 2001-2050 time periods; and for the B1 scenarios are for 2001-2050 periods.

---

[1] All ensemble scenarios of IPCC considered

10. Regional climate model performance of MM5 and REMO

Regional models share similar problems but differ in magnitude. Notable are MM5, MAR and REMO (Vizy and Cook, 2002; Gallée et al., (2004); Paeth et al., 2005). However, Schnitzler et al. (2001) suggest that integrating the interaction with vegetation cover and albedo considerably improves the simulation of rainfall over the Sahel in the global ECHAM4 model.

For the hydrological modeling for which these regional climate models are required, rainfall is a very important input parameter apart from temperature because it is the major driver of moisture input to the hydrological cycle. Hence it is important to ascertain that the annual, monthly and daily distributions of the data represent the amounts and the frequency statistics of the data, e.g., the exceedance of extremes are consistent with the long-term mean observed. It is important to assress the reliability of the MM5 and REMO future climate scenario for the evaluation of the impacts of climate change on water resources in the Volta Basin, for instance, the REMO-simulated and MM5-simulated mean rainfall for the time slice 1991-1997 were obtained from the basin weather stations and compared to the mean observed rainfall for the same area and periods (Figure 15) From the outcome, the comparison for the station, for example, show a good correlation between the observed and MM5-simulated and REMO-simulated monthly rainfall (Figure 15). Pearson correlation of gauged 1991-1997 and REMO 1991-1997 = 0.823; P-Value = 0.001; for gauged 1991-1997 and MM5 1991-1997 = 0.957; P-Value < 0.0001 On average, MM5 and REMO overestimate rainfall for this selected time slice of 1,203 mm and 1,322 mm per annum, respectively, against the measured 1,101 mm per annum. MM5 overestimates the rainfall from April through July and September and underestimates for August and October. REMO, on the other hand, overestimates rainfall for February through April, July, October and November and underestimates for August. The strongest overestimation for MM5 is for the month of July, while for REMO, this is March and April (Figure 16).

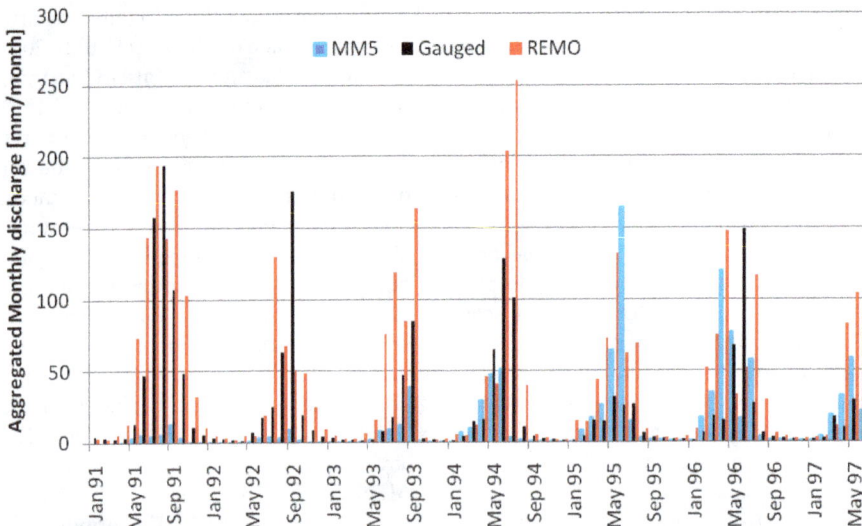

Fig. 15. MM5-simulated and REMO-simulated compared to observed monthly rainfall at Pwalugu (56,760 km²) catchment of the Volta Basin

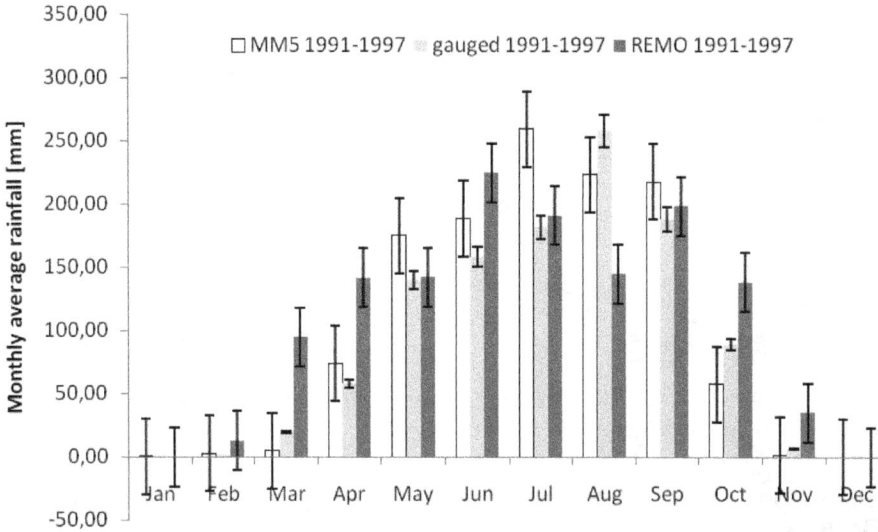

Fig. 16. MM5-simulated (mean over 7 years) and REMO-simulated (mean over 7 years) compared to observed (mean over 7 years) rainfall including standard deviation (error bars) rainfall at Pwalugu (56,760 km²) catchment of the Volta Basin

The resultant impacts on the hydrology are similar in most of the rainy seasons, apart from isolated extreme runoffs generated as a result of overestimations of some days and months of the season (Figures 17 and 18).

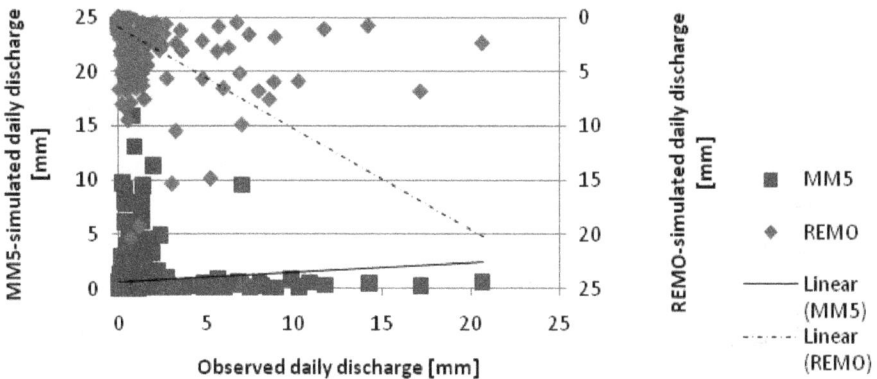

Fig. 17. MM5-simulated (over 4 years) and REMO-simulated (over 4 years) compared to observed (over 4 years) discharge with trend lines at Pwalugu (56,760 km²) catchment of the Volta Basin

Statistically, the Pearson correlation of MM5 and gauged = 0.181 with a P-Value = 0.108; Pearson correlation of REMO and gauged = 0.677 with a P-Value < 0.0001. The general trend for both MM5 and REMO for the period 1991-1997 is similar in pattern, and they simulate dry and wet years fairly well, with REMO better for hydrological simulations. Available daily gauged runoff exist data for 1994-1997 (Figures 17 and 18).

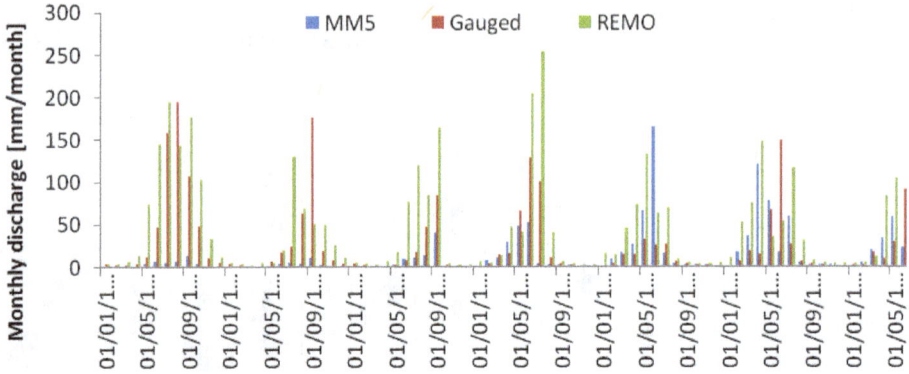

Fig. 18. MM5-simulated and REMO-simulated compared to aggregated observed discharge at Pwalugu (56,760 km²) catchment of the Volta Basin

11. Comparison of past, present and future hydrological dynamics of the Volta Basin

The 2001-2050 time slices representing the future period with increasing GHGs and changing land cover, according to the business-as-usual and a mitigation scenario from the IPCC SRES A1B and B1 scenarios (Nakicenovic and Swart 2000) are considered by REMO. The relative contribution of the land-use changes to total climate change in Africa was carried out using the A1B and B1 emission scenarios between 2001 and 2050. An annual growth rate of 2 %, land-use changes comprising of mainly prevailing cities and currently

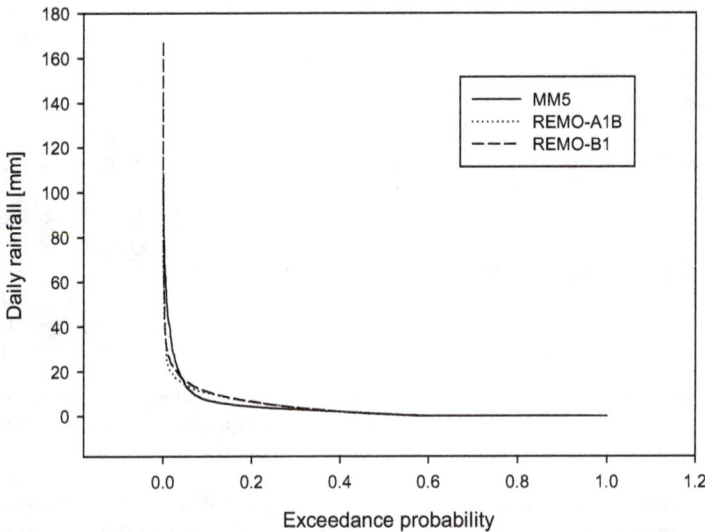

Fig. 19. Exceedance probability of daily rainfall simulated by MM5, REMO-A1B and REMO-B1 with average total rainfall of 1,287 mm, 1291 mm, and 1,391 mm, respectively, for the Pwalugu catchment of the Volta Basin for 2030-2039.

existing agricultural areas were some assumptions adopted (Paeth et al., 2005). MM5, on the other hand, did not consider land-use change scenarios and was based on IPCC's IS92a scenarios (May and Rockner, 2001), which are known to overestimate the CO_2 projections compared to the A1B and B1 emission scenarios.

A simulation by MM5 and REMO-B1 predicts high rainfall events compared to the scenarios of REMO-A1B. The resultant discharge in these scenario simulations shows a very high discharge for MM5 with a sharp gradient and larger amplitudes, closely followed by REMO-B1, and REMO-A1B. This is explained by the number of very low and rainy day event counts that are higher with MM5 than those of both scenarios of REMO (Figure 19).

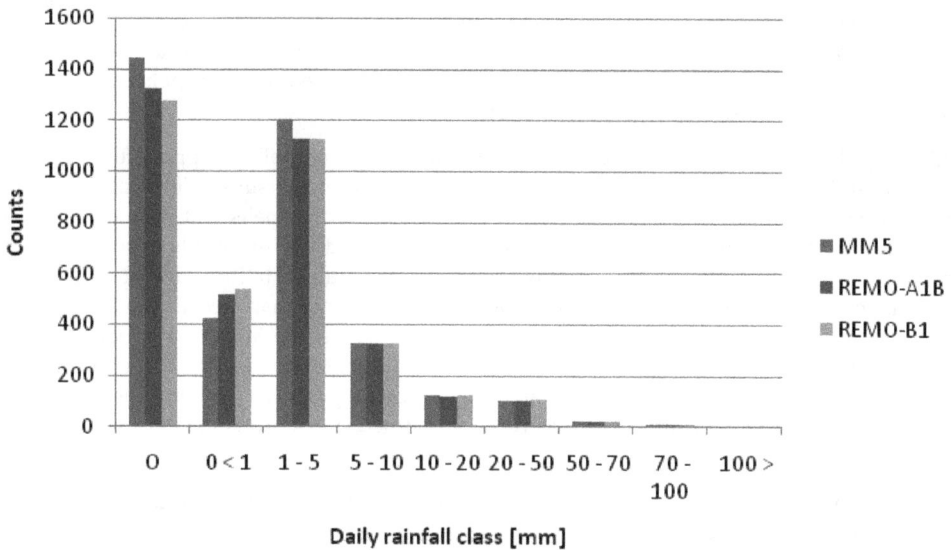

Daily rainfall class [mm]

Fig. 20. Rainfall distribution simulated by MM5, REMO-A1B and REMO-B1 with average total rainfall of 1,287 mm, 1291 mm, and 1,391 mm, respectively, for the Pwalugu catchment of the Volta Basin for 2030-2039.

12. Climate change projections and Policy implications

The regional climate models MM5 and REMO have demonstrated that they are able to simulate the observed main characteristics of African climate to some extent, with varied accuracy. For the resultant hydrological simulation with respect to discharge, the regional climate model REMO provides reliable and consistent high-resolution data for hydrological application of WaSiM-ETH for the Volta Basin Nevertheless, both climate models have divergent projections for the future climate and hence the hydrology of the basin.

There is evidence from models that the world's climate is changing and this change will impact regional development in various ways. According to climate scientists, increases in global temperatures are largely the result of increased greenhouse gas concentrations and that the continued increases in these concentrations will cause future climate warming.

Climate variability and change present risks relating to vulnerabilities that potentially undermine global efforts at poverty reduction and meeting the Millennium Development Goals (MDGs) of most countries. Semi-arid regions are vulnerable especially in Africa to this projected climate variability and change and the related impacts. These include projected temperature increases, changes in precipitation, projected change in cropping systems and food security etc.

Given the history of climatic variability, climate specialists predict a mix of droughts and floods of unusual magnitudes for West Africa that will threaten human security (IUCN, 2004; IPCC, 2007). Some climate change projections predict a decline in precipitation in the range of 0.5- 40% with an average of 10-20% by 2025. Others predict increases in precipitation. This will cause extreme climatic events such as floods and droughts - impacting on agriculture negatively (Houghton et al., 2001; Smith and Mendelsohn, 2006).in a similar case, a greater drought risks due to higher evapotranspiration and decrease in precipitation. In this case, it is high time every region will need to adapt policy implication to climate change by examining the nature of regional impact

Although there has been increased recognition of and global policy support for adaption, there is still the lack of policy commitment to adaptation at the sub-regional and national levels especially in developing countries. The incidence of dry spells and drought over the past two decades in the semi-arid region and the projected drying trends expose the sub-region to more drought related hazards and livelihood vulnerabilities among the human population. If the exposures to drying related hazards are not managed through concerted policy support, this potentially undermines the attainment of the Millennium Development Goals (MDGs) in the sub region. With increase in population; increasing demand for food and water use, coupled with poor water management practices and increasing risk of climate change, resultant impacts could reach undefined proportions for inhabitants of the semi arid regions.

Rainfall variability, land degradation and desertification are some of the factors that combine to make life extremely difficult in this part of the world.

Simi-arid regions which in recent times due to climate variability and change experienced massive losses of agricultural production and livestock; loss of human lives to hunger, malnutrition and diseases; massive displacements of people and shattered economies. Yet, most climate models predict that these regions will be drier in the 21st century. Even slight increases in rainfall are unlikely to reverse the situation since a hotter climate means that evapotranspiration will be more intense, exacerbating the already arid conditions.

13. References

Ajayi, A. E., (2004) Surface runoff and infiltration processes in the Volta Basin, West Africa: Observation and modeling, Ecology and Development Series No. 18.

Amisigo, B. A., (2005) Modelling riverflow in the Volta Basin of West Africa: A data-driven framework, PhD Thesis Ecology and Development Series No. 34.

Andreini M, van de Giesen N, van Edig A, Fosu M, and Andah W (2000) Volta Basin Water Balance. ZEF – Discussion Papers on Development Policy, No. 21, ZEF, Bonn

Blöschl,G. (1996): Scale and scaling in hydrology. Dissertation, TU Wien, 1-22.

Brown JD and Heuvelink GBM (2005) Assessing Uncertainty Propagation Through Physically based Models of Soil Water Flow and Solute Transport. Anderson, M.G. & J.J. McDonnell (Eds) Encyclopaedia of Hydrological Sciences. Wiley Blackwell. 3456 p

Busche H, Hiepe C, Diekkrüger B (2005) Modelling the effects of land use and climate change on hydrology and soil erosion in a sub-humid African Catchment. Proceedings of the 3rd International SWAT Conference 2005, pp434-443Carter TR and La Rovere EL (2001) Developing and applying scenarios. In: McCarthy, J.J., Canziani, O.F., Leary, N.A., Dokken,

FAO (2006) Food and Africulture Organization, Global forest resources assessment 2005. Progress towards sustainable forest management. FAO Forestry Paper, 147, 348pp

Gallée H Moufouma-Okia W Bechtold P Brasseur O Dupays I Marbaix P Messager C Ramel R and Lebel T (2004) A high-resolution simulation of a West African rainy season using a regional climate mode. J. Geophys. Res. 109, 10.1029/2003JD004020

Giorgi F (2006) Climate change hot-spots. Geophys. Res. Let., 33, doi:10.1029/2006GL025734

Green Cross International (2001) Trans-boundary Basin Sub-Projects: The Volta River Basin. Website: <www.gci.ch/Green- crossPrograms/waterres/pdf/WFP_Volta>.

Grell G A, Dudhia J, and Stauffer D R (1995) A description of the Fifth-Generation Penn State /NCAR Mesoscale Model (MM5). Tech. rept.

Hagemann S, Botzet M, Dümenil L and Machenhauer B (1999) Derivation of global GCM boundary conditions from 1 km land use satellite data. Max-Planck-Inst. f. Meteor., Report No. 289, Hamburg

Hayward D and Oguntoyinbo J (1987) Climatology of West Africa. Hutchindson.

Jacob D Van den Hurk BJJM Andrae U Elgered G Fortelius C Graham LP Jackson SD Karstens U Koepken C Lindau R Podzun R Rockel B Rubel F Sass BH Smith R and Yang X (2001) A comprehensive model intercomparison study investigating the water budget during the PIDCAP period. Meteorol. Atmos. Phys. 77, 19-44

Jung G (2006) Regional Climate Change and the Impact on Hydrology in the Volta Basin of West Africa. Dissertation Garmisch-Partenkirchen, German.

Kasei R (2009) Modelling Impact of Climate on Water Resources in the Volta Basin, West Africa. Series No. 69(Cuvillier Verlag Göttingen.)

Kunstmann H and Jung G (2005) Impact of regional climate change on water availability in the Volta basin of West Africa. In: Regional Hydrological Impacts of Climatic Variability and Change. IAHS Publ. 295

Majewski D (1991) The Europa-Modell of the Deutscher Wetterdienst. Seminar Proceedings ECMWF 2, 147-191

Martin N (2005) Development of a water balance for the Atankwidi catchment, West Africa – A case study of groundwater recharge in a semi-arid climate. PhD

May W and Röckner E (2001) A time-slice experiment with the ECHAM4 AGCM at high resolution: the impact of horizontal resolution on annual mean climate change; Climate Dynamics; Vol. 17: pp. 407-420

Mearns LO, Giorgi F, Whetton P, Pabon D, Hulme M, and Lal M (2004) Guidelines for use of climate scenarios developed from regional climate model experiments, Tech. rep., Data Distribution Centre of the IPCC

Nakicenovic N and Swart R (Eds.) (2000) Emission Scenarios. 2000. Special Report of the
    Intergovernmental Panel on Climate Change. Cambridge University Press, U.K.,
    570pp
Nyarko, B. K., (2007) Floodplain wetland-river flow synergy in the White Volta River basin,
    Ghana, Ecology and Development Series No. 53.
Paeth H Born K Podzun R and Jacob D (2005) Regional dynamic downscaling over West
    Africa: Model evaluation and comparison of wet and dry years. Meteorol. Z., 14,
    349-367
Parker DE and Alexander LV (2002) Global and regional climate in 2001. Weather 57,328-340
Roeckner E Arpe K Bengtsson L Christoph M Claussen M Dümenil L Esch M Giorgetta M
    Schlese U and Schulzweida U (1996) The atmospheric general circulation model
    ECHAM-4: Model description and simulation of present-day climate. Max-Planck-
    Inst. f. Meteor., Report No. 218. Hamburg
Roeckner E et al. (2003) The atmospheric general circulation model ECHAM 5. PARTI:
    Model description. MPI Report, 349, 127pp
Schellekens J (2000) Hydrological processes in a humid tropical rainforest: a combined
    experimental and modelling approach. Dissertation, Vrije Universiteit Amsterdam,
    7-9
Sintondji L (2005) Modelling of process rainfall-runoff in the Upper Quémé catchmentarea
    (Terou) in a context of climate change: extrapolation from the local scale to a
    regional scale. PhD Thesis. Shaker Verlag, Aachen, Germany
Storch H and Stehr N (2006) Anthropogenic climate change - a reason for concern since the
    18th century and earlier. Geogr. Ann., 88 A (2): 107–113
Vizy EK and Cook KH (2002) Development and application of a meso-scale climate model
    for the tropics: Influence of sea surface temperature anomalies on the West African
    monsoon. J.Geophys. Res. 107, 10.1029/2001JD000686

4

Fuzzy Nonlinear Function Approximation (FNLLA) Model for River Flow Forecasting

P.C. Nayak[1], K.P. Sudheer[2] and S.K. Jain[3]

[1]*Deltaic Regional Centre, National Institute of Hydrology, Kakinada*
[2]*Dept of Civil Engineering, Indian Institute of Technology Madras,*
[3]*NEEPCO, Department of Water Resources Development and Management,*
Indian Institute of Technology, Roorkee,
India

1. Introduction

It is well understood that the limitations of hydrological measurement techniques warrants for modeling of hydrological processes in a basin. However, most hydrologic systems are extremely complex and modeling them with the available limited measurements is a difficult task. The basic purpose of a model is to simulate and predict the operation of the system that is unduly complex, and also to predict the effect of changes on this operation. It is well known, of various hydrological processes, the rainfall-runoff process is the most complex hydrologic phenomenon to comprehend due to tremendous spatial and temporal variability of basin characteristics and rainfall patterns, as well as a number of other variables associated in modeling the physical processes (Tokar and Markus, 2000). The transformation from rainfall to basin runoff involves many hydrologic components that are believed to be highly nonlinear, time varying, spatially distributed, and not easily described by simple models. The artificial neural network (ANN) and Fuzzy Inference System (FIS) approaches are becoming increasingly popular in the context rainfall-runoff modeling due to their various advantages. This Chapter discusses an effective integration of these two models in a different manner.

It is a common belief that the ANN (and FIS models to an extent) models of the rainfall runoff process are purely black models as they do not explain the process being modeled, but for a few recent studies (Wilby et al., 2003; Jain et al., 2004; Sudheer, 2005). However, it must be realized that the hydro-meteorological data that are employed in developing rainfall runoff models (ANN, FIS or conceptual) contain important information about the physical process being modeled, and this information gets embedded or captured inside the model. For instance, a flow hydrograph, which is normally used as the output variable in an ANN rainfall runoff model, consists of various components that result from different physical processes in a watershed.

For example, the rising limb of a runoff hydrograph is the result of the gradual release of water from various storage elements of a watershed due to gradual repletion of the storage due to the rainfall input. The rising limb of the hydrograph is influenced by varying infiltration capacities, watershed storage characteristics, and the nature of the input *i.e.* intensity and duration of the rainfall, and not so much by the climatic factors such as temperature and evapotranspiration etc. (Zhang and Govindaraju, 2000). On the other hand, the falling limb of a watershed is the result of the gradual release of water from various storages of the watershed after the rainfall input has stopped, and is influenced more by the storage characteristics of the watershed and climatic characteristics to some extent. Further, the falling limb of a flow hydrograph can be divided into three parts: initial portion just after the peak, middle portion, and the final portion. The initial portion of the falling limb of a flow hydrograph is influenced more by the quick-flow (or interflow), the middle portion of the falling limb is more dominated by the delayed surface flow, and the final portion of the falling limb (of smaller magnitudes) is dominated by the base flow. Hence it is apparent that a local approximation technique, which maps the changing dynamics using different functions, would be an effective way to model the rainfall runoff process. Local approximation refers to the concept of breaking up the domain into several small neighbourhood regions and analysing these separately. This argument is supported by the results of Sudheer (2005), wherein he proposes a procedure to extract knowledge from trained ANN river flow models.

Farmer and Sidorowitch (1987) have found that chaotic time series prediction using local approximation techniques is several orders of magnitude better. In their approach, the time series is first embedded in a state space using delay coordinates, and the underlying nonlinear mapping is inferred by a local approximation using only nearby states. This approach can be easily extended to higher order local polynomial approximations. Singer et al. (1992) derived the local approximation as a state dependant autoregressive modelling. However, this becomes complex with large data sets as the inefficient computation of nearby state search makes the implementation much harder. In order to overcome this limitation we proposed to simplify the signal representation (input) with a vector clustering procedure. That is, the local model fitting is based on statistically averaged prototypes instead of the original state vector samples. Also, the nearby state search can be significantly simplified with all prototypes organized according to a certain metric such as pattern similarity.

Most of the applications of local approximation technique (e.g. FIS) employ linear relationship as an effective local approximation. However, the rainfall runoff process being highly nonlinear, a nonlinear local approximation would be a better approach. The objective of this Chapter is to illustrate that a nonlinear local approximation approach in modelling the rainfall-runoff process may offer better accuracy in the context of river flow forecasting. More specifically, the chapter discusses about subdividing the data into subspaces and evaluates the limitations of linear and nonlinear local approximations. The proposed approach is illustrated through a real world case study on two river basins. Both the applications are developed for river flow forecasting, one on a daily time step and the other on an hourly time step.

2. Theoretical considerations of local approximation

Consider modeling a river flow time series such that $y_t = f(x_t, \lambda)$, where it is required to forecast the value of flow (y_t) and x_t is the input vector to the model at time t. Generally, the modeler uses a set of n 'candidate' examples of the form (x_i, y_i), $i=1,2,...,n$, and finds an optimal set of parameter vector (λ) by calibrating an appropriate model. The inputs to the model typically are the previous values of the time series and the output will be the forecast value. The model is normally trained and tested on training and testing sets extracted from the historical time series. In addition to previous time series values, one can utilize the values or forecasts of other time series (or external variables) that have a correlation or causal relationship with the series to be forecasted as inputs. For a river flow forecasting problem, such exogenous time series could be the rainfall or evaporation over the basin. Each additional input unit in a model adds another dimension to the space in which the data cases reside, thereby making the function to be mapped more complex. The model attempts to fit a response surface to these data.

For many applications in data driven modeling (regression, classification etc.) an estimate of expected response (output) is desired at or close to one fixed predictor (input) vector. This estimate should depend heavily on predictor vectors in the sample which are close to the given fixed predictor vector. The predictive relationship between the current state x_t and the next value of the time series can then be expressed as:

$$y(t+T) = f_T^0(x(t)) \qquad (1)$$

where T is the prediction time horizon. The problem of one step ahead (T=1) predictive modeling is to find the mapping $f_T^0: R^N$ to R^1. In local approximation, a local predictor is constructed based on the nearby neighbors of x_t, that is, fitting a polynomial to the pairs $(x(t_i), y(t_i+T))$ with $x(t_i)$ being the nearest neighbors of $x(t)$ for $t_i<t$. The original signal can also be viewed as an evolution of the state $x(t)$ of a dynamical system in R^N. The signal history (input samples) compose the map from state space of dimension N to a scalar space, the parameters of the mapping can be estimated by interpolating $f(x)$ from noisy signal samples. The local modeling is superior and simpler under the condition that the given dynamics is locally smooth and a long enough signal history is available (Singer et al., 1992). Under this condition, $f(x)$ can be approximated by the first few terms of multi dimensional Taylor series expansion, resulting in,

$$f(x) = a^T x + b \qquad (2)$$

Based on the fuzzy modeling approach originally developed by Takagi and Sugeno (1985), the global operation of a nonlinear process is divided into several local operating regions. Within each local region, R_i, a reduced order linear model is used to represent the processes behavior. Fuzzy sets are used to define the process operating conditions such that the dynamic model of a nonlinear p`rocess can be described in the following way:

R_i: IF operating condition i

$$\text{THEN } \hat{y}_i(t) = \sum_{j=1}^{no} a_{ij} y(t-j) + \sum_{j=1}^{ni} b_{ij} u(t-j) \ (i=1, 2, 3, ..., nr) \qquad (3)$$

The final model output is obtained by firing strength

$$\hat{y}(t) = [\sum_{i=1}^{nr} \mu_i \hat{y}_i(t)] / (\sum_{i=1}^{nr} \mu_i) \qquad (4)$$

where y is the process output, $\hat{y}$ is the model prediction, u is the process exogenous input, $\hat{y}_i$ is the prediction of the processes output in the ith operating region, nr is the number of fuzzy operating regions, ni and no are the time lags in the input and the output respectively, μ_i is the membership function for the ith model, a_{ij} and b_{ij} are the local linear parameters, and t represents the discrete time.

Operating regions of a process can be defined by one or several process variables. A number of fuzzy sets, such as 'low, medium and high' are defined for each of these process variables. An operating condition is constructed through logical combinations of these variables which are used to define the process operating regions and assigned the fuzzy sets: 'low, medium and high'.

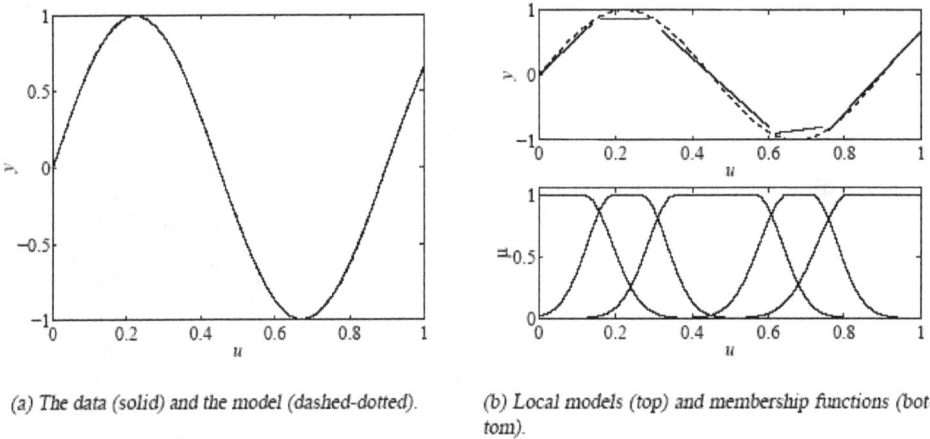

(a) The data (solid) and the model (dashed-dotted).

(b) Local models (top) and membership functions (bottom).

Fig. 1. Takagi-Sugeno fuzzy model computational procedure

The graphical representation for computational procedure for TS fuzzy model is presented for sine curve (Fig 1). From the figure it can be observed that if MFs can be developed for a given data set, then using the membership grade of each predictor variable, local linear models can be developed and their parameters can be estimated. The overlapping of different MFs and different local fuzzy regions are graphically presented in Fig 2. From the figure it is seen that different local linear models are developed from different fuzzy regions and fuzzy reasoning is applied to estimate the model output. Fuzzy regions are represented with different membership grade used in the fuzzy-if- then rules which is the crux of fuzzy modeling approach. Therefore, classification of different local regions in an input/output data set is very important in fuzzy modeling approach and fuzzy clustering technique is widely used for such purposes. In the clustering approach, classification is carried out using different distance measures. The degree of similarity can be calculated by using a suitable distance measure. Based on the similarity, data vectors are clustered such that the data within a cluster are as similar as possible, and data from different clusters are as dissimilar as possible.

Fig 3 gives an example of two clusters in $\mathbb{R}^2$ with prototypes v_1 and v_2. The partitioning of the data is expressed in the *fuzzy partition matrix* $\mathbf{U} = [\mu_{ij}]$ whose elements are the

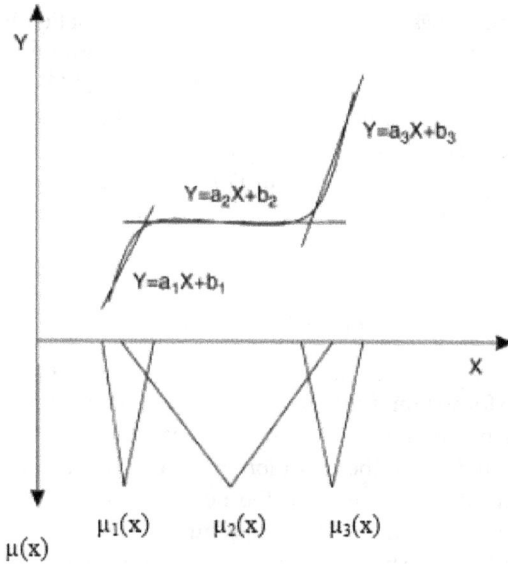

Fig. 2. Fuzzy local liner model developments

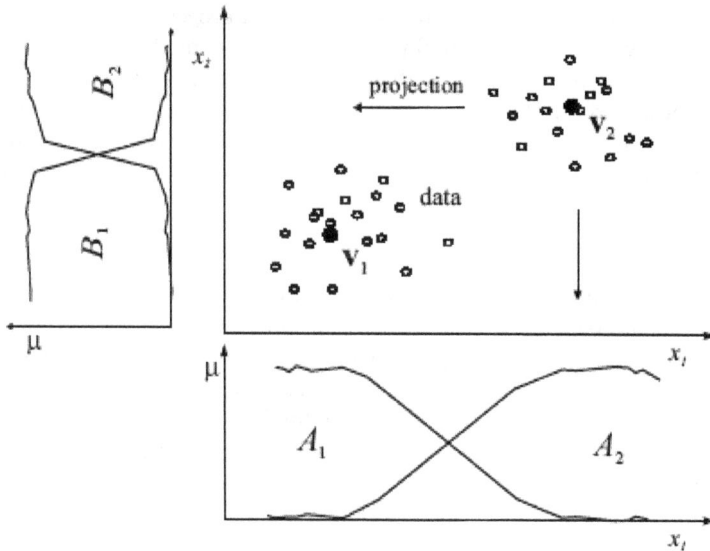

Fig. 3. Identification of membership functions through fuzzy clustering.

membership degrees of the data vectors x_k in the fuzzy clusters with prototypes v_j. The antecedent membership functions are then extracted by projecting the clusters onto the individual variables. The performance of models of this kind depends heavily on the definition of neighboring state space. Euclidian distance is the most commonly employed measure of closeness to find out the neighboring state space. However, it is not usually an appropriate measure of closeness due to the curse of dimensionality. For instance, consider that the predictor distribution is uniform on a ball of radius unity in D dimensional space. For an expected response at or close to origin of the ball, it would be reasonable to use only sample vectors inside a ball of radius $\rho < 1$, assuming Euclidean distance as a measure of closeness. But, since the probability of a sample vector lying in the smaller ball is ρ^D, it is necessary that sample sizes are exponential in D to get enough close vectors for accurate estimation. To avoid this curse, we might assume that $f_i(x)$ has a ridge approximation (i corresponds to the fuzzy region),

$$f_i(x) \cong \sum_{n=1}^{l} c_n g_n(a^T x) \tag{5}$$

in which g_n is a transformation function of the linear combiner of x; a^T, c_n are the parameters of the ridge function; l is the number of sub domains. Also note that the linear local approximation is valid under the condition that the dynamics of $f(x)$ is locally smooth. However, equation 5 typically accounts for the nonlinear local dynamics of $f(x)$. A close examination of equation 5 reveals that the ridge function is an ANN with single hidden layer having l nodes, and a linear transfer function on the output layer.

Hence it is apparent that if the state space is classified into sub-domains, and each of these domains is modeled independently by a neural network approach, which when combined together, the resulting model may provide a better global modeling of the nonlinear dynamics in the state space. It appears that this heuristic has not been addressed or confirmed by empirical trials. The present study illustrates this heuristic by comparing the performance of models developed using the proposed approach, local linear approximation and a global nonlinear approach.

3. Methodology

In general, the above discussed nonlinear local model fitting is composed of two steps: a set of nearby state searches over the signal history and model parameter fitting. For a given signal, this procedure results in a set of local model parameters which when combined together provide a single function over the entire space. Since the neighborhood search is performed over the whole signal history a lot of redundant computation results which in turn hinders effective implementation of this approach. These redundant computations can be avoided by classifying the state spaces into homogenous subspaces by means of an appropriate vector clustering technique.

4. Clustering for classification

The objective of cluster analysis is the classification of objects according to similarities among them, and organizing of data into groups. Clustering techniques are among the *unsupervised* methods, they do not use prior class identifiers. The main potential of clustering is to detect the underlying structure in data, not only for classification and pattern recognition, but for model reduction and optimization. Various definitions of a cluster can be formulated, depending on the objective of clustering. Generally, one may accept the view that a cluster is a group of objects that are more similar to one another than to members of other clusters. The term "similarity" should be understood as mathematical similarity, measured in some well-defined sense. In metric spaces, similarity is often defined by means of a *distance norm*. Distance can be measured among the data vectors themselves, or as a distance from a data vector to some prototypical object of the cluster.

Since clusters can formally be seen as subsets of the data set, one possible classification of clustering methods can be according to whether the subsets are fuzzy or crisp (hard). Hard clustering methods are based on classical set theory, and require that an object either does or does not belong to a cluster. Hard clustering in a data set X means partitioning the data into a specified number of mutually exclusive subsets of X. The number of subsets (clusters) is denoted by c. Fuzzy clustering methods allow objects to belong to several clusters simultaneously, with different degrees of membership. The data set X is thus partitioned into c fuzzy subsets. In many real situations, fuzzy clustering is more natural than hard clustering, as objects on the boundaries between several classes are not forced to fully belong to one of the classes, but rather are assigned membership degrees between 0 and 1 indicating their partial memberships. Fuzzy clustering can be used to obtain a partitioning the data where the transitions between the subsets are gradual rather than abrupt. Most analytical fuzzy clustering algorithm is fuzzy c-means (FCM) clustering. The FCM clustering algorithm is based on the minimization of an objective function called *C-means functional*. In the current study subtractive clustering algorithm is used for classification. Subtractive clustering method (Chiu, 1994) is an extension of the FCM and mountain clustering method (Yager and Filev, 1994), where the potential is calculated for the data rather than the grid points defined on the data space. As a result, clusters are elected from the system training data according to their potential. Subtractive clustering compared to mountain clustering has an advantage that there is no need to estimate a resolution for the grid.

5. Fuzzy non linear local approximation model

As discussed earlier a novel hybrid model is proposed herein which performs independent nonlinear local approximation, and combines each of them using the fuzzy framework. The architecture of the proposed model is presented in Fig 4. The method is based on the concept that the input space is divided into sub regions of similar dynamics using an appropriate clustering algorithm (subtractive clustering algorithm in this study), and modeling of each of these regions is carried out using nonlinear local function approximation (ANN in this study). The proposed model is termed as Fuzzy Non Linear Local Approximation (FNLLA) model, which is based on fuzzy concept and neural technique is applied for nonlinear local function approximation.

In Fig 4, the gating is done to identify the membership grade associated with any given input vector for each of the clusters. This input vector will be passed to each of the developed ANN, and the output from each ANN is combined by computing the weighted mean with the membership grade in each cluster.

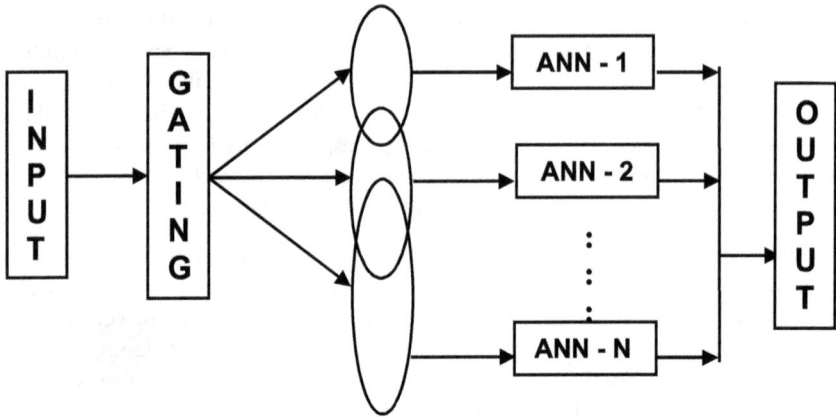

Fig. 4. Schematic representation of the proposed FNLLA model

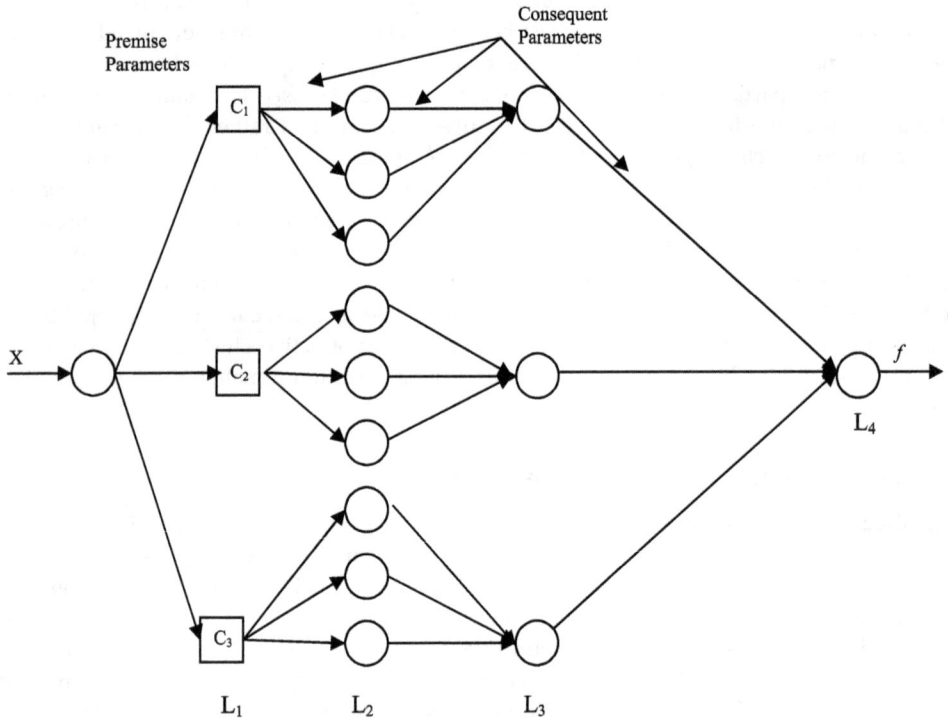

Fig. 5. Computational architecture of the proposed FNLLA model

The architecture of FNLLA is depicted in Fig 5, in which the fuzzy computing scheme is represented in an adaptive neural network structure. The consequent part of each of the fuzzy rule is a nonlinear function (ANN in this case). The computations are performed in 4 layers in FNLLA. In the layer 1, (L_1), the incoming input vector $(\mathbf{x})$ is passed to different sub regions $(C_1, C_2$ and $C_3)$ and the associated MFs are computed as:

$$\mu_j = e^{-\alpha \|x_j - x_i\|^2} \tag{6}$$

in which x_i is the ith input vector, x is the cluster centre and j the sub-region number, and α is a function of the cluster radius. In layer 2 (L_2), the weights of the hidden nodes of the ANN consequent models are estimated, and the output from each ANN consequent model is arrived at layer 3 (L_3). At layer 4 (L_4), each of these consequent outputs are combined to arrive at the final output.

6. Demonstrative case examples

The proposed model is illustrated through two examples by developing rainfall-runoff models: (a) for Kolar basin up to the Satrana gauging site in India, (b) the Kentucky basin, USA. The Kolar River is a tributary of the river Narmada that drains an area of about 1350 sq km before its confluence with Narmada near Neelkanth. In the present study the catchment area up to the Satrana gauging site is considered, which constitutes an area of 903.87 sq km (Fig 6). The 75.3 km long river course lies between north latitude 21⁰09′ to 23⁰17′ and east longitude 77⁰01′ to 77⁰29′. For the study, rainfall and runoff data on an hourly interval during the monsoon season (July, August, and September) for three years (1987–1989) are used. The rainfall data available were in the form of areal average values in the basin. The total available data has been divided into two sets, calibration set (data during the years 1987–1988) and validation set (data during the year 1989). Different models for lead times of up to 6 hours have been developed in the study.

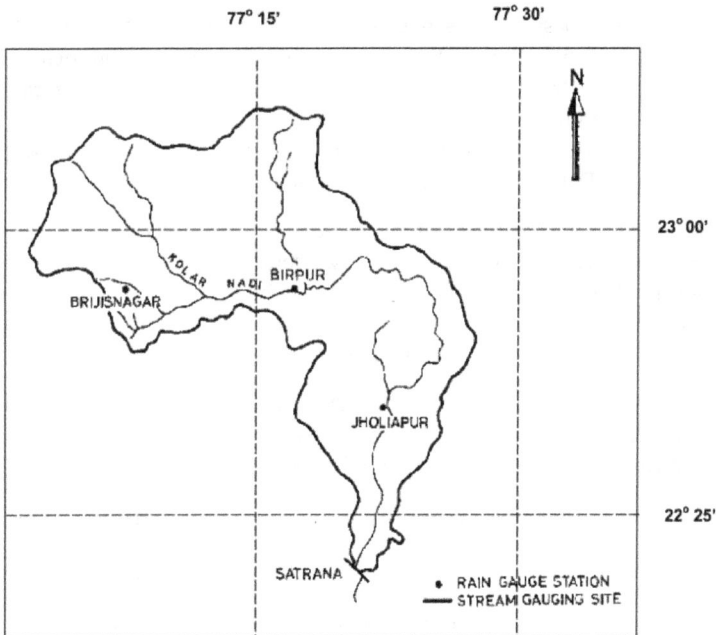

Fig. 6. Map of Kolar River basin, India

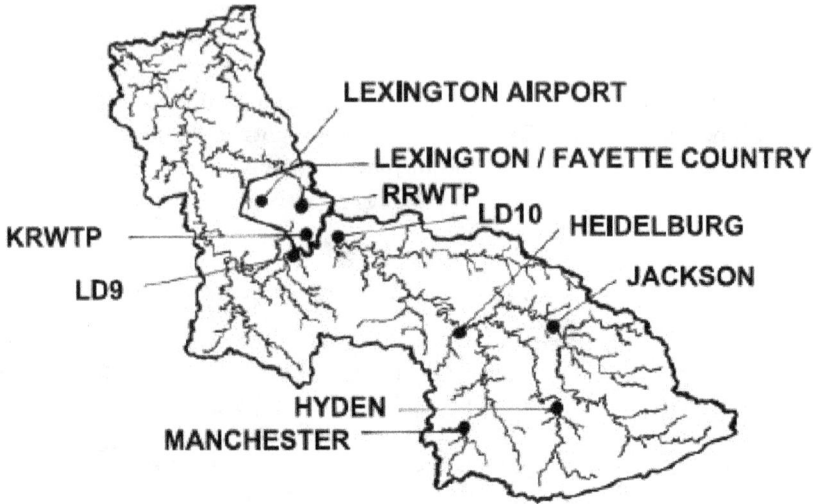

LD 9   = Lock and Dam 9
LD10   = Lock and Dam 10
KRWTP = Kentucky River Water Treatment Plant
RRWTP = Richmond Road Water Treatment Plant

Fig. 7. Study area map of Kentucky River basin in USA

The Kentucky River basin covers over 4.4 million acres of the state of Kentucky. Forty separate counties lie either completely or partially within the boundaries of the river basin. The Kentucky River is the sole water source for several water supply companies of the state. There is a series of fourteen locks and dams on the Kentucky River, which are owned and operated by the US Army Corps of Engineers. The drainage area of the Kentucky River at Lock and Dam 10 (LD10) near Winchester, Kentucky is approximately 6,300 km² (Fig 7).

The data used in the study presented in this paper include average daily streamflow (m³/s) from the Kentucky River at LD10, and daily average rainfall (mm) from five rain gauges (Manchester, Hyden, Jackson, Heidelberg, and Lexington Airport) scattered throughout the Kentucky River Basin. The total length of the available rainfall runoff data was 26 years (1960-1989 with data in some years missing).

The input vector identified, according to Sudheer et al. (2002) for modeling the river flow in Kolar, included a total number of 4 variables. Accordingly, the functional form of the model, in the case of Kolar, for rainfall runoff modeling is given by:

$$Q(t) = f[R(t-9), R(t-8), R(t-7), Q(t-1)] \tag{7}$$

where Q(t) and R(t) are river flow and rainfall respectively at any time t in hour.

The functional form for rainfall-runoff dynamic for Kentucky River basin is given by:

$$Q(t) = f[R(t), R(t-1), R(t-2), Q(t-2), Q(t-1)] \tag{8}$$

Different statistical indices that are employed to estimate the model performance include coefficient of correlation (CORR), efficiency (EFF), root mean square error (RMSE) and noise to signal ration (NS).

7. Results and discussions

7.1 Parameter Estimation in FNLLA

The optimal number of clusters in FNLLA has been obtained by varying cluster radius in the subtractive clustering algorithm, combined with the ANN model development. Different ANN models are developed for different clusters of input space changing hidden neuron from 2 to 10. Single hidden layer with sigmoid function nodes is used in the ANN. The sigmoid activation function is considered in the output layer also. A standard back propagation algorithm with adaptive learning rate and momentum factor has been employed to estimate the network parameters for different clusters. In order to have a true evaluation of the proposed nonlinear local approximation in fuzzy models, the result obtained for both the basins from FNLLA was compared with FIS, which performs a linear local approximation. In FIS model, subtractive clustering has been used for fuzzy model identification which includes optimal number of if-then-rule generation and consequent parameters are optimized using least square error (LSE) technique.

Kolar River basin		Kentucky River basin	
Cluster Radius	Number of clusters	Cluster Radius	Number of clusters
0.050 - 0.060	7	0.0010 - 0.0080	4
0.061 - 0.074	6	0.0081 - 0.0100	3
0.075 - 0.079	5	0.0110 - 0.2000	2
0.080 - 0.100	4	0.2100 - 1.0000	1
0.110 - 0.200	3		
0.210 - 0.300	2		
0.030 - 1.000	1		

Table 1. Partition of input space for Kolar and Kentucky basin

In the analysis, the radius of influence (r_a) of the cluster centre is fixed by various trials, which is the foremost interest for the current study. The value of r_a is varied from 0.001 to 1.0 with a step size of 0.01; at each stage number of clusters is estimated. From the current data set 1 to 7 clusters are found while changing the cluster radius from 0.05 to 1 for Kolar river basin and maximum 4 clusters are observed by changing radius from 0.001 to 1 for Kentucky basin. The number of clusters identified corresponding to various radiuses for both the basins are presented in Table 1. It is evident from Table 1 that as the cluster radius increases, the number of cluster decreases. Different ANN models are developed for data belonging to different clusters (effectively representing different ranges of flow), and as discussed earlier hidden neurons are by trial and error procedure. The stopping criteria for ANN model building was maximum efficiency.

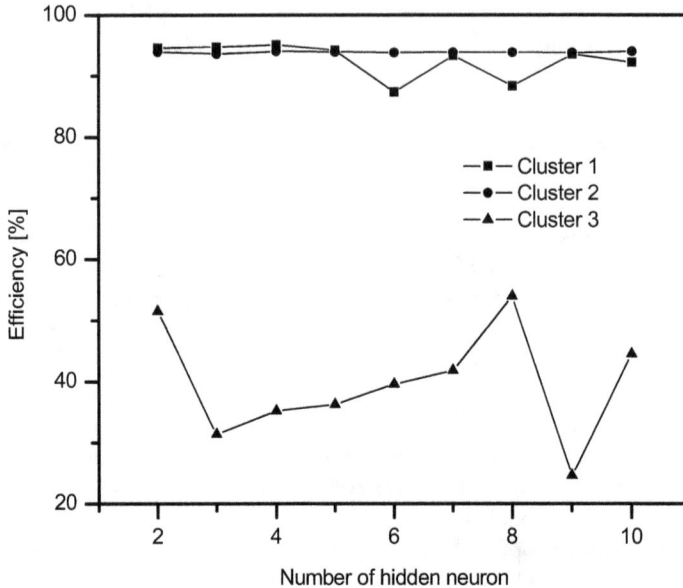

Fig. 8. Variation of Efficiency Plot for three clusters with respect to hidden neurons

It was observed that model performance was good when input space is classified into 3 clusters for both the basins. The efficiency plots with respect to different hidden neurons for 3 ranges of flows are presented in Fig.8 for Kolar basin. From the Fig 8 it is observed that the maximum efficiency is for the models having hidden neurons are 4, 4 and 8 for individual clusters for Kolar River basin. Similar procedure has been followed for optimization of model parameter for Kentucky River basin. The optimum hidden neuron obtained 3, 3 and 3 for Kentucky basin. The identified optimal clustering radius with number of if-then rules for Kolar and Kentucky basin are furnished in Table 2.

Model	Kolar River basin	Kentucky River basin
FIS	0.20 [3 rules]	0.008 [4 rules]
FNLLA	0.12 [3 ranges/rules]	0.010 [3 ranges/rules]

Table 2. Optimal model structure for three sub-domains for Kolar and Kentucky basin

The summary statistics of the flow data belonging to each of the identified clusters are presented in Table 3. It is clear from Table 3 that the clustering based on the input vector clearly identifies distinct clusters that have different nonlinear dynamics. This is evident from overlapping clusters; the cluster C1 contain flow range from 4.67 m³/s to 240.55 m³/s, and the cluster C2 contain flow range from 3.63 m³/s to 240.24 m³/s, in the case of Kentucky river basin data. A similar observation is found in the case of Kolar basin data too. It is worth mentioning that the consequent ANN models for each of these clusters preserve the summary statistics much effectively in both basins (see Table 3). Also, it is evident that the classification of the data into different clusters is according to the range of flow, though not exclusively forced by the FNLLA.

Statistic	C1		C2		C3	
	Observed	Computed	Observed	Computed	Observed	Computed
Kentucky River Basin						
Mean	13.45	13.19	32.25	32.13	239.16	239.08
Standard Deviation	13.08	12.05	25.83	20.79	281.21	272.51
Minimum	3.63	4.67	9.61	15.09	15.99	31.55
Maximum	240.24	240.55	297.43	266.96	2553.90	2293.20
Kolar River Basin						
Mean	4.09	4.07	12.47	12.43	142.73	138.69
Standard Deviation	1.94	1.74	4.55	4.27	308.65	290.72
Minimum	1.62	1.58	6.83	7.04	0.88	19.26
Maximum	29.03	28.86	125.57	124.84	2427.70	2062.40

Table 3. Summary statistics of the river flow in the identified sub-domain by FNLLA Model

Lead time	1-hour			3-hour			6-hour		
Cluster	C1	C2	C3	C1	C2	C3	C1	C2	C3
Calibration									
Correlation	0.82	0.98	0.91	0.64	0.90	0.82	0.94	0.77	0.91
Efficiency (%)	66.91	95.40	83.33	39.21	79.58	67.56	24.49	53.18	53.35
RMSE	2.62	66.18	0.79	5.88	139.41	1.92	11.03	210.71	3.43
Noise to Signal Ratio	0.58	0.21	0.41	0.78	0.45	0.57	0.39	0.69	0.41
Validation									
Correlation	0.93	0.98	0.45	0.63	0.87	0.48	0.91	0.75	0.74
Efficiency (%)	87.08	95.28	81.82	37.41	74.55	64.24	30.15	52.98	55.06
RMSE (m³/s)	1.43	53.97	1.50	4.03	125.44	2.43	9.45	170.65	3.15
Noise to Signal Ratio	0.36	0.22	0.96	0.79	0.51	0.93	0.49	0.69	0.43

Table 4. Cluster wise FNLLA performance for Kolar Basin

The results of the FNLLA were first analyzed for its effectiveness in capturing the nonlinear dynamics at local level. In order to achieve this, the performance indices were computed for each sub-domain for Kolar basin for the calibration and the validation period, and are presented in the Table 4. Note that in the study the FNLLA model classified 1975, 1782 and 593 patterns as low, medium and high flow respectively using the subtractive clustering algorithm. It is evident from the Table 4 that the nonlinear consequent models of the FNLLA are effective in capturing the nonlinear dynamics in each sub regions.

7.2 Performance of FNLLA at 1 step-ahead forecast

The values of various evaluation measures during calibration and validation period for FNLLA and FIS for both the basins for 1-hour lead forecast are summarized in the Table 5, from which it can be observed that both the models possess high value of correlation (0.96 and more) between the forecasted and the observed river flow at 1-hour ahead. The high value for the efficiency index indicates a very satisfactory model performance in capturing the nonlinear dynamics involved in the rainfall-runoff processes. Note that the FNLLA performs with higher efficiency in the case of Kolar basin compared to the FIS model. The value of RMSE varies from 24 m³/sec to 42 m³/sec for Kolar basin indicating a very good performance by both the model; the RMSE values are relatively less in Kentucky basin for FNLLA model. It is also noted that the NS ratio for the FNLLA model is less than that corresponding to the FIS model.

Statistical Indices	FNLLA		FIS	
	Calibration	Validation	Calibration	Validation
Kolar River Basin				
Correlation	0.98	0.99	0.98	0.98
Efficiency	95.84	96.72	95.91	96.61
RMSE (m³/s)	42.39	23.29	42.06	24.31
Noise to Signal ratio	0.20	0.11	0.34	0.12
Kentucky River Basin				
Correlation	0.98	0.97	0.96	0.96
Efficiency	95.68	93.87	92.17	91.68
RMSE (m³/s)	50.12	51.87	67.45	60.41
Noise to Signal ratio	0.21	0.25	0.28	0.29

Table 5. Statistical Indices for 1-hour and 1- day lead forecast for Kolar and Kentucky basin

The scatter plots of flows for 1-hour lead forecast validation period for both the models are presented in Fig 9 for Kolar basin (Fig 10 for Kentucky basin for 1 day lead). These plots give clear indication of the simulation ability of the developed model across the full range of flows. It is noted that most of the flows tend to fall close to the 45⁰ line (reduced scattering), showing a good agreement between observed and forecasted flows. From the plot it is observed that both the models are quite competent in forecasting river flow at 1 hour/day lead time. In general it is noted that the mapping of the low flow region is relatively better compared to high flow region.

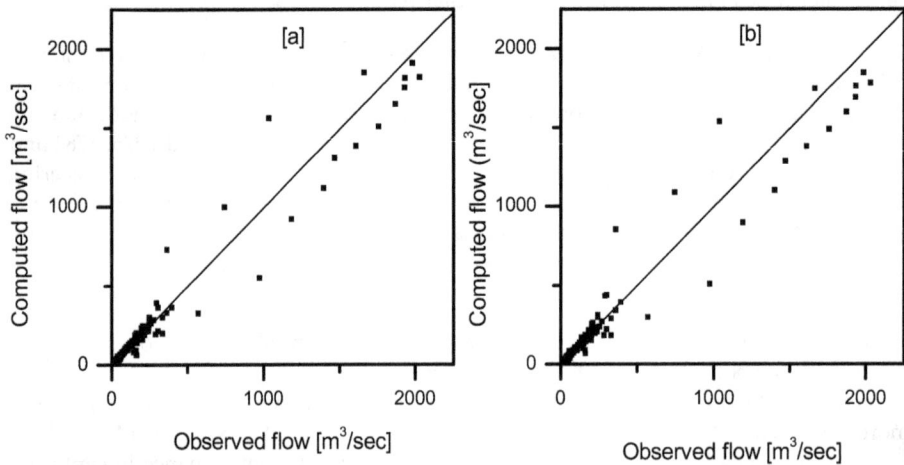

Fig. 9. Scatter plots for observed and computed flows by both models at 1 hour lead time for Kolar basin (a) FIS (b) FNLLA

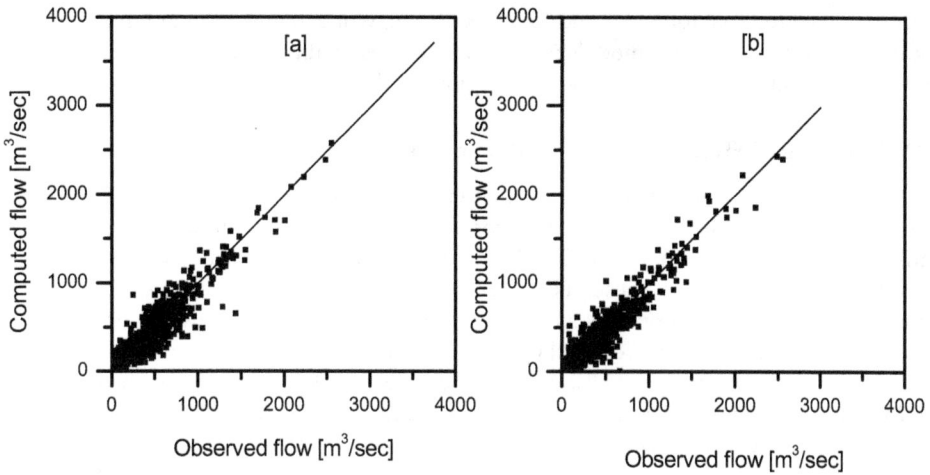

Fig. 10. Scatter plots for observed and computed flows (a) FIS (b) FNLLA model for 1 day lead forecast for Kentucky basin

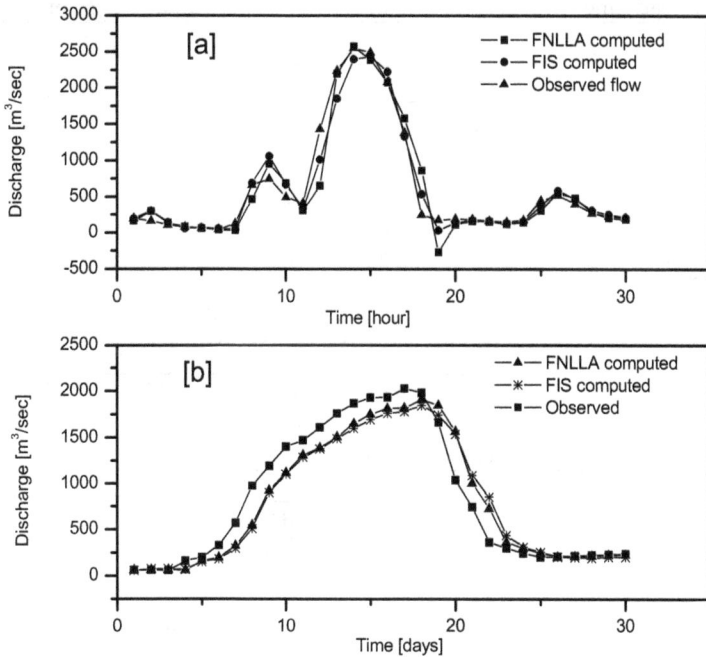

Fig. 11. Comparison plot between FIS and FNLLA model for a typical storm event for (a) Kolar basin (b) Kentucky basin

The forecasted hydrograph for a typical flood event (during validation period) for both the basins by both the models are presented along with its observed counterpart in Fig 11. It can be observed from Fig 11 that FNLLA is preserving the peak flows effectively than the FIS for

1 hour/day lead forecast for Kolar and Kentucky basin, while in low and medium ranges of flow the performance of both models is similar. This observation effectively brings out the capturing of nonlinear dynamics at the local regions of the input space.

7.3 Performance of FNLLA at higher lead time forecasts

Both the models are further evaluated for their effectiveness to forecast flows at higher lead times. The statistical indices for higher lead forecasting are presented in Table 6 for both the basins. The superior performance of the FNLLA compared to FIS is clearly visible at higher lead times from the results presented in Table 6. It is observed that even though the performance of FNLLA deteriorates as the lead time increases, it falls down to only 79.15% efficiency at 6 hours ahead, while FIS shows only 51.00% efficiency at the same lead time. A similar behavior is exhibited by FNLLA in the case of Kentucky basin also, as the efficiency is 77.23% at 3 days ahead compared to 46.92% in the case of FIS. A similar argument holds well in the case of other performance indices too.

	FNLLA			FIS		
Kolar River Basin						
Forecast Lead time	1 hour	3 hour	6 hour	1 hour	3 hour	6 hour
Calibration						
Efficiency (%)	95.84	81.52	79.15	95.91	57.79	51.00
RMSE (m³/s)	42.39	89.40	94.95	42.06	135.11	145.70
Noise to Signal ratio	0.20	0.43	0.923	0.34	0.65	1.95
Validation						
Efficiency (%)	96.72	79.95	77.73	96.61	50.54	46.92
RMSE (m³/s)	23.29	60.45	62.37	24.31	92.12	96.48
Noise to Signal ratio	0.11	0.35	0.39	0.12	0.77	0.83
Kentucky River Basin						
Forecast Lead time	1 day	2 day	3 day	1 day	2 day	3 day
Calibration						
Efficiency (%)	95.68	84.21	64.88	92.17	73.58	52.50
RMSE (m³/s)	50.12	95.82	142.9	67.45	123.93	166.20
Noise to Signal ratio	0.21	0.40	0.59	0.28	0.51	0.69
Validation						
Efficiency (%)	93.87	78.68	58.66	91.68	73.14	50.56
RMSE (m³/s)	51.87	96.68	134.63	60.41	108.53	147.24
Noise to Signal ratio	0.25	0.46	0.64	0.29	0.52	0.70

Table 6. Performance indices for FIS and FNLLA model at higher forecast lead times

8. Summary and conclusions

The objective of this chapter was to present and illustrate a nonlinear local approximation approach in modelling the rainfall-runoff process which offers better accuracy in the context of river flow forecasting. Based on the theoretical considerations of the fuzzy modelling in the state space (input-output), it is clear that if the state space is classified into sub-domains and each of these domains is modeled independently by a neural network approach which are combined together, it may provide a better global modeling of the nonlinear dynamics in the state space. In general, the proposed nonlinear local model fitting is composed of two steps: a set of nearby state searches over the signal history and model parameter fitting. For a given signal, this procedure results in a set of local model parameters which when combined together provide a single function over the entire space. Since the neighborhood search is performed over the whole signal history, a lot of redundant computations result which in turn hinders effective implementation of this approach. These redundant computations can be avoided by classifying the state spaces into homogenous subspaces by means of an appropriate vector clustering technique. The proposed model is termed as Fuzzy Non Linear Local Approximation (FNLLA) model, which is based on fuzzy concept and neural technique is applied for nonlinear local function approximation. The partition of the state space is achieved by subtractive clustering algorithm and nonlinear local function approximation is by ANN in the proposed method. The antecedent parameters of the model are simultaneously estimated during clustering, and the standard back propagation algorithm is employed for ANN parameter estimation.

The potential of FNLLA is illustrated using two case examples: (i) data pertaining to Kolar River basin, and (ii) data corresponding to Kentucky River basin. The optimal architecture of the FNLLA model, which is defined by the number of sub-regions in the data and the structure of each ANN, is arrived after a trial and error procedure. When the performance of the FNLLA is compared with that of a pure FIS, it is observed that the FNLLA certainly possesses the advantages of nonlinear mapping. Though both the models perform similar at 1 step-ahead forecasts, the FNLLA performs much better than FIS at higher lead times. Overall, the results of the study confirm the heuristic that a nonlinear local approximation is a better approach in fuzzy modeling especially when complex nonlinear functions are being mapped.

9. References

Chiu, S. (1994) Fuzzy Model Identification Based on Cluster Estimation. *Journal of Intelligent and Fuzzy Systems*, 2(3), 267-278,

Farmer, J.D. and J.J. Sidorowitch (1987) Predicting chaotic time series. *Physical. Review Letter*, 5(59), 845-848.

Jain, A., K.P. Sudheer, and S. Srinivasulu (2004) Identification of physical processes inherent in artificial neural rainfall-runoff models. *Hydrological Processes*, 18(3), 571-581.

Singer, A.C., G. Wornell, and A. Oppenheim (1992) Codebook prediction: a nonlinear signal modeling paradigm. *IEEE, In proc. Int. Conf. Acoustics, Speech & signal processing*, San Francisco, 5, 325-328.

Sudheer, K.P. (2005) Knowledge extraction from trained neural network river flow models. *Journal of Hydrologic Engineering*, ASCE, 10(4), 264-269.

Sudheer, K.P., A.K. Gosain, and K.S. Ramasastri (2002) A data-driven algorithm for constructing artificial neural network rainfall-runoff models. *Hydrological Processes*, 16, 1325-1330.

Tokar, S., and Markus, M. (2000). *Precipitation-runoff modeling using artificial neural networks and conceptual models*. Journal of hydrologic engineering, ASCE, 5(2): 156-161.

Wilby, R.L., R.J. Abrahart, and C.W. Dawson (2003) Detection of conceptual model rainfall–runoff processes inside an artificial neural network. *Hydrological Sciences Journal,* 48(2), 163–181.

Yager, R. and D Filev (1994) Generation of fuzzy rules by Mountain clustering. *Journal of Intelligent &Fuzzy Systems,* 2(3), 209-219.

Zhang, B. and R.S. Govindaraju (2000) Prediction of watershed runoff using Bayesian concepts and modular neural networks. *Water Resources Research,* 36(3), 753-762.

5

San Quintin Lagoon Hydrodynamics Case Study

Oscar Delgado-González, Fernando Marván-Gargollo,
Adán Mejía-Trejo and Eduardo Gil-Silva
Instituto de Investigaciones Oceanológicas,
Universidad Autónoma de Baja California,
Ensenada, Baja California,
México

1. Introduction

Hydrodynamics at coastal zone determine the way in which water flows and biochemical properties are exchanged. It is important to understand these processes in order to determine the potential effects caused by modifying the flow field as well as the mixing processes. For aquaculture purposes, the flow field distributes food to the organisms, thus by understanding the system hydrodynamic it is possible to determine suitable areas for aquaculture activities.

In México, approximately 2% of the 1.9×10^6 km² which constitutes the national territory corresponds to protected or semi protected coastal areas (Ortiz-Pérez and Lanza-Espino 2006) and is distributed over 11,000 km of coastline; however, norms and guidelines for commercial aquaculture development were not established until the early 1990s, so this activity is currently at an incipient stage in most coastal areas.

San Quintin Lagoon (SQL) is a coastal lagoon located on the northwestern coast of Baja California (Mexico) which is characterized as a semiarid region with a highly productive upwelling offshore. The lagoon covers an area of approximately 42 km² (Figure 1) and is divided in two basins named Bahia San Quintin (BSQ) and Bahia Falsa (BF). The mouth of the lagoon is 1 km wide and has maximum depth of 14 m. The lagoon shows several mud flats, some of them are exposed during low tides and others leave shallow areas which are optimal for aquaculture activities. The tidal regime is mixed semidiurnal with amplitudes as high as 1.6 m above MLLW (Mean Lower Low Water) during spring tides. The oceanic water entering the lagoon is nutrient rich due to upwelling processes associated with the north-south orientation of the coastline which mainly occur during spring and summer (Hernández-Ayón *et al.* 2004). The wind regime changes on a diurnal scale and is a breeze type with its maximum magnitude around 3pm with a dominant direction from the northwest. Due to the orientation of the lagoon and shallowness, the wind plays a very important role in circulation and generation of fetch limited waves. Inside the lagoon, the water column is well mixed (Millán-Núñez *et al.* 1982). The climate is Mediterranean and the lagoon shows as a negative estuary structure due to its net evaporation which makes the lagoon a hyper saline system throughout the year.

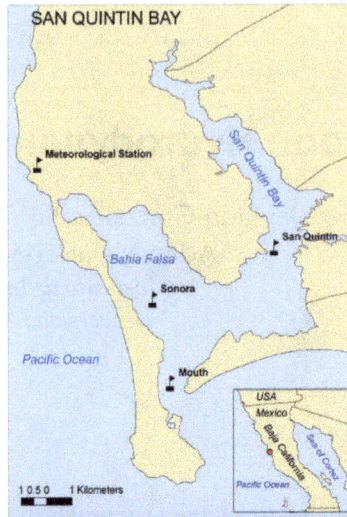

Fig. 1. Bahia San Quintin is a coastal lagoon on the northwest coast of Baja California México. The flags at the marine area represent the temperature sensor location, the one at the northwest continental location, marks the site location of the meteorological station.

During the 70s, studies were conducted at SQL to establish its viability for aquacultural activities (Lara-Lara and Álvarez-Borrego 1975, Álvarez-Borrego and Álvarez-Borrego 1982). These studies provided the first biophysical data and, among other findings, it was determined that the system is governed by mixed, semidiurnal tides.

In SQL, aquaculture activities have been conducted continuously for more than 30 years. The Pacific oyster *Crassostrea gigas* is extensively cultured in this area using a rack method. Even though most oyster farms use the same cultivation method, it is common to find differences of more than two months in the time at which organisms attain a commercial size throughout the lagoon (García-Esquivel *et al.* 2000).

The main factors involved in the selection of an area for aquaculture activities are: water, temperature, salinity, depth, and food availability (Héral and Deslous-Paoli, 1991). The conceptual model of factors and processes involved in the distribution of water properties by currents in BF (Delgado-González, 2010), gives an idea of the hydrodynamics and its importance in mixing and carrying food sources from one area to another. Phytoplankton spatial distribution at BF under neap and spring tides, considering oyster consumption, was obtained by using a 2D hydrodynamic model (Delgado-González, *et al.*, 2010) and the results were utilized to determine the most productive areas within BF.

The purpose of this chapter is to present the hydrodynamic regimes at BSQ using a 2D hydrodynamic numerical model that considers the wind effect as well as analyzing food input to the lagoon, residence times and waste production within the lagoon by using a lagrangean tracker method. It is necessary to understand the possible circulation patterns of coastal lagoons like BSQ in order to adequately manage their resources and their water quality, which could be affected as a consequence of the aquaculture activities taking place inside the lagoon.

The hydrodynamics regimes and waterborne particle transport was obtained using the Coastal Modeling System Flow (CMS) Flow in conjunction with the Particle Tracking Model (PTM). Both models are embedded within the SMS user interface. CMS is a two-dimensional finite volume numerical model developed by the US Army Corps of Engineers which is based on a rectangular grid allowing for grid refinement as well as telescoping grids. PTM (Particle Tracking Model) is a lagrangean particle tracker designed to trace the fate and pathways of sediment and other waterborne particles. It was designed jointly by the Coastal Inlets Research Program (CIRP) and the Dredging Operations and Engineering Research (DOER). Particles are transported by the flow field and it takes into consideration a logarithmic current profile for their advection.

2. Data collection

2.1 Tides

Tides in the north Pacific coast are the primary hydrodynamic forcing agent for coastal lagoons with a constant water exchange with the open ocean such as SQL. (Fischer *et al.*, 1979). The tides at SQL are mixed with a semidiurnal predominance (Figure 3). The tides exhibit a well-defined fortnightly modulation for which the range during spring tides is approximately 2.5 m and 1.0 m for neap tides. Using these values in combination with the surface area of the basin and the mouth cross-sectional area, for a linear-frictionless tidal wave, according to Stiegebrandt (1977), the amplitude of the tidal currents should be:

$$u_0 = aY\sigma/2A$$

where a is the tidal range, Y is the surface area of the basin (42×10^6 m^2), A is the cross-sectional area at the entrance (7500 m^2) and σ is the frequency of the main tidal forcing ($2\pi/12.42$ h). Thus, for spring tides u_0 should be approximately 0.98 m/s and for neap tides it should be around 0.31 m/s, which are similar to the observed values reported in this work.

The most intense tides occur during spring tides in December and January whilst the less exchange occurs during neap tides around the year (Delgado-Gonzalez, *et al.*, 2010). There is a 40 minute tidal lag between the lagoon mouth and the head in BSQ, located in the inner part of the right arm of SQB (Ocampo-Torres 1980).

2.2 Wind

Wind, in many places, is responsible for producing surface currents as well as vertical mixing under special conditions (Fischer et al., 1979). In the particular case of San Quintin the wind is generally strong and the direction does not change very much thus being also an important forcing agent for the lagoon hydrodynamics. Due to the lagoon configuration, wind is more important at BF than at BSQ mainly due to the primary axis of BF which coincides with the predominant wind direction and the fact that volcanoes shelter BSQ.

The wind data presented in this chapter was registered in the northwest area of BSQ (Figure 1). The wind hodograph shows the time of the day and magnitude for local and synoptic winds recorded at this station. This station also records air temperature and atmospheric pressure (Figure 2). The ellipse shape of the hodograph and its northern relative position from intersection of the West-East (u) and South-North (v) components can be used to identify the hourly characteristics for the local wind and the synoptic effect, respectively.

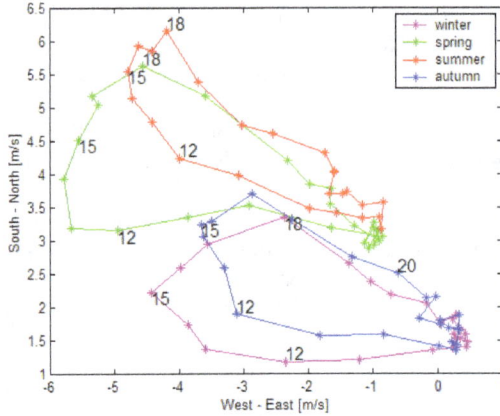

Fig. 2. Seasonal hodographs at La Chorera, Baja California. During spring and summer the effect of synoptic winds, associated with the Pacific semi-permanent high pressure center, moved the hodograph northwest from the origin. During both stations stronger winds (5-6 m/s) are recorded during afternoon hours and these winds can be associated with records of low oceanic water temperatures from the coastal upwelling waters.

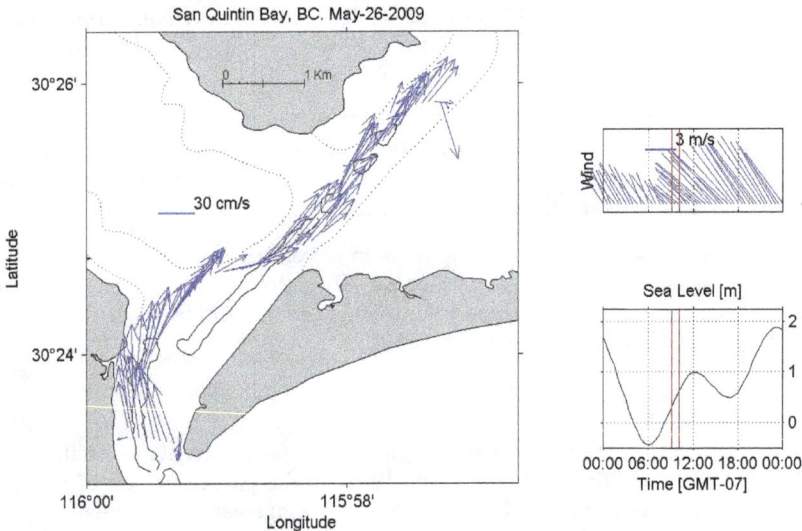

Fig. 3. ADP transects along the main channel during flooding tide. The red lines show the tide and wind conditions during the experiment.

2.3 Currents

Coastal lagoon hydrodynamics are mainly influenced by tides and winds as well as the lagoon internal configuration (Fischer et al., 1979). For SQB area there is limited currents data prior to 2004. Since 2004 there has been a growing interest from a scientific point of view as well as from the aquaculture companies to have a better understanding of the system circulation. To gain an idea of the hydrodynamics of San Quintin Lagoon, current data from ADP transects and drifters are shown next.

2.3.1 ADP transects

Several experiments were performed during different tidal and wind conditions. Figure 3 shows the depth averaged currents, flooding tide and wind conditions; it is possible to observe strong currents along the channels, a counter flow limited to the eastern part of the mouth and wind effects in the shallow areas. This experiment shows the complex circulation associated with the different forcing mechanisms; in some of them it is possible to recognize that the wind effect is mainly limited to the top layer (~1m).

2.3.2 Drifters

Field experiments using five drifters were done during 2010-2011 under ebb and flood tide conditions (Table 1).

Ebb tide						
HWL to LWL			Drifters			
Time (h)	Range (m)	Date	Initial time (T_o)	Ended time (T_f)	Distance (km)	Mean velocity (m s^{-1})
11:00-17:00	1.21	August 26, 2010	11:35	15:15	3.2	0.2
11:09-17:43	1.95	September 10, 2010	15:41	16:46	2.9	0.6
7:18-14:09	2.23	November 4, 2010	11:17	14:05	6.0	0.8
7:09-13:40	1.57	April 14, 2011	9:29	12:05	6.1	0.5
Flood tide						
LWL to HWL						
13:00-19:25	1.34	April 13, 2011	13:07	14:45	1.4	0.3
8:06-14:50	1.38	June 21, 2011	12:02	13:34	2.0	0.4

Table 1. The tidal condition during six field experiments with drifters at San Quintin Bay. During ebb tide the experiments conditions were from high water level (HWL) to low water level (LWL), and from LWL to HWL during flood tide.

For three of the experiments during ebb conditions, the drifters were located approximately 50 m apart from each other across the tidal channel; after being deployed, the ebb tide currents moved them to the central area of the tidal channel. In the August 26, 2010 experiment, the drifters were located approximately 100 m apart from each other along the central line of the tidal channel and their four hour trajectories remained along this central area.

Drifters from September 10th, pink color (Figure 4a), were deployed at the central zone of BSQ. The five drifters finished out of the lagoon and most of the time their trajectories remained at the central part of the main tidal channel. Just at the mouth area, the drifters were closer to the east part rather than to the central part of the channel and ended at the shoaling region of the tidal delta. Drifters from November 4th, 2010 and April 14th, 2011, black and gold color (Figure 4a), followed the central area of the channel until they reached the delta part, where they started to move away from each other following curved trajectories, which indicates the balance between the lagoon water and the oceanic waters. Drifters from August 26th, 2010, red color (Figure 4a) located at the inner part of BSQ moved slower, but followed the central area of the tidal channel.

Drifters from April 13th, 2011, red color (Figure 4b) were located outside the bay in order to determine if they would be suctioned inside. However the wind component was strong enough to move them away from the bay entrance. On June 21st, 2011, black color (Figure 3b), the drifters were deployed 4 hours after the flooding tide started to fill the lagoon, reducing the possibility of seeing the full expected trajectories and the drifters were forced by the wind stress to curve into the sand bar.

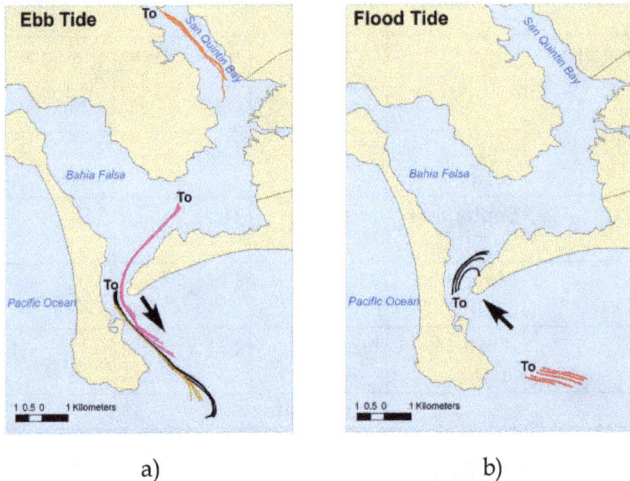

a)   b)

Fig. 4. During ebb tide conditions the mean velocities of drifters were 0.4m/s with maximum values around .8 m/s at the mouth area. During the experiments in flood conditions, the wind stress at the surface moved the drifters away when these were deployed outside San Quintin bay and to the coast when were deployed at the mouth. The initial time (To) is shown to represent the initial position in each experiment. The black arrow represents the outflow from and inflow to San Quintin Bay.

2.3.3 Water temperature

Water temperature at San Quintin Lagoon shows a clear seasonal signal, a clear variation between BF and BSQ and throughout a tidal cycle, and within its shallows areas the water temperature closely tracks air temperatures even with the constant effects of wind mixing processes.

Three HOBO Water Temp Pro v2 Data Logger sensors were deployed at BSQ during the 6 months (Figure 1). These sensors were located at aquaculture buoys and situated at a depth of approximately 1.5 m. The instrument recorded temperature at a rate of 15 minutes (Figure 5). The role of tides in the temperature oscillations during the August-November months and the upwelling index is well observed; it is possible to observe that during spring tides the water temperature inside the lagoon drops 3 or 4 °C, and most of the times this drop shows visual correlation with the upwelling index, and during neap tides the system increase its temperature. However, during the last days of November the spring tides reduce the temperature and the system does not return to the original warm temperatures, marking the beginning of winter inside the lagoon (Figure 5).

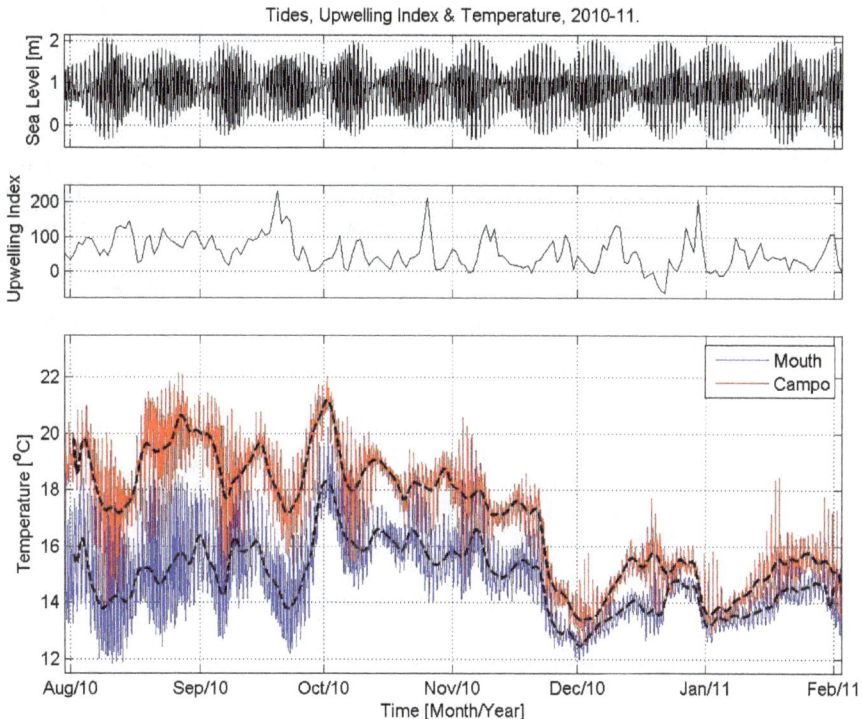

Fig. 5. The temperature time series at three stations in BSQ showed similar behavior, only two are presented. During summer months mean temperatures in the two inner is 19.0 °C ± 1.44 and 14.8 °C± 0.95 during winter months. The fortnightly tide has an important role in the main fluctuations observed.

In order to see the relationship between spring and neap tides and temperature changes, the first fifteen days of August and October (Figure 6). During neap tides, first 6 days of August (Figure 6a) the mean temperature is around 19 ° C and during the spring tide the mean temperature remains around 17.5 ° C, mainly because of the different hydrodynamics associated with these two tide conditions.

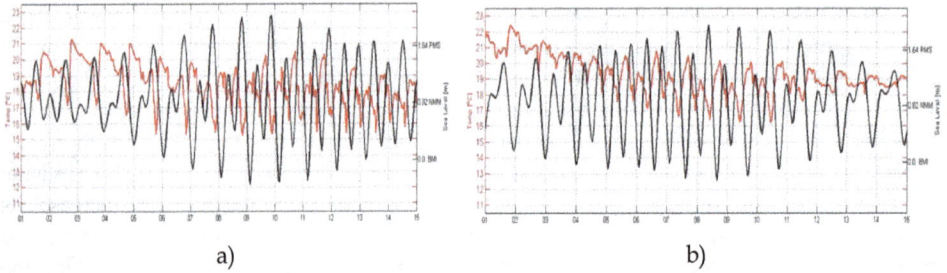

a)   b)

Fig. 6. First fifteen days of August a) and October b). It is possible to observe the relation between spring and neap tides with temperature. And also the inverse relationship between tide and temperature.

3. Numerical model

3.1 Hydrodynamics

The model domain was defined for the whole lagoon as well as an open water area which acts as a buffer zone between the lagoon and the forcing boundary (Figure 7). The grid is composed of 105,966 rectangular cells with size of 50 by 50 meters each and it has an azimuthal rotation of 333°. The domain measures 20.3 Km in the Y axis and 13 Km in the X axis. SQL is delimited by two sand bars which have some low altitude zones; for the modeling task these five areas have been considered as wetting and drying zones to allow for sand bar breaking and water exchange between the lagoon and the open ocean.

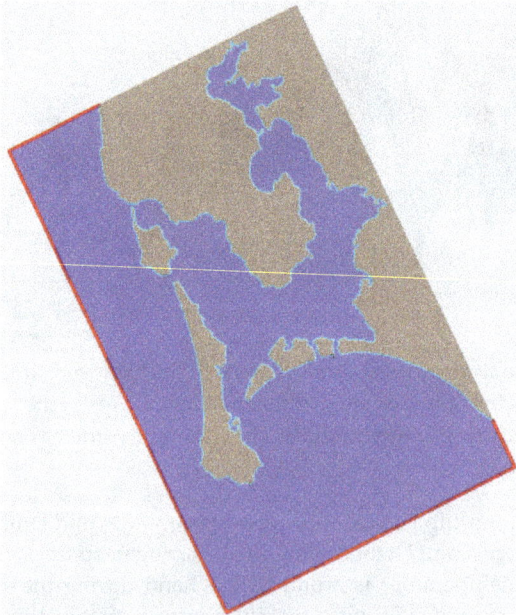

Fig. 7. The model domain used to calibrate the model at SQB.

The domain bathymetry was created using data reported by Flores (2006) and actualized with data collected 2009, 2010 and 2011 field work experiments; data set is referenced to mean low water. The resulting depths (Figure 8) clearly show the main channel as well as the bifurcation between BSQ and BF. The estuary delta is observed mainly at the east side of the estuary mouth. but a fan-shaped submerged sand bar is also visible surrounding the mouth. Inside the lagoon the orange colors represent the mud flat areas and it can be observed that the majority of the lagoon is shallow (between 0.5m and 1.5m).

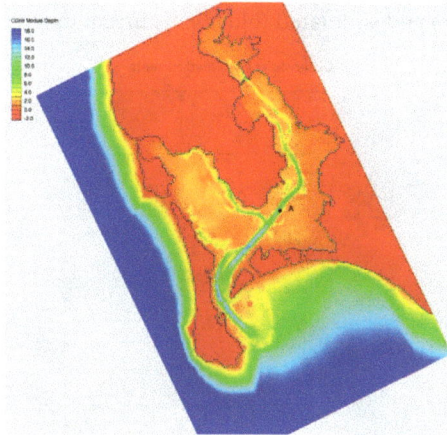

Fig. 8. Model bathymetry and calibration point A.

The main forcing processes within the lagoon are tide, most important, and wind (of secondary importance). At the open boundary (shown in Figure 7 as a red line) water surface elevation fluctuations due to tides are specified and all over the domain surface spatially constant wind condition is specified. To calibrate the model, wind and water surface elevation (WSE) data corresponding to a spring tide event from May 30, 2004 through June 10, 2004 was used (Figure 9). At the same time, Acoustic Doppler Profile (ADP) data from a fixed location (point **A** at Figure 8) was used to compare against model results. The comparison between observed and modeled current magnitude at point **A** (Figure 10), shows that a close relationship can be seen except at high magnitude currents where the model underestimates the currents by approximately 0.01m/s. Figure 11 shows a correlation plot between modeled and observed current magnitude and a R^2 factor of 0.87 was obtained.

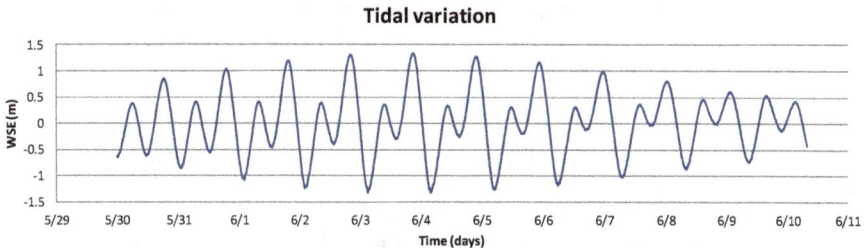

Fig. 9. Observed tidal period from May 30 to June 10, 2004.

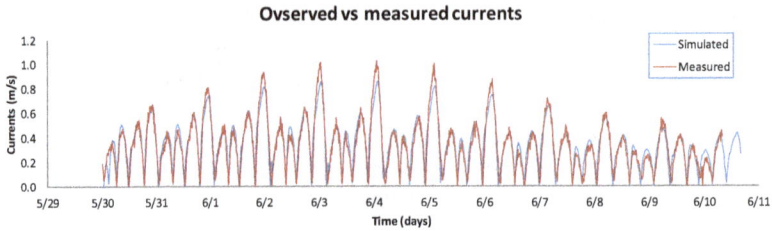

Fig. 10. Observed (red line) and simulated (blue line) current magnitude.

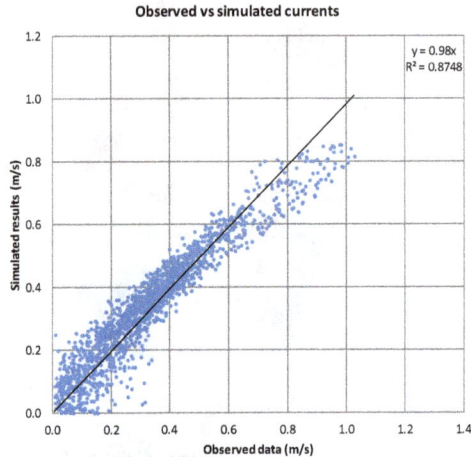

Fig. 11. Correlation between observed and simulated currents

Simulation results corresponding to spring and neap tide periods are presented in this section. The spring tide cycle corresponds to September 6th and 7th 2010 where the tidal range at the mouth was 2.25m. Figure 12 shows vector plots corresponding to ebbing and flooding moments for this particular tidal cycle. As the tide starts to descend, the flow field from the shallow areas converge at the main channels. At this point the most significant flow contribution to the main discharge originates from the west side of BF which is abundant in aquaculture activities. After approximately 1.5 hours, the flow field is well established at the BSQ Channel and Sonora Channel but is less evident at BF Channel. After four hours the strongest flow is mainly around the channels while the mud flats contribution is less significant. Finally after five hours the flow field is mostly observed at the channels. Throughout the ebbing period BSQ channel shows the most discharge. At the estuarine mouth, the flow follows the main channel and then forms a fan shaped flow field with the strongest currents observed at the southern section of the estuary entrance. As the tide changes, water starts to propagate inwards mainly through the southern section of the mouth in a funnel shape. Inside the lagoon the flow propagates mainly through the channels. For BF both channels contribute equally with the flow field propagating into the mud flats and a convergence of both water masses can be observed in the central area of this mud flat. After approximately 2.5 hours from low tide the water starts to propagate over the mud flat between Sonora Channel and BF Channel as well as over the east bank of the BSQ Channel. This pattern is then observed until the flooding process ends.

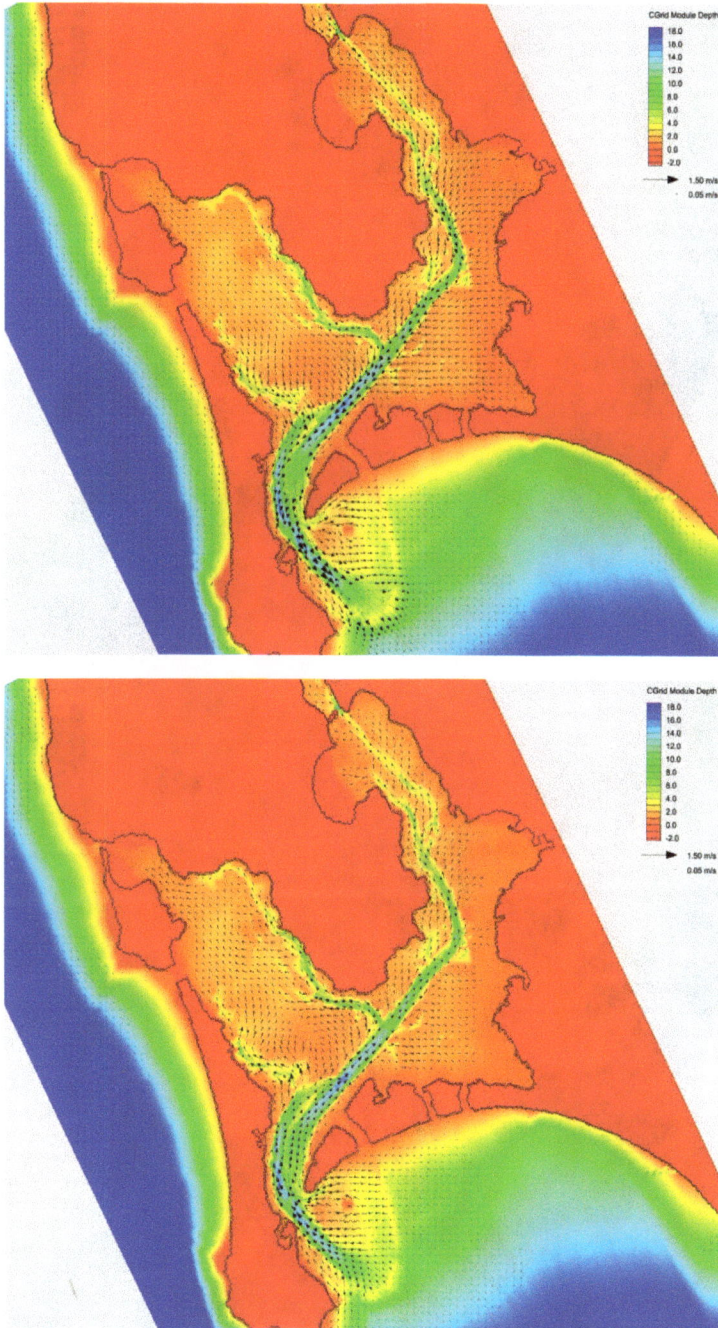

Fig. 12. (Top) Ebbing tide for a spring event, 3.5 hours after high water and (Bottom) flooding tide for a spring event, after 2.5 hours from low water.

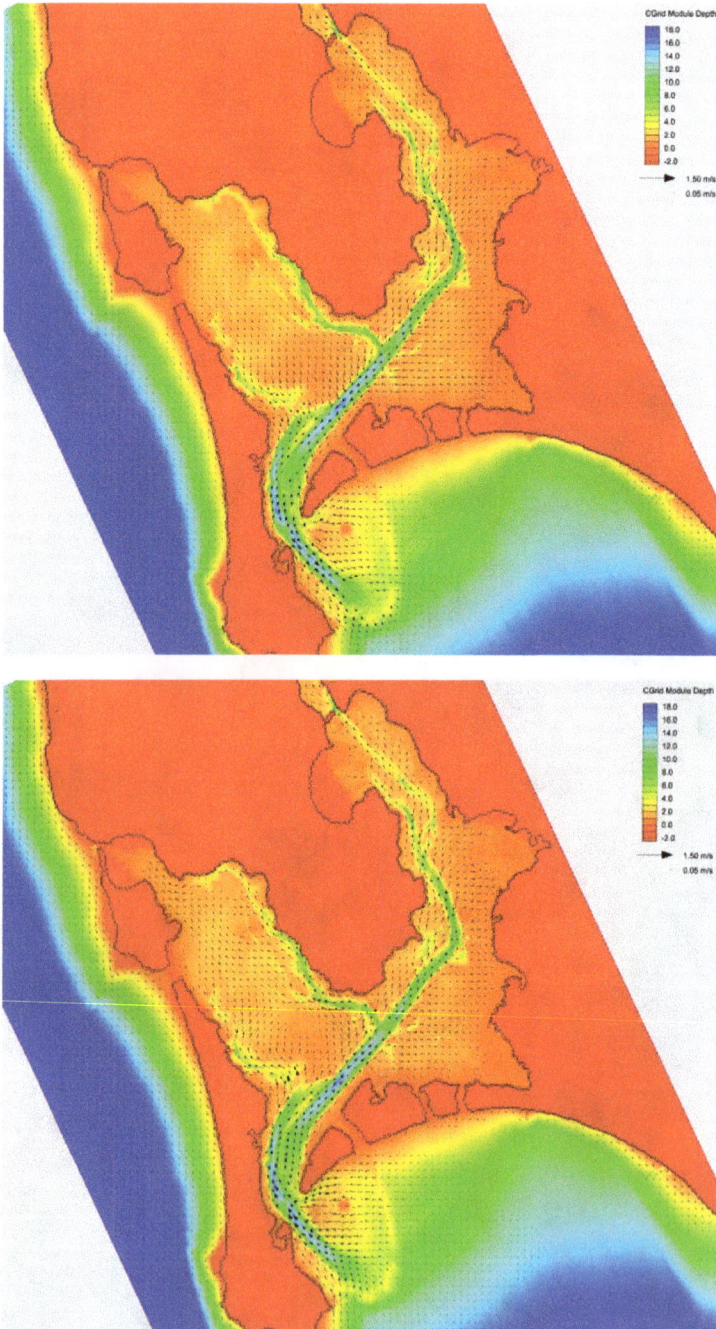

Fig. 13. (Top) Ebbing tide for a neap event, 1.75 hours after high water and (Bottom) flooding tide for a Neap event after 4 hours from low water.

The simulated neap tide corresponds to a range of approximately 1.25m. Figure 13 shows vector plots corresponding to an ebbing moment as well as a flooding moment. For the ebbing period the BF channel (similarly to spring tides) plays the most important discharge role. For the flooding time frame, it is observed that the channels distribute most of the inflow water rather than over the banks and through the mud flats.

3.2 Particle tracking

The particle tracking model (PTM) utilizes the computed flow field for the advection of simulated particles. Particle sources are set as point inputs where a discharge rate is set. The particles can be specified as neutrally buoyant or with a settling velocity. For this case since we will be focusing on diluted substances we specified the particles as neutrally buoyant.

The particle tracking model (PTM) was calibrated using drifter buoys. Since the model is bi-dimensional, the particles are transported using the vertically averaged flow thus, drifter velocities would be under estimated by the model (i.e surface currents measured by the buoys are higher than the average velocity as the average current is obtained through a logarithmic profile). Figure 14 shows a drifter experiment which took place on August 26th 2010 during an ebbing tide. This figure shows the drifters trajectory (black lines) as well as the simulated trajectories (purple lines)a bathymetric representation where the experiments were performed overlaid by the buoy trajectories (black lines) and the modeled trajectories (purple lines). As it can be observed the modeled trajectories are underestimated in comparison with the buoys. However, the path reflected by the model is similar to the observed path where an initial dispersion is noticed and then the flow field concentrates the particles into the main channel which then are re dispersed at the end of the trajectory. The best fit with the observed dispersion rate was obtained using a 0.6 m²/s coefficient.

3.3 Flushing time

Flushing time was calculated by setting different point sources throughout the domain and liberating a set of particles at time zero. Then after a certain time frame, an evaluation was performed to determine the amount of particles remaining inside the lagoon as well as their location with reference to their source. The analysis was done for a spring tide event where particles were released for two consecutive hours after high tide. Figure 15 show different time frames corresponding to an initial low water and subsequent tidal cycles at 1, 3 and 5 tidal cycles.

For the initial ebbing process most of the particles released from source **4** moved outside the lagoon into the delta area, particles from sources **1** to **3** are mostly aligned within the main channel, particles from source **5** are also transported through the Sonora Channel and particles from source **6** stayed within the upper section of BF. After one tidal cycle (**T1**) most of source **4** and **5** particles are outside the lagoon as well as some particles from source **3**, particles from sources **1** to **3** are somewhat mixed and stay on the main channel while particles from source **6** migrate to the eastward section of BF (where the BF channel is). After three tidal cycles (**T3**) most of the particles have been mixed and dispersed between BSQ, the main channel and outside the lagoon except particles from source **6** which are dispersed within the BF basin; overall approximately 2/3 of particles from sources **1-5** moved out of the system. After **5** tidal cycles (**T5**) only a few particles from sources **1-5** remain within the lagoon while most of particles from source **6** are still within BF basin. This experiment

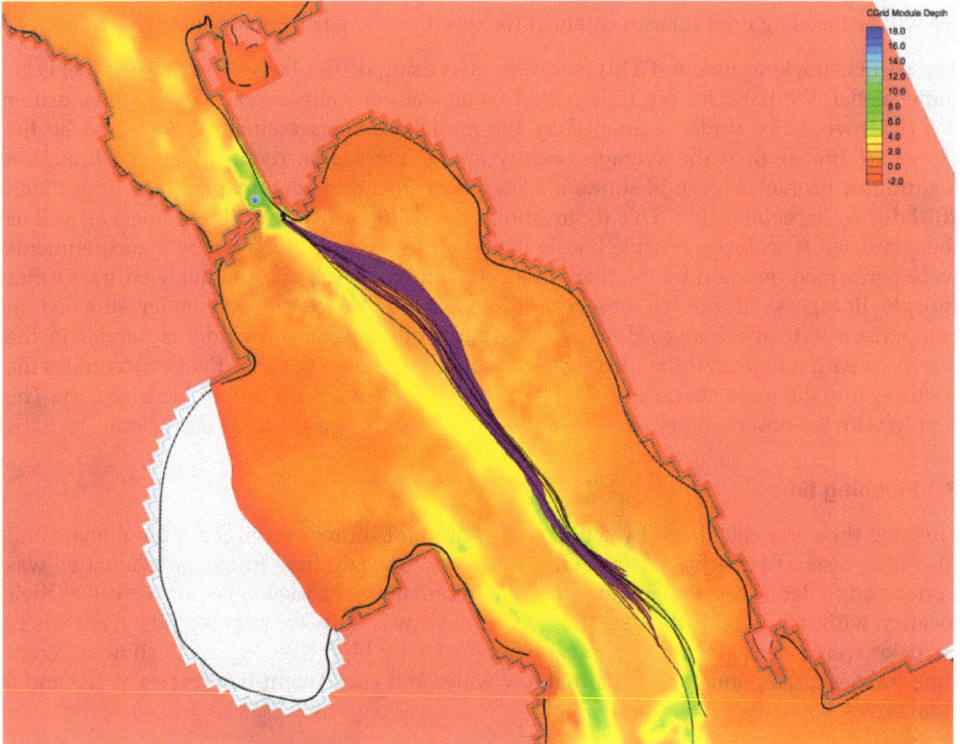

Fig. 14. Particle and buoy trajectories for August 26th ebbing tide.

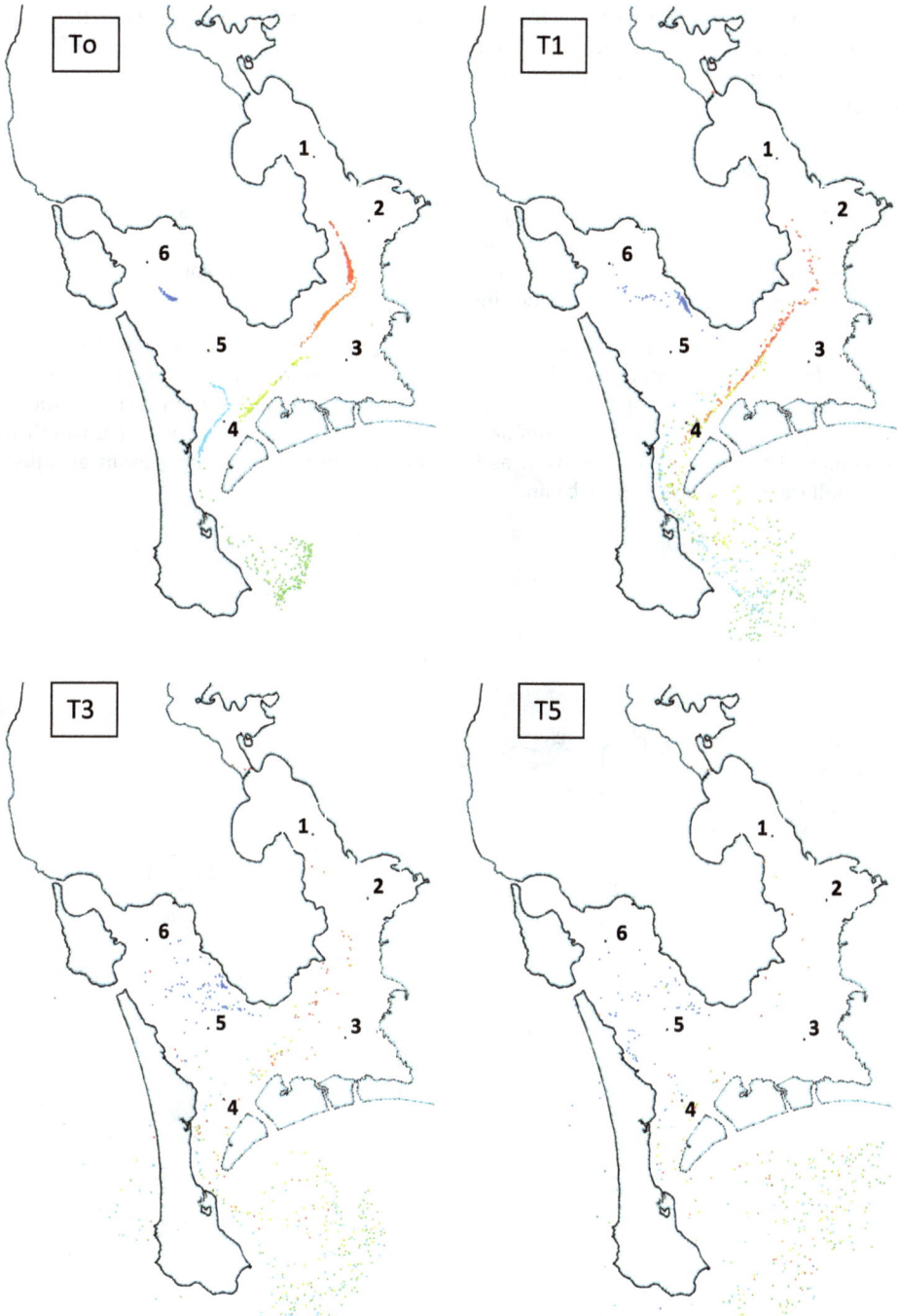

Fig. 15. Residence time for different zones within San Quintin Lagoon

shows that the flushing time from BSQ is much faster than for BF, especially the upper section of this basin. Also the Sonora Channel plays an important role in flushing BF as can be observed from source **5** and the main channel connecting to BSQ flushes the eastern mud flats represented by sources **2** and **3**.

3.4 Detritus inflow

Due to the fact that aquaculture is a very important activity inside BF basin, the model was set to simulate detritus inflow from open water into the lagoon. To do so, four point sources were specified just beyond the mouth delta. Figure 16 shows these sources as well as the advection of such particles during a flooding time frame.

From Figure 16a it can be observed that sources **1** and **2** are mainly transported into BF through the Sonora Channel, at the left side of BF, as well as over the central mud flat and a smaller fraction from this sources goes into SQB. For source **3** there is not much transported into the lagoon as the bulk stays outside and finally source **4** is transported through the main channel mostly into BSQ as well as the eastern mudflats from BSQ basin and just a very small fraction enters into BF basin.

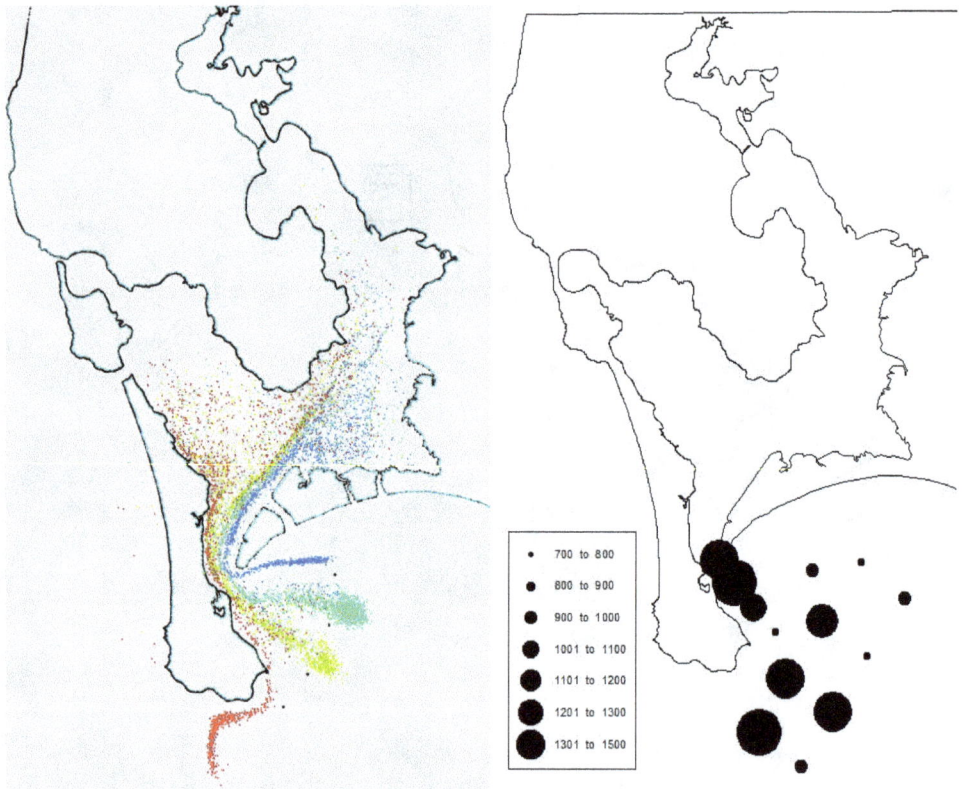

•	700 to 800
•	800 to 900
●	900 to 1000
●	1001 to 1100
●	1101 to 1200
●	1201 to 1300
●	1301 to 1500

Fig. 16. a) Modeled detritus inflow point sources and b) chlorophyll levels during a flooding period.

Figure 16b shows chlorophyll data taken with a Turner design fluorometer in June 21st 2011 during a flooding time period. This plot shows higher concentrations on the western side of the mouth thus, this chlorophyll rich water will be introduced into BF basin and utilized by the cultivated oysters.

Another relevant finding from this experiment is that the upper sections of BF basin and SQB basin do not show many particles which suggests low quality water which for BF basin is asseverated as the flushing of residues produced on this area is also poor.

4. Acknowledgements

This work is a contribution of project by UABC grants 14avaConvocatoria Interna and by F-PROMEP-36 SEP-23-006 grants.

5. References

Alvarez-Borrego, J. y Alvarez-Borrego S. 1982. Temporal and spatial variability of temperature in two coastal lagoons. CalCOFI Reports XXIII:, 188-197.

Delgado-González, O.E. 2010. Desarrollo y aplicación de una herramienta de gestiónpara el aprovechamiento acuícola en Bahía San Quintín,Baja California. PhD Thesis. Universidad Politénica de Cataluña, Barcelona, España. 182 p.

Delgado-González, O.E., Jiménez, J.A., Fermán-Almada, J.L., Marván-Gargollo, F. , Mejía-Trejo, A. ,García-Esquivel, Z. 2010. La profundidad e hidrodinámica como herramientas para la selección de espacios acuícolas en la zona costera. Ciencias Marinas 36(3). 249-265.

Fischer, H.B. List, E.J., Koh, R.C.Y., Imberger, J. and Brooks, N.H. 1979. Mixing in inland and coastal waters. Academic Press, Boston, 483 pp.

Flores-Vidal, Xavier, 2006. Circulación Residual en Bahía San Quintín, B.C., México. CICESE. MSc. 91pp.

García-Esquivel, Z., González-Gómez M. A., Ley-Lou F. y Mejía-Trejo A. 2000. Microgeographic differences in growth, mortality, and biochemical composition of cultures pacific oysters (*Crassostrea gigas*) from San Quintin Bay, Mexico. Journal of Shellfish Research 19 2. 789-797.

Héral, M. y Deslous-Paoli J.M. 1991. Oyster culture in european countries. CRC Press, Inc. 154-186.

Hernández-Ayón, M, Galindo-Bect M.S., Camacho-Ibar V.F., García-Esquivel Z., González-Gómez M.A. y Ley-Lou F. 2004. Dinámica de los nutrientes en el brazo oeste de la Bahía de San Quintín, Baja California, México, durante y después de El Niño 1997/1998. Ciencias Marinas 30 1A. 129-142.

Lara-Lara, J.R. y Álvarez-Borrego S. 1975. Ciclo anual de clorofilas y producción orgánica primara en Bahía San Quintín, Baja California. Ciencias Marinas 2 1. 77-97.

Millán-Núñez, R., Alvarez-Borrego S. y Nelson D.M. 1982. Effects of physical phenomena on the distribution of nutrients and phytoplankton productivity in a coastal lagoon. Estuarine, Coastal and Shelf Science 15. 317-335.

Ortiz-Pérez, M.A. and D.L. Lanza-Espino, 2006. *Diferenciación del espacio costero de México: Un inventario regional*. México: Serie deTextos Universitarios, Instituto de Geografía-UNAM, México.137p.

Ocampo-Torres, F.J. 1980. Análisis y predicción de velocidad mediante un modelo unidimesional en Bahía San Quintín, B.C, México. Universidad Autónoma de Baja California BSc 90 pp.

Stigebrandt, A. 1977. On the effect of barotropic current fluctuations on the two-layer transport capacity of a constriction. Journal of Physical Oceanography 7, 118-122.

6

Consequences of Land Use Changes on Hydrological Functioning

Luc Descroix and Okechukwu Amogu
IRD / UJF-Grenoble 1 / CNRS / G-INP, LTHE UMR 5564, LTHE,
Laboratoire d'études des Transferts en Hydrologie et Environnement,
France

1. Introduction

Land use and land cover changes are well known to have a strong influence on the water cycle and its evolutions. Vegetation cover is the main component of the land surface that explains the behaviour of rainwater transforming into runoff, infiltration or evapo-transpiration. Land use changes such as afforestation or deforestation, increase in cropping or grazing areas, likely modify this water balance and land surface behaviour. However, urbanisation and sometimes some geopolitical events, could significantly influence the water cycle.

Human activities that change the land cover can considerably change the water cycle within a basin; for example, the present-day situation of the Aral Sea basin in Central Asia where intensive irrigation resulted in a near cessation of the water inflow from the Syr Daria and the Amu Darya rivers. In the Mediterranean basin and the Sahel, the successive removal of vegetation by forest clearing and overexploitation increased drought risks. Inversely, the increase in frequency of flooding in the Ganges basin can be explained by deforestation (Kuchment, 2008). Furthermore, dams and reservoirs dedicated to irrigation increase evaporation and decrease the transport of sediments.

In the following sections, we will describe the main changes observed in the water cycle in different areas. Firstly, we will compare the hydrodynamic changes in two different mountainous areas subject to very different socio-economic changes in the recent decades and centuries. As "water towers", mountains are the natural reservoir and providing areas for most of the semi-arid and arid regions in the World with the land use and land cover changes in these areas causing changes in water supply in the downstream plains. Secondly, we will compare two different areas in West Africa in the current period in order to highlight the role of soil hydrological functioning from the primary scale to the regional scale. The non-linearity of hydrological processes in this area gives rise to the "Sahelian paradox" as the anthropic impact exceeds the influence of the climatic inter decadal signal. LUC/LCC (land use change/land cover change) not only lead to consequences in hydrological balances, but also in sedimentary balance. Land cover has a direct influence on soil erosion and conservation. Both erosion and runoff determine the river transport capacity, whose evolution appears as geomorphic evidences of the historical behaviour of the basins. In a third section, we will describe the consequence of land use changes in erosion, transport and sedimentation processes; finally, other causal factors of water cycle modifications will be discussed in the last part of this chapter.

2. Consequences

Mountainous areas are commonly characterized by forests, whose removal reduces infiltration and improves the conditions of overland flow (Kuchment, 2008). As a consequence, flood runoff and peak discharges may significantly increase. On the contrary, the main effect of afforestation on the water cycle is an increase in transpiration and interception of precipitation, which in turn results in a decrease of the volume of total discharge. Simultaneously, the higher infiltration capacity of forest soils increases the opportunity for groundwater recharge, and the flow of small rivers tends to be more sustained.

In this section, the hydrological behaviours of two different mountains are compared.

The French Southern Alps constitute a mountain area in stage of resilience. This region suffered up till the end of the 19th century from severe land degradation due to deforestation, crops extension and overgrazing. Large areas and hillslopes were degraded and numerous "torrents" appeared. Floods and inundations in the alpine valleys and in the downstream flood plains were frequent and devastative. Flood flows increased and on the contrary, base flows were reduced, making water management very difficult. A strong increase in soil losses in the upstream part of the basins led to river beds widening and rising. As in other mountainous areas of Southern Europe, a hard policy of land reclamation was initiated in the second half of the 19th century (French 2nd Empire and 3rd Republic). Reforestation and torrent correction were strongly developed (Table 1); then, socio-economic evolution and three wars (against Prussia in 1870, and the 1st and 2nd World Wars) depleted these areas of their inhabitants (Descroix & Gautier, 2002; Descroix & Mathys, 2003).

Department	total area (km^2)	Forested area (%)					1992 forested area in 1992 (km^2)
		1770-1780	1878	1914	1948	1992	
Alpes Hte Provence	6954	10	18	26	31	36	2500
Hautes Alpes	5534	11	19	27	29	34	1860
Drôme	6560	15	23	29	31	43	2800

Table 1. Evolution of the forested area in the three French Southern Alps districts since 1770

Bed rock outcrop	soil losses in mm.yr^{-1}	number of sites documented
Black marls	9.4	13
Blue marls	8.6	4
Grey marls	8	1
Molasses	1.4	1
Glacial fields	16	1
Oligocene clays	30	1

Table 2. Measured erosion depth in the main fragile outcrops in Southern French Alps

The most important sediment providing material in the French Southern Alps is the black marls (Table 2), which outcrops exceed 1000 km^2; but some other fields are also very sediment productive, as indicated in Table 2. For centuries, these very degraded marly terrains were providing stream flow and suspended sediment load to the Alpine rivers.

Nowadays, the forested area is more than three times that of the mid 19th century, and erosion is no more a concern for the population neither is the flood, and inundation has become very uncommon in the last few decades. Flooding and sedimentation are now very sparse and they are no more a constraint to development either for agriculture or for other socio-economic activities. The Durance river valley was completely developed during the 20th century and a series of dams were built along the river in order to provide electric power as well as water to the irrigated district of the lower Provence and the Comtat Venaissin and for the cities of Marseille and Toulon.

Inversely, in Northern Mexico, the hillslopes of the **Western Sierra Madre** have for some decades been going through a great erosion stage. This is due to a generalised over-exploitation of the environment. This area has been characterized at least since the 1970s by a general overgrazing and forest clearing in the upper part of the basins (Descroix et al., 2002, Viramontes & Descroix, 2003). Both human originated processes cause an increase in bare soils areas, leading to soil compaction and a significant decrease in soil hydraulic conductivity and water infiltration. Other consequences are a decrease in soil water holding capacity and an increase in runoff. Instead of gullies, this area is characterised overall by sheet erosion and sheet runoff, all the soil surface being degraded: fine soil particles are removed by drop detachment and transported by runoff. This decrease in soil permeability is demonstrated by data collected in Table 3 and Table 4. At the end of the 20th century, deforestation was severe and more than 50% of the extended pine and oak forests of the upper parts of the Sierra had already been cleared, with the exploited wood volume being commonly three times higher than the authorized volume (Viramontes & Descroix, 2003). Measurements made at the 50 m^2 plot scale show that runoff was increased by one order of magnitude and erosion by two orders of magnitude after the tree cutting (Table 3). In the first years following clearing, the remaining litter partially protects the soil.

Plot characteristics	runoff coefficient (%)	soil losses (g.m^{-2})
Non cleared plot	2.8	1.1
Cleared plot with remaining litter	8.5	30
Cleared plot without litter	23	133

Table 3. Influence of clearing on runoff and erosion measured at the plot scale (50 m^2) in a forest environment in the Western Sierra Madre (Northern Mexico)

In the same upper part of the Sierra Madre, soil physical characteristics were measured on grassland areas, in order to compare their behaviour in non grazed or normally grazed areas on the one hand, and in spread overgrazed sectors on the other hand. Obviously, overgrazing causes a strong increase in runoff (five times higher) and erosion (one order of magnitude higher), and a significant reduction in hydraulic conductivity and soil porosity and inversely a rise in soil bulk density (Table 4). The three latter parameters are part of the explaining factors of the two former ones. They are due to cattle trampling which provokes soil compaction and a closing of the pores, most of them becoming inactive (Descroix et al., 2002).

Soil characteristics	runoff coefficient %	soil losses g.m^{-2}	hydraulic conductivity mm.h^{-1}	bulk density g.cm^{-3}	Porosity %
Non grazed	8	7	77	1.21	49
Overgrazed	43	90	24	1.55	35

Table 4. Soil characteristics on non-grazed and overgrazed areas in a grassland environment. Runoff and erosion measured on 50 m² and 1 m² plots (10 repetitions per class); hydraulic conductivity measured with disk infiltrometer, bulk density by the "pool method", porosity using a mercury porosimeter.

Thus, we observed in the last 3 decades the following features at the basin scale (those of Rio Ramos and Rio Sextin, two tributaries of the Rio Nazas, whose catchment areas are respectively 7100 and 4700 km²) (Fig.1):

- a reduction of the basin lag time to precipitations
- a reduction of the total duration of river flow
- an exaggeration of the flood peak
- an increase in flood flow
- and a decrease in base flow

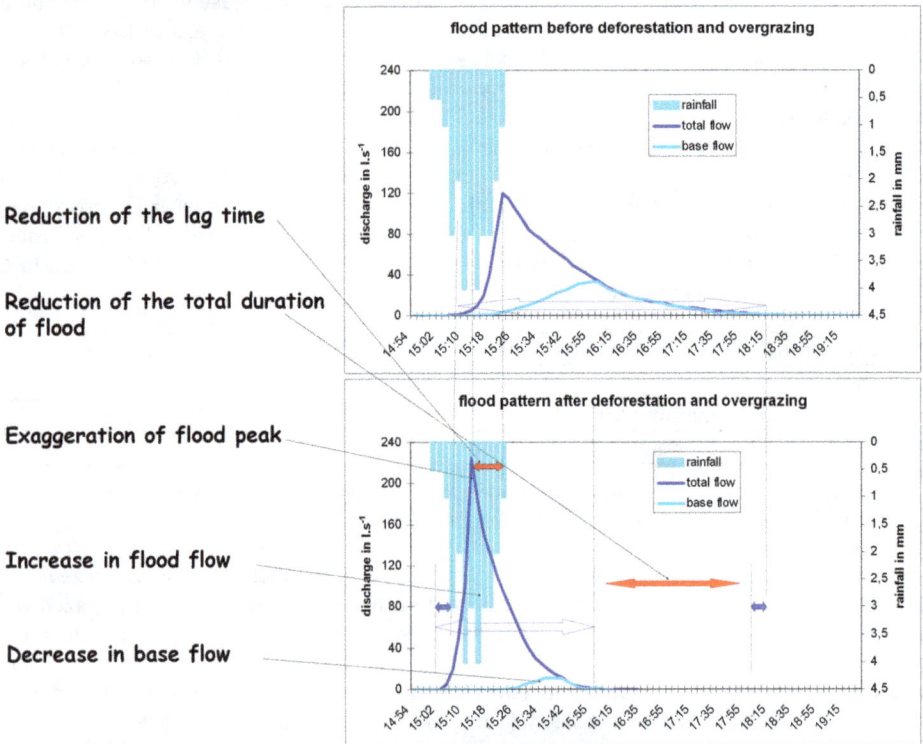

Fig. 1. Observed hydrological changes linked to land use changes by the evolution of hydrographs in sub-basins of some km² into the Ramos and Sextin Rivers basins

All these facts are the results of a reduction of the water holding capacity of the soils and the basins.

These consequences make water management more difficult, because there is less water in the rivers during the dry season, at the moment where water is most required for agriculture, grazing and other needs. Simultaneously, the increase of the discharge, the peak flow and the flooding risk during the rainy season also makes water management more difficult due to both the necessity to increase the volume of reservoirs to ensure the same volume of supplied water and the severe increase in silting up which could cause other problems as dam sedimentation, and local difficulties for irrigation and navigation.

This will also have consequences in the sediment balance (see part 4).

Land cover highly impacts the water cycle; this is even more marked in mountainous areas where there is commonly more rainfall, and steep slopes favourable to runoff and erosion. Mountains being natural water towers, land use and land cover changes could have there consequences in the water balance of extended areas. Processes described in fig.1 applied at the scale of great water providing areas could lead considerable changes in water supply and oblige to great water management modification. In sub-Saharan Africa, two main water towers (the Guinean mountains and the Ethiopian Highlands) provide water to hundreds of millions inhabitants, with significant proportion of them living in arid and semi-arid areas. An increase in both flood and drought hazards is observed from some decades and this is a threat for the downstream located societies. This is also a geopolitical challenge, and water scarcity could be in the future the cause of severe conflicts.

3. The sahelian paradox

In the West African Sahel, it was observed in recent decades that during the Great Drought that occured in this region (this has not yet come to an end, although rainfall amounts are partially increasing again; see Ali and Lebel, 2009 and Lebel and Ali, 2009), Sahelian rivers showed a significant increase in runoff in spite of the reduction of the annual rainfall amount. Since this cannot be due to rainfall, there is a consensus to estimate that this is a consequence of land use change (Albergel, 1987; Descroix et al., 2009; Amogu et al., 2010). West Africa suffered a great drought from 1968 to 1995. The Sahelian area (semi-arid part of West Africa, with annual rainfall lower than 700 mm) was more particularly affected with a decrease in rainfall ranging from 20 to 35% during the 1970s and the 1980s compared with the 1950-1968 period. The drought has not really ended, only attenuated in the eastern half of the Sahel, and rainfall remains very low in its western part (Ali and Lebel, 2009; Lebel and Ali, 2009).

The large rivers of West Africa (Senegal, Gambia, Niger, Chari among others) experienced an expected decrease in discharge since the beginning of the Drought (1968); for the period of 1970–2000, the decrease in the mean annual discharge of the region's largest rivers (fig.2), namely the Senegal and Niger rivers, was in proportion almost twice as much as the decrease in rainfall. Similar trends have been observed for smaller river systems while, in contrast, other studies have indicated a runoff increase in some Sahelian catchments. Albergel (1987) remarked a paradoxical fact: although the rainfall was decreasing from 1968, runoff increased in experimental Sahelian catchments in Burkina Faso; in the Sudanian catchments of the same country, more classically, the discharges were decreasing. This is even more paradoxical because rainfall decreased stronger northward, in the Sahelian area, than in the Sudanian one (Fig. 3). Amani and Nguetora (2002) highlighted an increase in runoff of some right bank tributaries of the Niger River, and they observed that their flood began one month earlier during the 1980s than during the 1960s; they also remarked that in the Niger River at Niamey station, sometimes in the recent years (at the moment) the first flood peak was higher than the "guinean flood" peak, a process that never occurred before 1984. Mahé et al. (2003) observed an increase in discharge of the Nakambé River, one of the upper Sahelian tributaries of the Volta River; Mahé and Paturel (2009) noticed an increase in runoff coefficients and in discharges in the Black and White Gorgol basins in the western Sahel (Mauritania). This is also observed in the south Sahelian Sokoto basin in Nigeria (Mahé et al., 2011). Most of the Sahel seems to be concerned by this "Sahelian paradox". Amogu et al (2010) proposed a regionalisation of hydrological processes, with a big difference between Sudanian areas where runoff decreased strongly and Sahelian (and, in recent years, north-Sudanian) areas where runoff increased significantly. They also determined that this increase in runoff was observed at all the spatial scales.

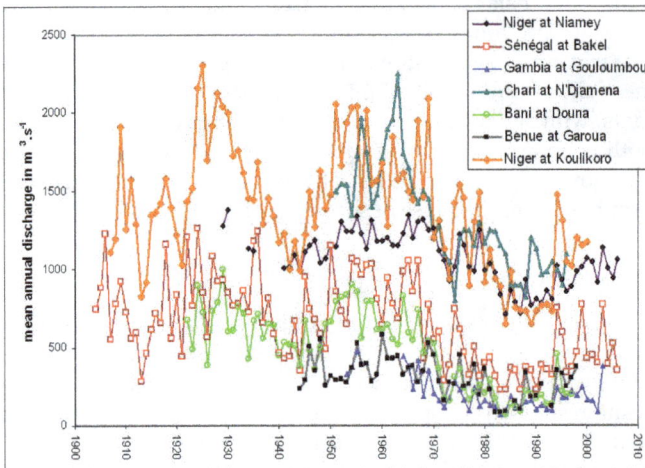

Fig. 2. The effect of the Great Drought on the discharge of the main West African rivers: a general expected decrease in runoff after 1968

Until the end of the 1990s, all the water supplied to the Niger River downstream from Gao came from the right bank tributaries. On the left bank, endorheism is the main hydrological mode, and the observed increase in runoff led to an increase in the number, the volume and the duration of ponds. As the ponds are the main groundwater recharge points, this caused a significant increase in groundwater recharge and a rise in water table level in the western part of the Iullemeden sedimentary basin, during the drought; this was named the "Niamey's paradox" by Leduc et al (1997) and described by Séguis et al. (2004). Also in this case, the increase in runoff is supposed to be due to land use change, particularly land clearing (Leblanc et al., 2008). Finally, if the drought is the most cited process in the Sahel, "floods in drylands" are not to be considered as an oxymoron, and they are unfortunately not well known, in spite of their current rise in number and strength (Tschakert and al., 2010).

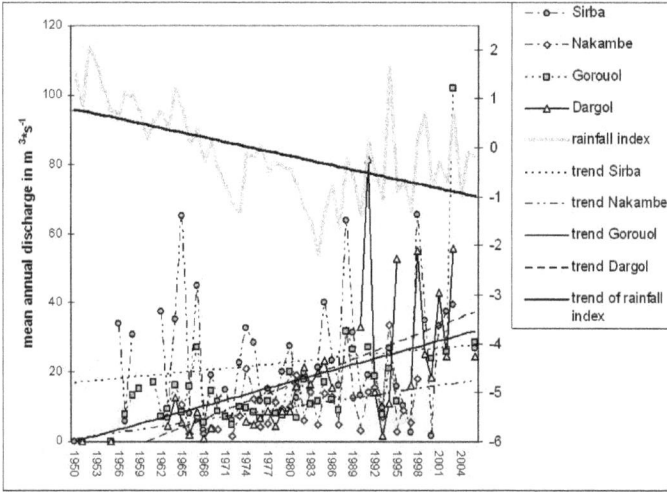

Fig. 3. The unexpected increase in discharge of gauged Sahelian rivers during the Great Drought

Most of the changes in the water cycle characteristics in West Africa result from the long drought period (observed since 1968) and the reduction of the soil water holding capacity. However, recent studies (Mahé et Paturel, 2009, Descroix et al., 2009) have shown that anthropic factors better explain the spatial and temporal variability of the processes than the climatic factors.

As a matter of fact, as in all geo-climatic environments, a combination of natural and human parameters leads to changes in processes, which explains the resource evolution and possibly, as a feed back effect, the climatic evolution; but the last hypothesis remains to be founded on tangible evidence. The combined action of the very clear climatic and anthropic trends led to the following hydrological consequences (Albergel, 1987, Mahé, 1993, Amani et Nguetora, 2002, Mahé et al, 2003, Mahé et al, 2005, Andersen et al., 2005, Mahé, 2009, Mahé et al, 2009, Amogu, 2009, Descroix et al., 2009, Amogu et al., 2010; Descroix et al. 2012):

- An increase in runoff and river discharge in the Sahelian region (fig.3), linked to the partial removal of vegetation cover and to the soil crusting in increasingly widespread areas, causing a reduction of the soil water holding capacity; runoff becomes mostly of the "hortonian" type (produced when rainfall intensity exceeds the soil infiltration capacity) (Horton, 1933); as this increase in runoff occurred during the great drought, it is called the "hydrological Sahelian paradox";

In the Sudanian region, where the rainfall deficit was lower, an inverse trend has long been observed, and runoff and discharge were logically and expectedly decreasing with the reduction in rainfall. Hydrological processes remain there mostly "hewlettian" (Hewlett, 1961, Cappus, 1960), i.e. runoff occurs only when the soil is saturated; as soils are mainly deep, the reduction of rainfall firstly affected the part of water which was dedicated to runoff (fig.4). The increase in runoff is not observed in the Sudanian area (where rainfall amount ranges from 700 to 1300 mm per year), where the reduction in rainfall was not so strong (10-15%) due to lower land use changes and a reduced impact of these changes. With soils being deeper in this area, the important role of soil and groundwater in runoff yield is highlighted. In this "hewlettian" context, a decrease in rainfall firstly reduces the part dedicated to runoff.

- In the north of the Sudanian area, the evolution of discharges is intermediate between the two previous ones described above; the discharge seems to be unchanged (Fig. 5). This can be explained by the development of "horton-type" processes due to land clearing and associated soil crusting, in an area which remained relatively void of inhabitants for a long time (due to sanitary and historical reasons).

- A decrease in the discharge of the main regional rivers (Gambia, Senegal, Niger, Volta, Chari) (fig.3); their main water sources originate in the sudano-guinean area more than in the sahelian region, although their basins extend into the sahelian area (Niger, Sénégal) ; therefore, the increase in sahelian runoff does not compensate the reduction in Sudan-originated flows. Thus, the discharge of the Niger River at Koulikoro (in Mali, upstream from the Niger Inner Delta –NID-) was halved during the last three decades of 20th century (Olivry, 2002; Andersen et al., 2005).

- In certain endorheic areas of the Sahel, the increase in runoff caused for some decades a rise in the number of ponds, of their volume and their duration; while these ponds were defined as the main recharge area for groundwater, a significant rise in water table level was observed (Leduc et al., 1997; Leblanc et al., 2008). Recently it was determined that the sandy deposits linked to the acceleration of soil erosion (gully bottoms, spreading areas, dejection fans, etc) are new recharge areas in extension (Descroix et al., 2012).

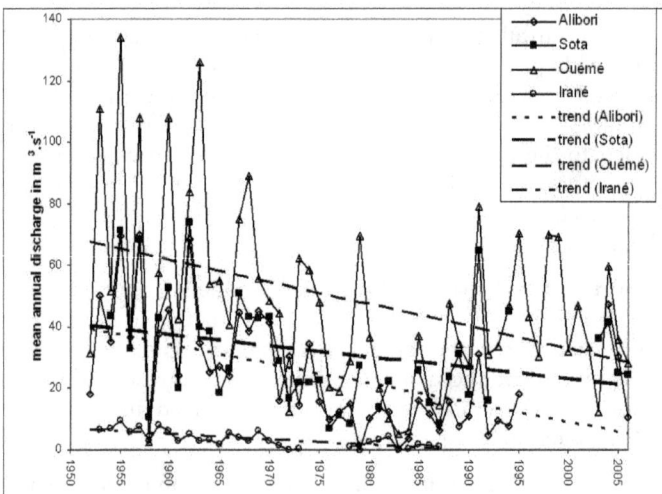

Fig. 4. The expected decrease in discharge of gauged Sudanian rivers during the Great Drought

During the end of summer 2010, the surroundings of the Niger River in the Niamey area suffered a severe flood. We show here, that although there was a partial recovery after the severe droughts of the 80's and 90's, rainfall was not the main factor of this exceptional flood: 2010 annual rainfall was below the 50's 60's values, and neither rainfall distribution during the year, nor the size of the rainy events changed significantly during the last decades. Rather, the hydrological behavior of the right bank tributaries of the Niger River shows a severe increase in runoff since the 80's (Fig.3), which is still ongoing for two of them, and resulted in changes in the hydrograph of the Niger River during the rainy season, lasting from June to August in Niamey (Fig. 6). Otherwise, during the last decade (2000-2009), the flood occurred 2 or 3 weeks earlier at the Niamey station than in previous decades. During the last two decades, the end of the main flood has come relatively early in spite of the increase in discharge, continuing to occur 2 months earlier than in the 1950s and 1960s (before the drought) (Fig. 6). We show that this is the result of an increase in bare and crusted soils as a consequence of human pressure and/or of non reversible effects of the drought. This questions the idea of the re-greening Sahel arising from recent remote sensing studies. Of particular interest to policymakers is the effect of human pressure (and particularly intensive cropping and grazing) on hydrological regimes of Sahelian rivers, which could be partly mitigated by either the use of updated urbanization rules of by promoting re-greening experiments which had already some success in this area.

The hydrograph of Niger River downstream from the Niger Inner Delta (northern Mali) is bi-modal, being constituted by (Fig. 6):

-   a first, local and sahelian flood, fed by the discharge of the sahelian right bank tributaries of the Niger River, mainly the Gorouol, the Dargol and the Sirba, as well as some small direct tributaries from both the right and the left banks; it is also called the "red" flood due to the important suspended load of lateritic origin; it lasts from July to late September;

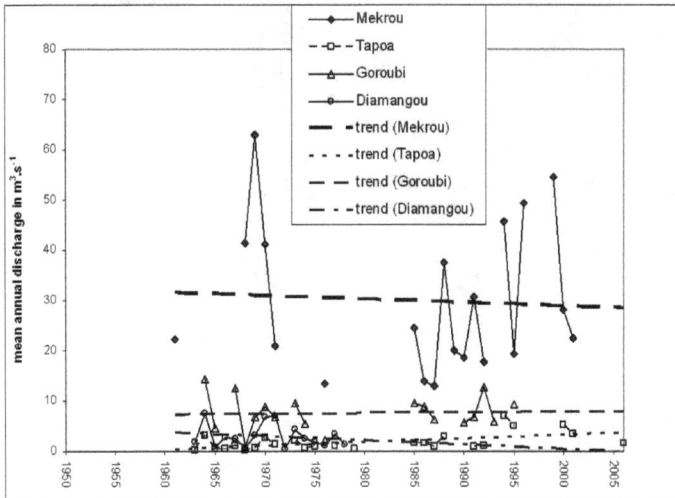

Fig. 5. The intermediate behaviour of northern Sudanian basins: towards a Hortonian process ?

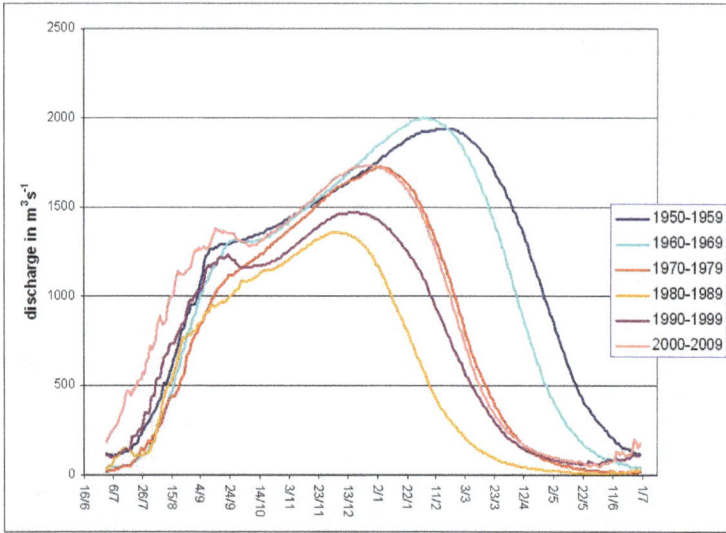

Fig. 6. Evolution of the hydrograph of Niger River at Niamey station representing the evolution of discharge and regime

Fig. 7. Inundation of the right bank part of the city of Niamey (Niger) by the Niger River in August 2010

- a second, sudano-guinean flood occurring at the beginning of October and lasting 4 to 6 months, is fed by water coming from the mountains of Guinea by the Upper Niger and its tributaries; it is called the "black" flood due to weak sediment transport (suspended load is blocked and stored in the Niger Inner Delta).

During the red flood of 2010, the water reached its highest recorded level, the flooded area was 3.1 km² in the city of Niamey, some 5 000 people lost their homes and crops and the food shortage resulting from drought of the two previous years provoked an increased havoc on local population life (Fig. 7) (Descroix et al., 2011).

4. Influence on sedimentation

In this section, the consequence of land use changes on erosion, transport and deposition of sediment will be analysed through the pattern of river beds evolution. As it was shown in section 2, a major land management policy was driven in the Southern French mountains at the end of the 19th century (as in other European Mediterranean areas, and, later, in French North African dependencies). The aim was land reclamation in order to control soil degradation as an origin of flooding and sedimentation in downstream floodplains. At the small basin scale, the results were very effective and efficient, since even very degraded hillslopes were reforested and protected against erosion (see fig. 8).

Fig. 8. The Brusquet catchment (French Southern Alps) in 1880 (left) and in 1990 (right) (*photos IRSTEA (formerly CEMAGREF)*)

This led to important changes in the downstream hydrography, the decrease in bed load and suspended load linked to the afforestation and flood hazard reduction causing a strong evolution of river beds in most of the major rivers in the southern European mountains (Descroix and Gautier, 2002). The river beds were affected by:

- a reduction in river bed width;
- a river bed entrenchment;
- an invasion of the river bed by riparian vegetation.

These processes appeared some decades after the beginning of reforestation which was initiated in the second half of the 19th century and the rural abandonment strongly accelerated the increase in forested areas during the 20th century. During the 1970s, the first evidence of the sediment load deficit appeared with the beginning of the river bed entrenchment, accelerated by gravel extraction in most of the French Alps rivers. Although this activity was forbidden at the beginning of the 1980s, the river entrenchment remains severe due to the remaining sediment transport deficit linked to the natural afforestation, preventing soil material from being detached and transported. In many cases, the river bed entrenchment reached several meters, up to 3 meters in the Buech river valley (Descroix and Gautier, 2002), locally more than 10 m in northern French alpine rivers (Peiry, 1988). In the last decades, 90% of the total length of French Alpine rivers have been affected by this sedimentary déficit (Peiry et al., 1994). This could lead to the destruction of bridges and embankments, and led the authorities to drive new land management policies applied to river beds: new embankment, protection of bridges, etc (Fig. 9).

In Northern Mexico, the strong overgrazing and deforestation cited in section 2 are linked to speculation; it is easy to earn much money without taking into consideration the land conservation; selling wood to sawmills, and calves to the US market became good businesses during the last decades. In the 1980s and the 1990s, Durango state (Northern Mexico) produced 5 millions m^3 of wood per year instead of the 1.7 million m^3 authorized by federal forestry institutions (Descroix et al., 2001). During the 1990s and the beginning of the 2000s, a general trend of overgrazing was observed in the whole Northwestern part of Mexico (Chihuahua, Sonora, Sinaloa and Durango States) (Descroix, 2004).

Fig. 9. The Céans river bed in 2004 (left) and in 2011 (right) (Southern French Alps). The bed of the Céans was entrenching strongly in the last decades, due to a great decrease of erosion and sediment transport. The river banks were falling regularly in the entrenched bed (left), obliging the authorities in building new embankments to protect surrounding areas from destruction (right).

This led to the on-going land degradation described in section 2. This causes a severe increase in suspended and bed load sediment in river beds, which in turn leads to the widening and enhancement of these beds (fig. 10a): downstream, this provokes the silting up of numerous dams located at the boundary of the coastal Sinaloa and Sonora plain (Fig. 10b); the latter being arid or semi arid, it needs water from the Western Sierra Madre to supply the irrigated districts. Clearly, land overexploitation in the upper part of the basins is threatening the agricultural and economic development of the relatively rich coastal plain, as well as the endorheic basin of the Nazas River and the Conchos basin, the main Mexican tributary of the Rio Bravo/Rio Grande River.

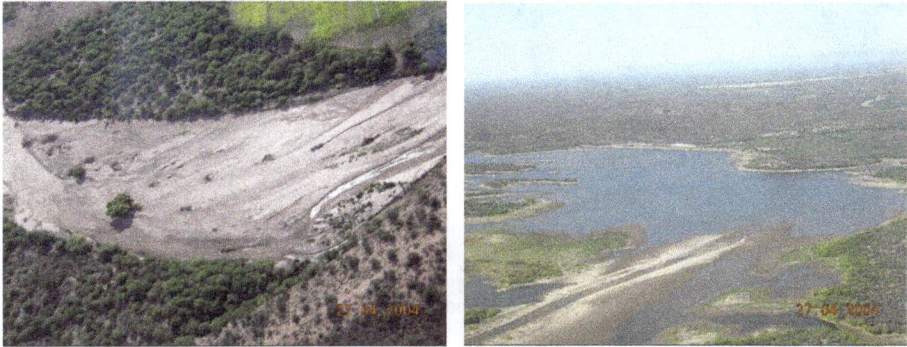

Fig. 10. Evidence of silting up in the Matapé River bed: 10a (left) a general widening and enhancement of river bed is observed; 10b (right) downstream, sedimentation is filling the dams reservoirs, threatening the irrigation downstream in the Coastal, semi-arid, plain.

In section 3, we saw that West Africa is suffering a strong land use change due to both a 40 year long drought and a severe demographic pressure (the highest demographic growth in the World since the beginning of the 1990s).

What are the consequences of this pressure on the regional sedimentary balance?

There is a strong opposition in the sedimentary balance of the Niger River upstream vs downstream from the Niger Inner Delta.

- In the Bamako area (upsteam reach) a strong sedimentary deficit has been observed due to sand resource strong overexploitation; furthermore, part of the sediment load is being stored in the Selingué dam (Sankarani River, one of the main tributaries of Niger Upper reach) (Ferry and Mietton, 2011). This is recent and was not observed previously (Picouet, 1999).
- Inversely, in the downstream reach, a strong silting up of the Niger's riverbed has been observed, linked to both the severe increase in soil erosion in the Sahelian basin and the significant decrease in Niger river discharge (-50% after 1968). (Amogu, 2009, Amogu et al., 2010) (Fig.11).

The sedimentary deficit of the Niger riverbed upstream from Bamako is also evidence of the impact of urbanisation on the hydrological behaviour of river basins.

Like the Niger River, some other great river basins in the world are affected by changes in sedimentary balance; the following examples show the importance of taking into account this balance in developing land management policies:

Fig. 11. Evidence of the silting up of Niger River in its middle reach: at left, sandy deposit in the river bed at Niamey (Niger); at right: the new Kourtéré (formed in 1998) alluvial cone invading the Niger River bed in its right bank, just upstream from Niamey (*photo E. Gautier*)

- The Nile River: all the sediment is stored in the Upper Aswan dam (total capacity 169 billion m³), thereby retaining sediment that could fertilize the lower part of the valley, where this fertile sediment would ensure good harvests in a time of demographic growth. This sediment retention also caused a strong decrease in the volume of fisheries in the mouth of the Nile River, due to the reduction of nutrients, retained in the dam. In addition, the overuse of Nile River water for irrigation causes a salinization of the lower part of the river, the low stream flow allowing sea water to enter the delta and the lower river bed.
- As the Nile, the Volta River in West Africa, is equipped with one of the biggest dams in the world (Akosombo dam, 160 billions m³ stored); but with all the alluvium being blocked in the dam, the sediment balance became negative some decades ago as a feed back effect. The littoral area east of the Volta River's mouth in the Gulf of Guinea is suffering a strong erosion and the shore is receding, in some places more than 30 m per year, obliging fishermen in the coastal areas of Togo and Benin to regularly move their villages back;
- The Yellow river in northern China (Huang He) is known to have the highest suspended and bed load in the world, mostly due to the erosion of very fertile but very fragile loess plateau soil. The combination of silting up in the riverbed and the overexploitation of stream water flow, evidenced by the building of some dams (San Men, Liujiaxia, Qingtongxia among others) have caused the river to not reach the sea for periods lasting several months per year since the beginning of the 2000s (Lasserre & Descroix, 2011).

In these three cases, an increase in the silting up in the cited dams is observed.

5. Specific impacts of agriculture and urbanisation

In this section, we will study other human factors that influence the modification of hydrological behaviour, such as agriculture, irrigation, drainage, urbanisation and socio-economics, in several parts of the World (mainly in tropical areas).

As has been shown in section 4, rural activities (mostly cropping, livestock grazing and forest clearing) are commonly the cause of an increase in soil losses in the hillslopes, due to the consequent decrease in vegetation cover which causes a soil direct exposure to wind and water erosion. Even if the crops can reach or exceed the same biomass as the removed natural vegetation, this is the case only at the end of the cropping season: evapotranspiration, soil coverage, and protection against wind and water erosion and against runoff remains low during the first weeks and months of the crop cycle. Seasonal influence also impacts deforestation, and clearing has little effect on flow under moderate climates during winter and early spring, because in these seasons, evapotranspiration is very low (Kuchment, 2008).

The impact of irrigation on the water cycle is particularly significant in arid areas, but it is also considerable in moderate climates; it is common for runoff and evaporation from irrigated areas to increase significantly (Kuchment, 2008). In many dry areas, a rise in water table can occur because of water filtration from reservoirs, leakage from water distributing systems or inappropriate practices; this may cause waterlogging and soil salinisation (Kuprianov and Shiklomanov, 1973; Kuchment, 2008). In order to avoid water logging, drainage is applied in many regions of the world; the first effect of drainage is the lowering of water table; as a result, evapotranspiration may drop considerably; it increases flow velocities and runoff can rise by 20-30%. The effect of agriculture and forestry on the water cycle depends on physiographic and climatic conditions. Ploughing usually breaks up overland flow and increases infiltration (Kuchment, 2008). Tillage and the activity of plant roots system modify the structure of the upper soil layer and change the vertical permeability and the soil water retention capacity. Extension of vegetation cover increases the interception of precipitation and increases evapotranspiration.

Urbanisation is well known to disturb natural hydrology, mostly due to the reduction in infiltration capabilities and in the obstruction of stream flows. The increasing urbanisation of watersheds increases the amount of impervious areas; thereby increasing the volumes of stream flow and shortening the time of rain water concentration. The consequence is an increase in maximum flow discharges thereby increasing the flooding risk (Kuprianov and Shiklomanov, 1973). Natural land that used to soak up runoff is replaced by roads and large areas of pavement. Due to the increased pavement area, there is more storm runoff and erosion because there is less vegetation to impede the flow of water as it runs down hills; more land erosion and more sediment is washed into streams, less water will soak into the ground. This means that the water table will have less water to recharge it; it can cause subsidence of the urban area, which is commonly exaggerated by the overuse of ground water for urban needs. Mexico City pumps almost 1.5 billion $m^3.yr^{-1}$ from its water table, provoking a subsidence of several cm per year. Otherwise, large areas of Bangkok became liable to flooding in the recent decades due to the subsidence caused by the overuse of groundwater.

Thus, urbanisation causes an increase in both the runoff coefficient and the stream velocity due to the increasing proportion of impervious surfaces (factories, roofs, roads, car parks, pavements, etc). It also causes strong modifications in water quality, rain water draining these impervious areas and loading urban and industrial dust and wastes through the rivers and the water tables.

As has been shown in section 3, urbanisation needs gravel and sand for construction and river beds have long been considered as abundant supplying areas for these materials. However, these reserves are not inexhaustible; uncontrolled extraction of material may affect the sediment balance and the hydraulic equilibrium of the river, and sometimes that of surrounding lands, bridges and embankments. These consequences are amplified in mountainous areas due to the steep slopes.

A case study of consequences of mountain urbanisation: the La Ravoire torrent (Northen French Alps) appearance in March 1981.

At the end of March 1981 a major ecological catastrophe occurred in the French northern Alps. At the snowmelt, the 31st March, the Ravoire brook (Upper Isère valley, in Savoie region) suddenly reawakened (Périnet, 1982) (Fig.12).

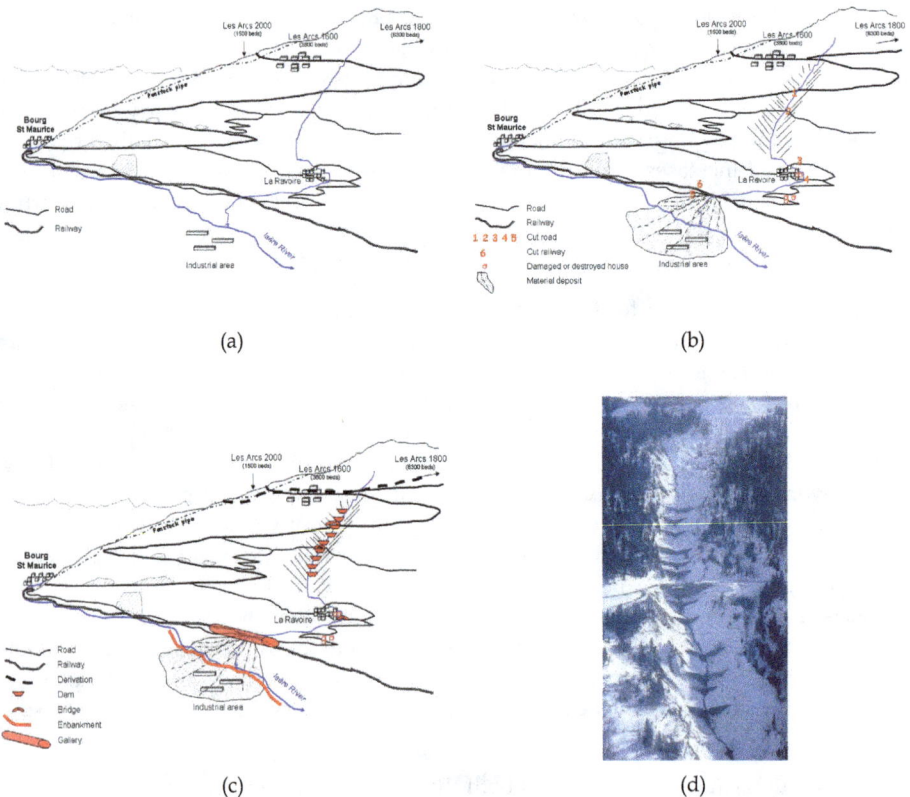

(a)   (b)

(c)   (d)

Fig. 12. a) Situation of the Les Arcs hillslope before the 31/3/1981; b) Situation after the 31/3/1981: damages caused by the rainy event; c) Main restoration measures; d) the series of dams built to prevent further deepening of the torrent bed (photo F. Périnet)

The banks were destabilized and collapsed in great bundles and the brook became a torrent, carrying enormous volumes of mud mixed with snow, trees and rocks. Within five days, its bed had deepened by more than 10 meters; 300 000 m^3 of rocks and material hurtled down from the mountain, covering more than 5 hectares of grassland at foothill, blocking the Isère River and pushing its course towards its right bank, in the industrial area of Bourg Saint Maurice. Significant damages were recorded in this area with some buildings destroyed; on the left bank of Isère River, where the Ravoire brook flows, the railway and five roads (including the access road to Les Arcs ski station) were cut, five houses partially or completely destroyed. In a first assistance operation, a single way road was established to provide access to the 15 000 stranded tourists trying to leave the ski station, the torrent water was diverted towards the penstock pipe of the Malgovert hydropower station (located near Bourg St Maurice in the bottom of the Isère valley). Then works for the torrent course reclamation, the housing protection and the building of new bridges were undertaken as well as a gallery for the railroad (Fig.12).

6. Geopolitics

Geopolitical events can significantly impact the water cycle. The main concern is the strong land use changes caused by massive migrations due to wars or local conflicts.

"The Indochinese section of the Mekong Basin has been subjected to major environmental disturbances over the last half century" (Lacombe et al., 2010). The Vietnam War is invoked as a central explanation for the extensive deforestation in specific areas while conflict induced exoduses caused the abandonment of cultivated lands, followed by forest regeneration. Although the socio-economic consequences of these episodes have been analysed, their hydrological impacts remained unknown until the study of Lacombe et al. (2010). They investigated hydrological changes in two catchments of the lower Mekong Basin that were either heavily bombed (in southern Laos) or depopulated (in northern Laos) (Fig.13). The first one is located on the Ho Chi Minh trail; one third of the latitudinal extent of the Ho Chi Minh trail in Laos is included in this southern catchment and it was a target which the United States Air Forces (USAF) made a concerted effort to neutralize. During this conflict, the USAF practised deliberate massive removal of vegetation as a military tactic to deny cover and land to opposition forces. A total of between 690 000 and 2 948 000 tons of high explosive ordnance were exploded over this southern catchment between October 1965 and September 1973 (Lacombe et al., 2010). In the northern basin, while air operations of the Vietnam War caused losses of forest, continuing ground battles in northern Laos over the period 1953–1975 induced significant emigration. Between 730 000 and 1 million people (one fourth of the country's population as per 1970) were displaced over this period (Taillard, 1989; Goudineau, 1997; cited by Lacombe et al., 2010). This migration flow was sustained after the end of the war. As an immediate consequence of the communist takeover more than 300 000 people fled to Thailand between 1975 and 1985, escaping the new regime. This massive exodus particularly affected the northern provinces of Laos, traditionally inhabited by Hmong ethnic groups (Central Intelligence Agency, 1970, cited by Lacombe et al., 2010) which served the US army during the Vietnam War and were forced to flee the new communist regime after the war ended. The abandonment of large mountainous areas, which were traditionally subject to shifting or permanent cultivation, most likely induced a shift toward forest regeneration (Lacombe et al., 2010). The analysis (Lacombe et al., 2010) is based on the widely and independently recognized fact that vegetation, via evapo-transpiration, is a central driver of groundwater recharge and stream flow production. The analysis of the most complete Vietnam War air mission database and of available hydro-meteorological data over the period 1960–2004 reveals a sharp runoff increase in the southern catchment when bombing climaxed in the early 1970s while no hydrological change is observed in the northern catchment over the same period (Lacombe et al., 2010). From 1995 onwards, the northern and southern catchment's runoff productions are significantly lower and higher than in the pre-war conditions, respectively. Although causalities could not be ascertained because of data limitations, these short- and long-term hydrological shifts were found to be consistent, in terms of occurrence, spatial distribution and magnitude, with the expected changes in the vegetation cover, either denser in the north (in response to abandonment of cultivated lands) or sparser in the south (as a result of bomb-induced deforestation and soil degradations) (Lacombe et al., 2010). Furthermore, as a consequence, some areas near Vientiane (the capital of Laos) and overall near Luang Prabang (a city located 200 km north of Vientiane) were cleared in past years including on steep slopes (Chaplot et al., 2007), causing the appearance of gullies (Fig.14).

Fig. 13. Location of the two sub basins studied by Lacombe et al. 2010

In other areas in the world, geopolitics could lead to severe land degradation: in the Republic of South Africa, the apartheid regime created some Bantustans, also known as black African homelands or simply homelands), which were territories set aside for black inhabitants, not recognized by the international community. Due to population density and land management, soil erosion was strong in some of these areas such as the Kwazulu. The same kind of process occurred in northern Pakistan during the USSR-Afghanistan war; a great number of Afghans took refuge in Pakistan. Finally, in Hispaniola Island (West

Indies), which includes two countries, Haïti on the western side and the Dominican Republic to the east, the rural areas of the Republic of Haïti, are much more densely populated than the ones of the Dominican Republic resulting in severe land degradation (deforestation to harvest crops, grass and wood). The fact that the population is much poorer in Haiti should also be noted. In these three examples, there is a strong gradient in land exploitation between two areas, noticeable in aerial pictures and satellite scenes. Overall, this provokes severe damages in the overexploited side due to cyclones in Haïti and to monsoon rains in Pakistan and in South Africa.

Fig. 14. Gullies appearing in rain fed rice fields in central Laos, 10 km eastward from Luang Prabang. These areas were settled on fields recently cleared by peasants coming from the eastern Laotian mountains near the Vietnam boundary, judged insecure by the government which encouraged people to migrate within easily controlled areas (*photo NAFRI*) .

7. Conclusion

These last two types of human impacts (urbanisation and geopolitics) demonstrate the great influence of societies, socio-economic conditions, the strategic context and the migration processes, on the land cover and thus on the water cycle. The poor understanding of the influence of these factors could lead to exaggerations recorded in history, such as the sentencing of goats as "the main enemy of Mediterranean forests and soils" by the French governments during the 19th century, while most of the land degradation problem could result from the lack of forest policy during the period just after the French Revolution (from 1797 to 1804) and the forest exploitation during previous centuries in order to build the ships of the French Royal Marine. The Little Ice Age (from the end of Middle Age to the end of 19th century) and its low winter temperatures (mostly during the 19th century) was also a natural constraint for the vegetation. It is always more appropriate to link these severe crises of water cycle to a combination of natural and human factors.

8. Acknowledgements

We are grateful to the French research Institute for Development (IRD), the Niger Basin Authority (NBA), and the AMMA program which funded the field experiments in Mexico, in France and in Niger. This study was also partially funded by the French ANR projects ECLIS (Contribution of livestock to the reduction of rural population vulnerability and to the promotion of their adaptability to climate and society changes in Sub-Saharan Africa) and ESCAPE (Environmental and Social Changes in Past, present and future)

9. References

Albergel, J. (1987) Sécheresse, désertification et ressources en eau de surface : application aux petits bassins du Burkina Faso. In *The Influence of Climate Change and Climatic Variability on the Hydrologic Regime and Water Resources*; IAHS publication N° 168, Wallingford, UK, pp. 355-365.

Ali, A. & Lebel, T. (2009) The Sahelian standardized rainfall index revisited. *Int. J. Climatol.* 29, 1705-1714.

Amani, A. & Nguetora, M. (2002) Evidence d'une modification du régime hydrologique du fleuve Niger à Niamey. In *FRIEND 2002 Regional Hydrology: Bridging the Gap between Research and Practice*, Proceedings of the Friend Conference, Cape Town, South Africa, 18-22 March, 2002; Van Lannen, H., Demuth, S., Eds.; IAHS publication N°274, Wallingford, UK, pp. 449-456.

Amogu, O. (2009). The land degradation in the sahel and its consequences on the sediment balance of the Mid Niger River PhD thesis, university Grenoble 1, 420 p.

Amogu O., Descroix L., Yéro K.S., Le Breton E., Mamadou I., Ali A., Vischel T., Bader J.-C., Moussa I.B., Gautier E., Boubkraoui S..& Belleudy P. (2010) Increasing River Flows in the Sahel?. *Water*, 2(2):170-199.

Andersen, I., Dione, O., Jarosewich-Holder, M.& Olivry J.C. (2005) *The Niger River Basin: a Vision for Sustainable Management*, Golitzen, K.G., Ed., The World Bank: Washington, DC, USA.

Cappus, P. (1960). Etude des lois de l'écoulement. Application au calcul et à la prévision des débits. Bassin expérimental d'Alrance. La Houille Blanche, No. A. Grenoble, France, pp. 521–529.

Chaplot, V., Khampaseuth, X., Valentin, C., & Le Bissonnais, Y. (2007). Interrill erosion in the sloping lands of northern Laos submitted to shifting cultivation. *Earth Surface Processes and Landforms*, **32**, 415-428.

Descroix, L., Viramontes, D., Vauclin, M. , Gonzalez Barrios, J.L. and Esteves, M. (2001). Influence of surface features and vegetation on runoff and soil erosion in the western Sierra Madre (Durango, North West of Mexico). *Catena*. 43-2 :115-135.

Descroix, L., Gonzalez Barrios, J.L., Vandervaere, J.P., Viramontes, D. & Bollery, A. (2002). An experimental analysis of hydrodynamic behaviour on soils and hillslopes in a subtropical mountainous environment (Western Sierra Madre, Mexico). *Journal of Hydrology*, 266:1-14.

Descroix, L. & Gautier, E. (2002). Water erosion in the French Southern Alps : climatic and human mechanisms. *Catena*, 50 : 53-85.

Descroix, L. & Mathys, N. (2003). Processes, spatio-temporal factors and measurements of current erosion in French Southern Alps : a review. *Earth Surface Processes and Landforms*. 28 ; 993-1011.

Descroix L. (2004) : Sintesis de las observaciones de Luc Descroix en cuanto a los Términos de Referencia para el « Estudio de factibilidad del proyecto de Restauración y Conservación de suelo e incremento de la cobertura vegetal para aumentar la recarga de los acuíferos de la cuenca del Río Sonora ». Comisión Nacional del Agua, Subdirección General Técnica, World Meteorological Organisation. Research and Technical Report, 54 p.

Descroix, L., Mahé, G., Lebel, T., Favreau, G., Galle, S., Gautier, E., Olivry, J-C., Albergel, J., Amogu, O., Cappelaere, B., Dessouassi, R., Diedhiou, A., Le Breton, E., Mamadou, I. & Sighomnou, D. (2009). Spatio-Temporal Variability of Hydrological Regimes Around the Boundaries between Sahelian and Sudanian Areas of West Africa: A Synthesis. *J. Hydrol*, 375, 90-102.

Descroix, L., Laurent, J-P., Boubkraoui, S., Ibrahim, B., Cappelaere, B., Favreau, G., Mamadou, I., Le Breton, E., Quantin, G. & Boulain, N. (2012). Experimental evidence of deep infiltration under sandy flats and gullies in the Sahel. J. Hydrol. (2012), doi:10.1016/j.jhydrol.2011.11.019

Descroix, L., Genthon, P., Amogu, O., Sighomnou, D., Rajot, J-L., Vauclin, M. (2011). Recent hydrological changes of Sahelian rivers: the case of the 2010 red floods of the Niger River at Niamey. AGU conference, San Francisco, CA, USA

Ferry, L. & Mietton, M. (2011). L'équilibre du fleuve Niger perturbé. Sciences au Sud, 60, 3.

Hewlett, J.D. (1961). Soil moisture as a source of base flow from steep mountain watershed. US forest Service, Southeastern Forest Experiment Station, Asheville, North Caroline.

Horton, R.E. (1933) The role of infiltration in the hydrologic cycle. EOS. American Geophysical Union Transactions 14, 44–460.

Kuchment, L.S. (2008). Runoff generation (genesis, models, prediction), Water problems institute of RAN, 394p. (in Russian).

Kuprianov V.V. & Shiklomanov, I.A. (1973). Influence d el'urbanisation, de l'industrie er de l'agriculture sur le cycle hydrologique. Acts of the meeting « Hydrological problems in Europe » Bern, Switzerland, August 1973, UNESCO & WMO.

Lacombe, G. Pierret,A. Hoanh, C.T., Sengtaheuanghoung, O. & Noble, A.D. (2010). Conflict, migration and land-cover changes in Indochina: a hydrological assessment. *Ecohydrol.* 3, 382–391 (2010) Published online 1 October 2010 in Wiley Online Library (wileyonlinelibrary.com) DOI: 10.1002/eco.166.

Lasserre, F. & Descroix, L. (2011). *Eaux et Territoires.* third edition, Presses Universitaires du Québec, ISBN 978-2-7605-2602-0, Sainte Foy, Québec, Canada.

Lebel, T. & Ali, A. (2009) Recent trends in the Central and Western Sahel rainfall regime (1990–2007). *J. Hydrol. 375*, 90-102.

Leblanc, M., Favreau, G., Massuel, S., Tweed, S., Loireau, M. & Cappelaere, B. (2008) Land clearance and hydrological change in the Sahel: SW Niger. *Glo Pla Cha 61*, 49-62.

Leduc, C., Bromley, J. & Shroeter, P. (1997) Water table fluctuation and recharge in semi-arid climate: some results of the HAPEX Sahel hydrodynamic survey (Niger). *J. Hydrol. 188-189*, 123-138.

Mahé, G. (1993) Les écoulements fluviaux sur la façade Atlantique de l'Afrique. Etude des éléments du bilan hydrique et variabilité interannuelle. Analyse de situations hydroclimatiques moyennes et extrêmes. Etudes et Thèses, Orstom, Paris.

Mahé, G., Leduc, C., Amani, A., Paturel, J-E., Girard, S., Servat, E. & Dezetter, A. (2003) Augmentation récente du ruissellement de surface en région soudano sahélienne et impact sur les ressources en eau. In *Hydrology of the Mediterranean and Semi-Arid Regions, proceedings of an international symposium. Montpellier (France)*, 2003/04/1-4, Servat E., Najem W. , Leduc C., Shakeel A. (Ed.) :, Wallingford, UK, IAHS, 2003, publication n° 278, p. 215-222..

Mahé, G., Paturel, J.E., Servat, E., Conway, D. & Dezetter, A. (2005) Impact of land use change on soil water holding capacity and river modelling of the Nakambe River in Burkina-Faso. *J. Hydrol. 300*, 33-43.

Mahé, G. (2009) Surface/groundwater interactions in the Bani and Nakambe rivers, tributaries of the Niger and Volta basins. West Africa. *Hydrol. Sci. J.*, *54*, 704-712.

Mahé, G. & Paturel, J-E. (2009) 1896-2006 Sahelian annual rainfall variability and runoff increase of Sahelian rivers. *C.R. Geosciences*, *341*, 538-546.

Mahé, G., Lienou, G., Bamba, F., paturel, J-E., Adeaga, O., Descroix, L., Mariko, A., Olivry, J-C., Sangaré, S., Ogilvie, A., and Clanet, J-C. (2011). The Niger River and climate change over 100 years. *Hydro-climatology : variability and change ; proceedings of symposium J-H02 held during IUGG 2011, Melbourne, Australia, july 2011*, IAHS Publ n°344, 2011.

Olivry, J-C. (2002). Synthèse des connaissances hydrologiques et potentiel en 862 ressources en eau du fleuve Niger. World Bank, Niger Basin Authority, 863 provisional report, 160 p, Niamey.

Peiry, J-L. (1988). Approche géographique spatio-temporelle des sediments d'un cours d'eau intra-montagnard: l'exemple de la plaine alluviale de l'Arve (Haute Savoie, France) ; PhD thesis, Lyon 3 University, 378pp.

Peiry, J-L., Salvador, P.G. & Nouguier, F. (1994). L'incision des rivières dans les Alpes du nord: état de la question. Revue de Géographie de Lyon, 69 (1), 47-56.

Périnet, F. (1982). Stations de sport d'hiver ; réflexionc à propos d'un accident. Revue Forestière Française, 34 (5), 99-111.

Picouet, C. (1999). Géodynamique d'un hydrosphère tropical peu anthropisé : Le Bassin supérieur du Niger et son delta intérieur. PhD Thesis, University Montpellier 2, 469 p.

Séguis, L., Cappelaere, B., Milési, G., Peugeot, C., Massuel, S. & Favreau. G. (2004) Simulated impacts of climate change and land-clearing on runoff from a small Sahelian catchment. *Hydrol. Process. 18*, 3401-3413.

Tschakert, P., Sagoe, R., Ofori-Darko, G. & Nii-Codjoe, S. (2010) Floods in the Sahel: an analysis of anomalies, memory and anticipatory learning. *Climatic Change*, 103: 471-502.

Viramontes, D. and Descroix, L. (2003). Variability of overland flow in an endoreic basin of northern Mexico : the hydrological consequences of environment degradation. *Hydrological Processes*, 17 :1291-1306.

7

Unsteady 1D Flow Model of Natural Rivers with Vegetated Floodplain – An Application to Analysis of Influence of Land Use on Flood Wave Propagation in the Lower Biebrza Basin

Dorota Miroslaw-Swiatek
Warsaw University of Live Sciences - SGGW
Poland

1. Introduction

The present studies on renaturalization of river valleys, river beds as well as quantitative estimation of water demand of protected hydrogenic habitats exposed to flooding, in many cases require application of flooding flow models. The state of fluviogenic ecosystems is dependent mostly on the conditions of their hydrological alimentation by flooding waters. The possibility of particular swamp vegetation occurrence on those areas is related to the presence of annual floods of certain duration and depth. So there is a significant relationship between the plant communities' pattern and persisting hydrological conditions. The most important hydrological characteristics, conditioning of the growth and the development of swamp vegetation, are, first of all (Oswit, 1991; Hooijer, 1996; Zalewski et al., 1997): inundation surface, mean inundation depth, the frequency and the duration of inundation. That indicates that there is a very strong relationship between a vegetation structure and water conditions. On the other hand, floodplain vegetation significantly affects flood extent in the valley. Hydrodynamic models coupled with GIS techniques make it possible to obtain necessary data for the determination of abovementioned hydrological characteristics. They are also tools which facilitate estimation of the influence of different river valley management methods on hydraulic conditions of water flow. They can be applied in the process of decision-making for: projects, investments and operational activities in the field of flood protection and environmental impact assessment.

The valley of the Biebrza River, in particular the Lower Basin of Biebrza, is an exceptionally valuable natural object with inundation of surface waters occurring annually as a vital site-shaping factor – this builds specific habitat conditions, crucial for the unique character of that area. This area is an extremely valuable wetland site of global significance protected by Ramsar Convention and annual flooding influences formation of a unique character of the study site. A zonal system of various plants reflects water conditions of the Lower Biebrza Basin (Oswit, 1991) and causes spatial division of the resistance to water flow (Swiatek et al., 2008).

The hydrodynamic model, providing a correct description of flooding water flow in the area of the Biebrza Valley, could be used as research tool for executing effective policy of natural values protection within Biebrza National Park. The influence of the different floodplain land use, described by changes in vegetation structure, on flood wave propagation in the Lower Biebrza River Basin (LBRB) was analyzed as an example of developed model applying.

Mathematical modelling of river flow with water levels not exceeding a bank elevation is widespread and described by many authors (Cunge et al., 1980, Szymkiewicz, 2010). Applications of hydraulic models, particularly for lowland river valley flooding with water overflowing the main channel, flooding adjacent areas, and flowing in the floodplains covered with various vegetation, are not very common. The models used usually apply simplified approaches to the influence of land use in river valleys on flow conditions. Flow resistance, in rivers as well as in floodplains covered with vegetation, is usually described with generalized Manning's coefficient values subject to plants vegetative alterations. In another simplified approach widely used, the complexity of water flow on floodplains, which is very hard to be sufficiently represented, is considered in the form of so-called 'dead' areas of the cross-section (storage zones), which do not conserve momentum, but take part only in storing water in the river valley. In this approach, floodplain geometry is accounted for in only one of St Venant's equations – a continuity equation – and the momentum equation reduces that to hydraulic parameters within the main channel geometry (Cunge et al., 1980).

The advantage of one-dimensional models is their computational effectiveness for large-scale areas. These models provide fairly good results in those cases, when the assumption of one-dimensional flow is valid, that is for river valleys with no extensive floodplains (Horrit & Bates, 2002). On account of their applicational usefulness, these models are constantly developed and upgraded.

The developed model, as opposed to many existing commercial models, withdraws from a simplified description of flow resistance expressed by the spatially differentiated Manning's coefficient and the use of the Darcy-Weisbach relationship. It also enables introduction of water mass and momentum exchange process between the main channel and floodplains, and parts of a cross section covered with vegetation and those with no vegetation. To this end, additional flow resistance along imaginary vertical boundaries between the main channel and floodplains has been introduced (Pasche & Rouve, 1985). Thus the developed model enables accounting for unsteady flow calculations flow resistance resulting from both vegetation covering a cross section and momentum exchange between the main channel and floodplains, proposed in the Pasche approach (Pasche ,1984). The 1D unsteady river flow model with vegetated floodplain was proposed by Helmio (2002). In this model the flow resistance for the riverbed and the floodplains was calculated according to the method by Nuding (1991). That method is poorly documented in literature, especially in respect of the determination of applied parameters. The method of Pasche was also applied in SPRUNNER model (Laks & Kałuża, 2007), where it is used for the determination of active zone of flow, extended by the interaction of the main channel with floodplains.

In this paper an unsteady 1D flow model was developed for a river with vegetated floodplains. The basic form of the non-linear St Venant equations combined with retention effects of the vegetated areas on flood wave conveyance were used in the model. In this approach friction caused by vegetation and additional resistance caused by interaction between the main channel and vegetated areas (the Pasche method) were taken into account. The model developed in this paper, on the contrary to the other models mentioned above, enables also considering of the flow resistance, resulting from any configuration of vegetation cover in a compound channel, in the calculations of unsteady water flow. The model was applied to the Lower Biebrza River, situated in the north-eastern part of Poland, flowing through the last extensive, fairly undisturbed river-marginal peatland in Europe. The roughness height coefficients of vegetation (reeds, sedges, grasses, bushes as well as trees and scrubs) in floodplains were determined by field monitoring. Water level observations collected by an automatic monitoring system were used in calibration and validation process The model was calibrated and validated for flood events in 2006 and 2007 year. Elaborated model will be used as a tool for assessment of various agricultural practices with regard to the effective management for protection of the unique wetland site of the Biebrza National Park

2. Numerical model development

2.1 The unsteady flow model

In practice unsteady flow in natural rivers is usually treated as one-dimensional flow, and is based on St Venant equations. St Venant equations consist of the dynamic equation and the continuity equation. When water discharge and water level are dependent variables, these equations are written respectively as:

$$\frac{\partial Q}{\partial t} + \frac{\partial}{\partial x}\left(\frac{\beta Q^2}{A}\right) + gA\frac{\partial h}{\partial x} + A\frac{|Q|Q}{K^2} + q\frac{Q}{A} = 0, \tag{1a}$$

$$B\frac{\partial h}{\partial t} + \frac{\partial Q}{\partial x} = q \tag{1b}$$

where: Q = discharge [m³/s]; h = water level [m]; x = distance [m]; t = time [s]; A = cross area of flow [m²]; B = width of water surface [m]; K = conveyance factor [m³/s]; g = gravitational acceleration [m/s²]; q = lateral inflow [m³/s/m]; β = momentum correction factor.

The conveyance factor K is expressed as

$$K = A\left(\frac{8gR}{\lambda}\right)^{1/2} \tag{2}$$

where: R = hydraulic radius [m]; λ = average dimensionless friction factor.

The equation (1) requires also determining of boundary conditions as well as initial conditions. Boundary conditions refer to hydraulic properties at the upstream and downstream ends of a river. The model developed in this paper enables only a subcritical flow description. The upstream boundary condition is determined as a discharge hydrograph $Q(t)$. A water level hydrograph $h(t)$, discharge hydrograph $Q(t)$, rating curve $Q(h)$ or friction slope S_f can be used as the downstream boundary condition.

The initial condition refers to the state of flow in the river when the simulation starts. In the model presented in this paper a steady flow in the channel is used as the initial condition (Swiatek et al., 2006). The finite element method in the Galerkin formulation was used to solve a pair of Eqs. (1) (Szymkiewicz, 2010). This approach leads to the non-linear ordinary differential equations, in which the time-weighted finite difference method is used in the approximation of the time derivative. This method forms a system of algebraic non-linear equations which are numerically solved by iteration method.

The total conveyance (Eq. (2)) for a compound cross-section is obtained by summing the subdivision conveyances of the channel and floodplains:

$$K = K_{lf} + K_{ch} + K_{rf} \tag{3}$$

where: K = total conveyance; K_{lf} = total conveyance of the left floodplain; K_{ch} = conveyances of the channel; K_{rf} = total conveyance of the right floodplain.

The total floodplains conveyances are calculated according to the vegetation distribution with the Pasches's method used for computing the total Darcy-Weisbach friction factor λ. The total conveyance K was introduced to the St Vetnant Eq. (1) computation. It was calculated for each cross-section and water level in the iterative method of solving a system of algebraic non-linear equations (Swiatek, 2008).

2.2 Determination of friction factors

Hydraulic calculations of the flow in natural rivers with floodplain require methods which include natural vegetation structure of the river waterside zones and the floodplain. The values of the resistance coefficients of the floodplain vegetation still belong to rareness. Presently the most often used are roughness coefficients for Manning's equation, but the choice of the sufficient one among the tabular values is subjective. Another method is determination of plant characteristics elaborated by Rouvè (1987). The division on high, medium and low vegetation (proposed by Bretschneider and Schulz 1985) is used in such calculations. High vegetation is considered here as higher than water flow depth (trees and shrubs) and in small degree go under hydrodynamic water pressure, medium vegetation as approximately equal to water depth (mostly shrubs) and low vegetation which refers mostly to sedge and grass communities (Fig.1). Established criterion is ambiguous and in fact the same vegetation can be ranked into different types in view of the natural water levels variability.

The basis of hydraulic calculations of river flow including its natural vegetation structure of high and medium vegetation is assumption that water flow resistances are the same as resistances which occur when water overflow regularly distributed vegetation with averaged geometric parameters (DVWK, 1991; Kubrak & Nachlik, 2003).

Parameters which describe vegetation of the floodplain and are used in calculations are an average tree diameter or shrubs branches d_p and distances between them in the direction of the water flow a_x and transversal to it a_y (Fig.2). Named parameters are determined on the basis of field measurements in the area of the water flow.

The resistance of flow caused by the roughness of low vegetation occurring on the scarp, the channel bed and floodplain is calculated from the formula given by Colebrook-White:

$$\frac{1}{\sqrt{\lambda_s}} = -2.03 \log\left(\frac{2.51}{\mathrm{Re}\sqrt{\lambda_s}} + \frac{k_s}{14.84\ R} \right) \tag{4}$$

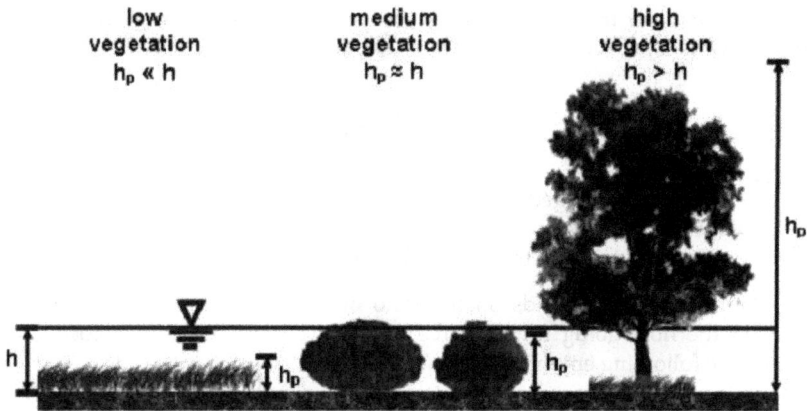

Fig. 1. Vegetation classification proposed by Bretschneider and Schulz (1985)

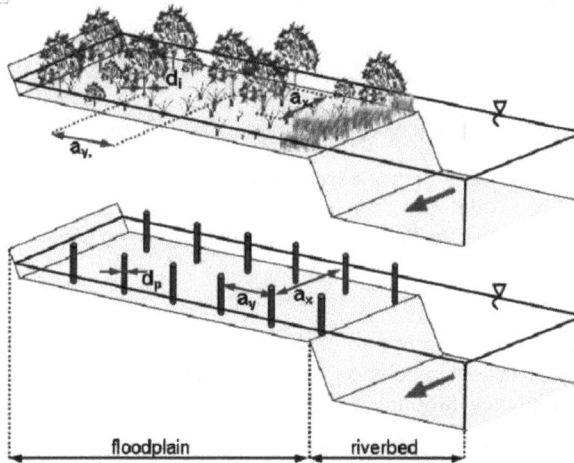

Fig. 2. The high vegetation geometric characteristics

where: λ_s = friction factor of low vegetation or not overgrown part of the cross- section [-]; k_s = roughness height of the overgrown (low vegetation) or not overgrown cross-section [m]; Re = the Reynolds number for part of the cross-section.

As it results from the Colebrook-White law, the friction factors of flow depend on the Reynolds number and on relative roughness k_s/R. The influence of the Reynolds number on friction factors decreases along the increase of its value and of relative roughness of channel sides. In natural channels the influence of the Reynolds number for values higher than 25000 may be neglected without any harm to the precision of calculations. Therefore Rickert (1988) recommends application of the equation (4) for practical calculations, in the following form:

$$\frac{1}{\sqrt{\lambda_s}} = -2.03\log\left(\frac{k_s}{14.84\ R}\right) \tag{5}$$

The friction factor λ_v for trees and bushes (submerged part of high vegetation) is calculated from the following expression (Pasche & Rouve, 1985):

$$\lambda_v = \frac{4h_z d_p}{a_x a_y} C_{WR} \tag{6}$$

where: h_z = height of submerged part of trees [m] (Fig.3); d_p = trees diameter [m]; a_x, a_y = distance between plants along the flow and transversal; C_{WR} = dimensionless drag coefficient for submerged part of the trees or bushes.

The drag coefficient C_{WR} depends on the ratio of the V_i flowing velocity to the average velocity V_v of the flow going through tree overgrown areas (relative velocity V_r), and is described by the following empirical formula (Rickert, 1986):

$$C_{WR} = \left(1.1 + 2.3\frac{d_p}{a_y}\right)(V_r)^2 + 2\left(\frac{1}{1 - d_p / a_y} - 1\right) \tag{7}$$

$$V_r^2 = \left(\frac{V_i}{V_v}\right)^2 = 0.6 + 0.5\log\left(\frac{a_x}{a_y}\right) \tag{8}$$

The expression (7) is valid when 0.05<d_p/a_y<3 and 0.2<a_x/a_y<2.

The resistance of flow in parts of cross-sections overgrown by high vegetation depends on both vegetation and bed roughness. The friction factor for this area, according to the concept issued by Einstein and Banks (Indlekofer, 1981), is the following sum:

$$\lambda = \lambda_s + \lambda_v \tag{9}$$

where: λ = average friction factor in part of the cross-section [-]; λ_s = friction factor caused by channel bed roughness or low vegetation[-]; λ_v = friction factor for non-submerged and non-flexible vegetation (high vegetation, Fig.3.) [-].

Physically, composite roughness along the wetted perimeter of the compound cross section modifies velocity distribution in the cross section (Fig.3). A detailed examination of the effects of varying wall roughness and cross sectional geometry would require a three-dimensional analysis of the flow. Pasche (1984) proposed one-dimensional analysis of steady flow in a compound cross-section of the lowland river based on the Darcy-Weisbach formula. According to the observed velocity distribution a compound river cross section is divided into sections with vertical imaginary walls with roughness height of channel k_T and friction factor λ_T between the main channel and neighboring floodplains (Fig.4). The heights of these boundaries (h_{T1}, h_{T2}) are taken into consideration in calculations of the wetted perimeter of the main channel. Mean velocity in each section (in Fig.4. signed by symbols T1 and T2 for vertical imaginary walls and numbers 1, 3, 2 for scarps and channel bed) for a channel is calculated from the Darcy-Weisbach equation, which results from the momentum balance in part 'i' of the cross-section:

$$v_i = \sqrt{\frac{8gR_i S_f}{\lambda_i}} \tag{10}$$

Unsteady 1D Flow Model of Natural Rivers with Vegetated Floodplain – An Application to Analysis of Influence of
Land Use on Flood Wave Propagation in the Lower Biebrza Basin

151

where: S_f = hydraulic slope [-]; v_i = average velocity in the 'ith' sub-domain of the main channel; λ_i = friction factor of the 'ith' sub-domain.

The average friction factor in the whole main channel λ_g is calculated with consideration of the friction factors in every sub-section of the main channel (Fig.4):

$$\lambda_g = \frac{\lambda_{T1}h_{T1} + \sum_{i=1}^{3}\lambda_i wp_i + \lambda_{T2}h_{T2}}{h_{T1} + \sum_{i=1}^{3}wp_i + h_{T2}} \qquad (11)$$

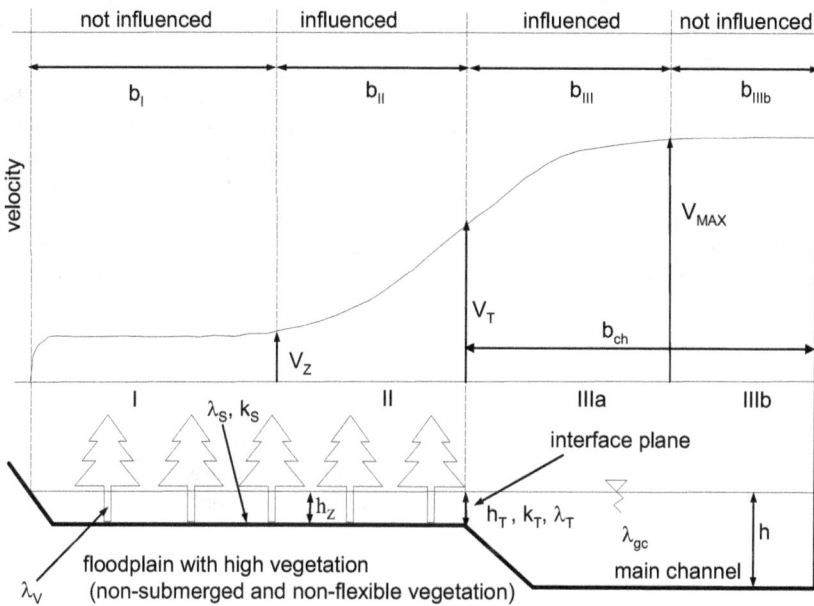

Fig. 3. Scheme of compound channel with floodplain with high vegetation and velocity distribution

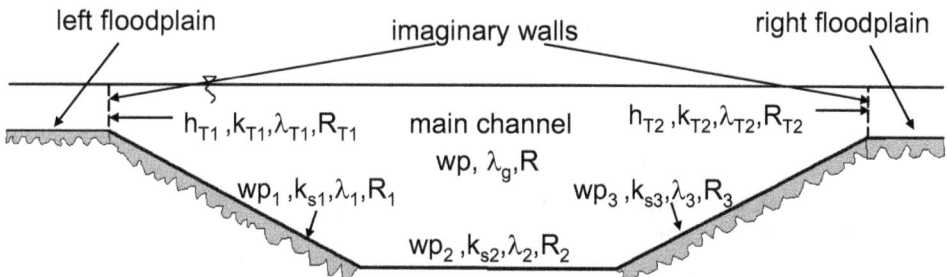

Fig. 4. Scheme of main channel sub-sections with different hydraulic parameters

The friction factor λ_i is calculated from the formula (5), in which hydraulic radius R_i and R_{T1}, R_{T2} related to different roughness factors in cross-sections is determined according to the concept by Einstein (Kubrak, 2003), assuming the equality of average velocity (4) in the main channel v and average velocity v_i in the every sub-domain i:

$$\sqrt{\frac{8gR_iS_f}{\lambda_i}} = \sqrt{\frac{8gRS_f}{\lambda_g}} \tag{12}$$

When floodplain is covered by low vegetation, the interface of the floodplain from the main channel is treated as the rough wall with roughness height k_{T1}, k_{T2}, that have the same values like left and right scarp of the main channel. The calculation of the hydraulic radius for different roughness factors in cross-sections is conducted using the iterative method and basing on the formula (12).

The way of calculation of flow intensity in the floodplain with trees is more sophisticated and based on the theory proposed by Passche and Rouve (1985). According to the observed velocity distribution (Fig.3) the compound channel is divided into four sections. Due to the horizontal mass and momentum exchange between areas with steep velocity gradients the flow is reduced in the main channel (sub section IIIa) and accelerated on the floodplain (sub section II). At calculation of the discharge in the main channel, the interface plane of the floodplain from the main channel is treated as the rough wall with roughness of channel k_T and friction factor λ_T (Fig.3). The friction coefficient λ_T in interface plane is introduced to determine the flow reduction in the main channel along the width bIII, while the discharge on the floodplain is increased along the width bII.

In reality the resistances to flow by the interface are made by intensive cyclical impulses of mass and momentum in transverse direction to the main flow, as well as by associated high turbulence strains and whirls on the surface of water in round-the-tree flows. The roughness height in the interfaces (Fig.3) is calculated from the formula (Pasche, 1985):

$$k_T = 0.854 \; R_T \; \Omega \left(\frac{2\,b_{II}}{b_{ch}} \right)^{1.07} \tag{13}$$

where: R_T = hydraulic radius concerning the resistance by the interface [m]; bII = width of the floodplain zone of influence of the area with trees [m]; b_{ch} = width of the main channel [m].

The width of the area of floodplain influenced on flow in main channel is calculated from the formula:

$$b_{II} = \frac{h_T}{\lambda_v (0.068 \; e^{0.56\,C_T} - 0.056)} \tag{14}$$

where: h_T = depth of interfaces plane between main channel and floodplain [m]; c_T = dimensionless velocity in interfaces.

The friction factor λ_v of the high vegetation (trees or bushes) is calculated from expression (6). The dimensionless velocity in interfaces plane is calculated from the formula:

$$C_T = -3.27 \log \Omega + 2.85 \tag{15}$$

where: Ω = vegetation parameter.

The coefficient Ω is calculated from the following form:

$$\Omega = \left(0.07 \frac{a_{NL}}{a_x}\right)^{3.29} + \left(\frac{a_{NB}}{a_y}\right)^{0.95} \tag{16}$$

$$a_{NL} = 128.87 \, C_W d_p \left(1 + \frac{g \, a_{NL} \, S_f}{V_T^2 / 2}\right)^{-2.143} \tag{17}$$

$$a_{NB} = 0.24 \, a_{NL}^{0.59} (C_W d_p)^{0.41} \tag{18}$$

where: a_{NL}, a_{NB} = the length and the width of the Karmann path formed at a single plant [m]; C_W = drag coefficient; V_T = velocity in the interface plane between the main channel and its floodplain.

The drag coefficient C_W is determined for a single tree at ideal two-dimensional flow. Its variability for different forms of the turbulent flow was given by Lindner (1982), and in natural channels, where Reynolds number for a single plant varies between $8 \cdot 10^3$-$1 \cdot 10^5$ has value 1.2.

The velocity V_T in the interface plane between the main channel and its floodplain is calculated from the following formula:

$$V_T = C_T \sqrt{\frac{\lambda_v}{8}} v \tag{19}$$

The velocity V_T calculated from the equation (19) is usually different from the one assumed at the beginning of a_{NL} calculation (17), therefore the whole set of calculations should be repeated, accepting the calculated value V_T according to formula (19) as the next approximation.

The method described above has the following limitations (Schumacher, 1995):

- friction factor of the imaginary boundary must be larger than the friction factor of the bottom of the main channel;
- distance between plants along the flow a_x must be smaller than the length of the Karmann path a_{NL} formed at a single plant ;
- ratio of the main channel and the floodplain width must be less than 40.

In the elaborated model, when the first or the second of the above conditions is not valid, in the formula (11) in place of λ_{T1} , λ_{T2} the friction factors of the flow left, right scarp λ_i is used. The third condition is usually fulfilled for natural lowland rivers with floodplain.

The next problem, which can occur in calculation when Colebrook-White equation (5) is used, is related to the value of relative roughness k_s/R. When the value of k_s/R is approximately 14.84, the friction factor λ_s is indeterminate, and in the case of relative roughness a little bit smaller than this number it has a huge non–physical value. Therefore, to avoid these difficulties the limited value of λ_s is introduced in the calculation in the model.

3. Application: the lower Biebrza river basin

3.1 Research area

The Valley of the Biebrza River is situated in the north-eastern part of Poland (Fig.5). The research focused on the Lower Biebrza Basin (LBB), which is located in the southern part of the Biebrza Valley, from the village of Osowiec to the mouth of Biebrza to the Narew River. The length of the lower basin amounts to 30 km, while its width ranges from 12 to 15 km. The largest surface in the area of that basin is occupied by a flood terrace, which includes vast, flat peatlands and gently folded, loamy near-riverbed zone of the width from 1 to 2 km. The riverbed of Biebrza has a winding course in the lower basin, it forms numerous meanders, side reaches and ox-bows, through which water flows during floods. The length of the riverbed varies between nearly 20m to 35m. The riverbed is significantly emerged from the valley, and in the southern part, close to the mouth to the Narew River, the riverbed is sharply incised in the bottom of the valley. The most important tributaries to the Biebrza River in the lower basin are: the right-side tributaries - the Klimaszewnica and the Wissa Rivers, and the left-side tributary: the Kosodka River. The Valley of the Lower Biebrza has a characteristic, zonal pattern of plant communities, which entirely reflects hydrological conditions dominating in the study area.

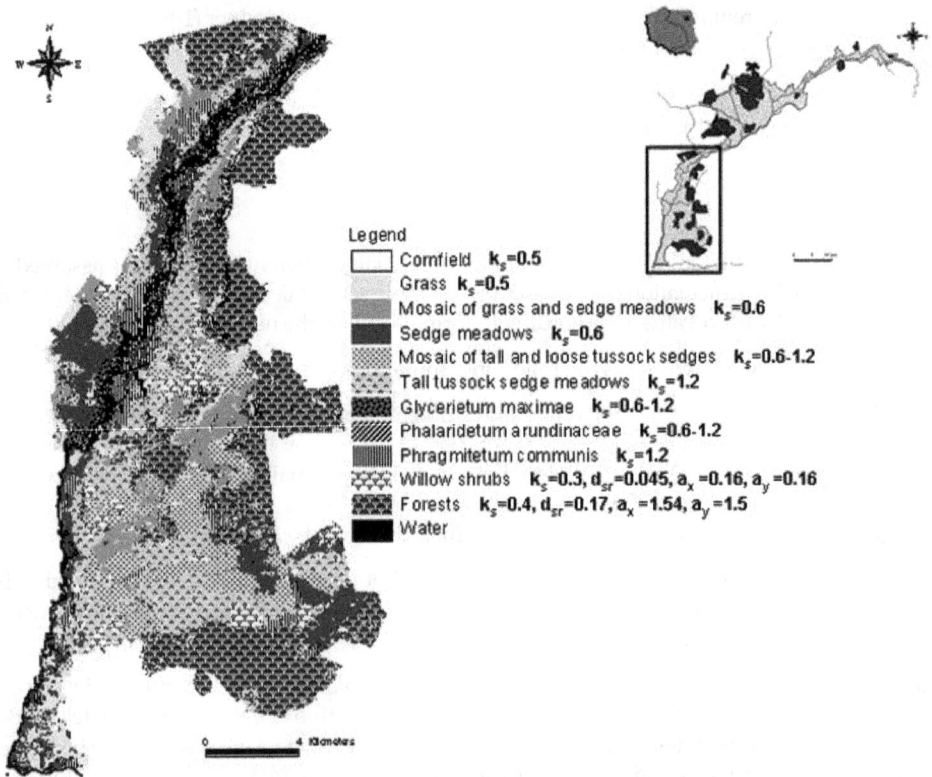

Legend
☐ Cornfield k_s=0.5
Grass k_s=0.5
Mosaic of grass and sedge meadows k_s=0.6
Sedge meadows k_s=0.6
Mosaic of tall and loose tussock sedges k_s=0.6-1.2
Tall tussock sedge meadows k_s=1.2
Glycerietum maximae k_s=0.6-1.2
Phalaridetum arundinaceae k_s=0.6-1.2
Phragmitetum communis k_s=1.2
Willow shrubs k_s=0.3, d_{sr}=0.045, a_x=0.16, a_y=0.16
Forests k_s=0.4, d_{sr}=0.17, a_x=1.54, a_y=1.5
Water

Fig. 5. Location of the Lower Biebrza Basin and plant characteristics used in this study

According to the previous investigation by Oświt (1991) five vegetation zones are distinguished: reed communities Phragmition, sedge-reed communities Magnocaricion, sedge-moss communities Scheuzchzerio caricetea fuscae, willow and birch shrubs and also the communities of alder carr and birch-alder forest Alnetea glutinosae. The most valuable ecosystems here are not only natural areas of peatlands but also large areas of open semi-meadows, which are the result of extensive agricultural use. The hydrogenic dependent habitats in the Biebrza marshes run on a stable ground water inundation or river flooding which occurs regularly and reinforce these representative water ecosystems every year. Changes of water conditions and discontinuance of extensive agriculture causes transformation of meadows and pastures into tall herb vegetation, reed and in the end the succession of shrubs and forest upon non-forest ecosystems of peatlands which can be observed in some places in the Biebrza River valley.

3.2 Hydraulic and topographical data

The developed hydrodynamic model was used in river and floodplain flow simulations in the LBB. A river water flow model was elaborated for the reach BD1 –Burzyn, where the geometry of the riverbed and the floodplains were described by 47 cross-sections (Fig. 6). The upstream boundary condition was defined in the form of flow hydrograph for BD1 cross-section as the sum of flows for Osowiec gauge at the Biebrza River and for Przechody gauge at the Rudzki Channel. The downstream boundary condition was defined by the rating curve for Burzyn gauge at the Biebrza River (cross-section BD17).The alimentation by the main tributary – the Wissa River – was described as a point lateral inflow, located at the cross-section BD11, applying the discharge data for Czachy gauge. Available hydrological data for gauges: Osowiec, Przechody, Czachy and Burzyn proves that in the period of annual spring floods the outflow from the study area (recorded at Burzyn gauge) is much higher than the sum of inflows from Osowiec, Przechody and Czachy, which indicates a significant alimentation from a sub-catchment (Fig. 7).

In the elaborated hydrodynamic model, on account of insufficient recognition of hydrological alimentation by ground and snowmelt waters, the inflow from a sub-catchment was described with the use of the following, simplified formula:

$$Q_{lateral}(t) = Q_{out}(t + \Delta t) - Q_{in}(t) \tag{20}$$

where: $Q_{lateral}(t)$ – sub-catchment discharge at time t $[m^3/s]$; $Q_{out}(t)$ – discharge at Burzyn at time t $[m^3/s]$; $Q_{in}(t)$ – inflow discharge (at BD1 and Czachy gauge) at time t $[m^3/s]$, Δt – time lag [s].

Next, the $Q_{lateral}$ inflow was uniformly imposed along the river section. The value of Δt was adapted as a result of numerical experiment during a calibration process. The developed solution of sub-catchment alimentation is obviously a rough approximation of the actual situation, as it does not account for the spatial variation of the lateral inflow. A more realistic approach to this issue needs to be elaborated in further study.

The analyzed reach of the Biebrza River has a length of 41.5 km. The Rudzki Channel flows into the Biebrza River about 8 km upstream of the opening cross-section BD1, however, the Wissa River, which is the right-side tributary, inflows to Biebrza at the kilometer 14.5 (Fig. 6). The river bed elevations were measured by manual sounding, the neighbourhood of

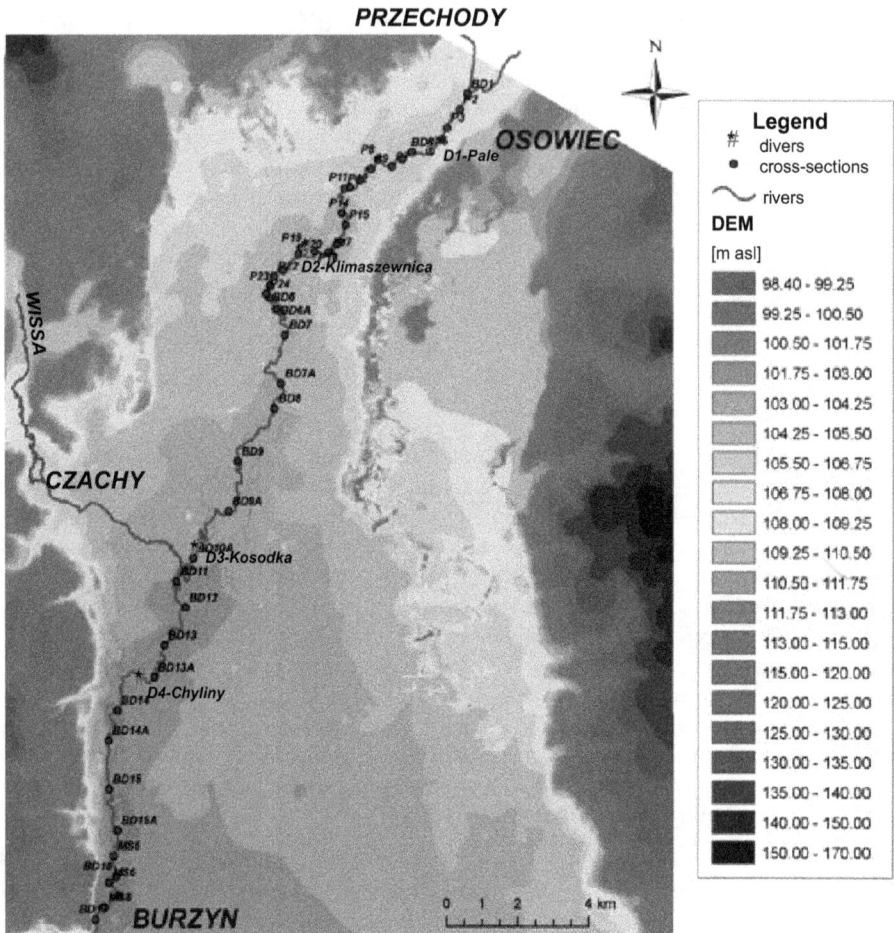

Fig. 6. Digital Elevation Model of the LBB with location of the cross – sections used in hydrodynamic model

river banks was leveled and points situated in floodplain were calculated from Digital Elevation Model and topographic maps for each cross section. Close to the river, the cross sections were measured in the field by coupling the traditional leveling survey with Differential GPS (DGPS) techniques. The topography of the floodplains was determined by the analysis of the existing Digital Elevation Model (Swiatek and Chormanski, 2007) (Fig. 6). The DEM was elaborated with the use of ArcGis software. The elevation model represents the shape of the valley in the form of a raster map of 25 m resolution. The main sources of data for DEM development were contour lines digitized on a 1:25 000 scale topographical map and also elevation points measured by GPS technique at transects perpendicular to the river. The elaborated DEM is characterized by mean square error not exceeding 0.35 m. (Chormanski, 2003).

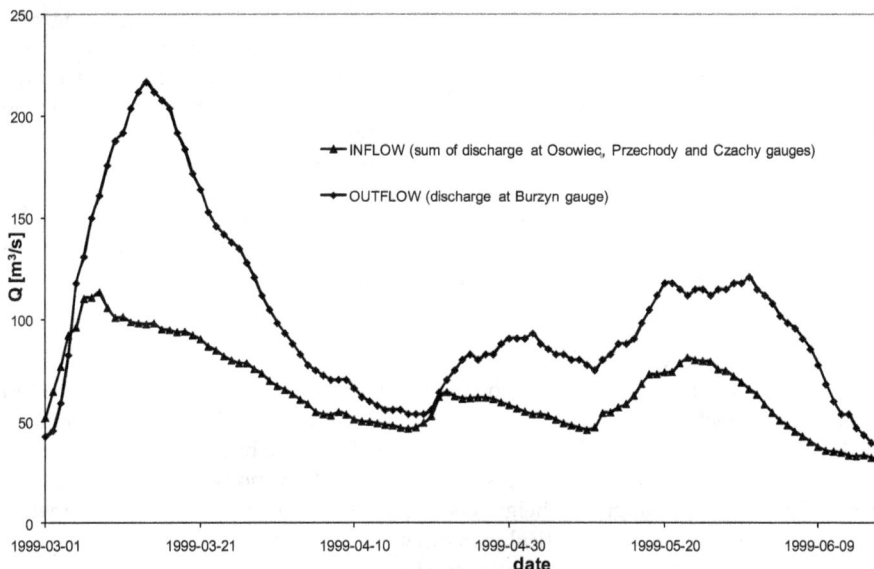

Fig. 7. The comparison of inflow with outflow for Biebrza River in the period 01 March 1999
– 17 June 1999 time period (Q– flow [m3/s])

3.3 Floodplain vegetation data

The vegetation parameters of Biebrza floodplains used in calculations were determined on
the basis of vegetation inventory in the Lower Biebrza Basin, vegetation map and PHARE
aerial photos. Every part of a cross-section has one bottom roughness parameter, related to
the type of plants overgrowing in this site of a floodplain. An example of vegetation
distribution in selected BD3 cross-section is shown in Fig.8.

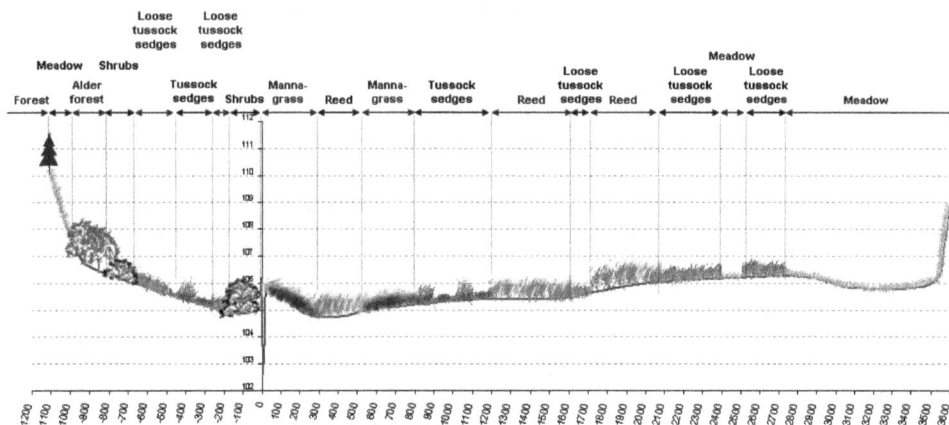

Fig. 8. Example of vegetation distribution in LBB floodplain (cross-section BD3)

The methodology of plant characteristics determination proposed by German engineers (DVWK, 1991) was applied in this study for high and medium vegetation in every cross-section established in the Lower Biebrza Basin. Water flow resistance in flowing round of high vegetation such as bushes or trees is calculated on the basis of supplementary average vegetation diameter d_p and supplementary distances between plants in direction of water flow a_x, and transversal to it a_y (equation 6). Resistances of totally flooded vegetation are characterized by roughness height of flexible and stiff vegetation k_s. It was determined on the basis of values given by Ritterbach (1991) for the Wupper River Valley in Germany and field measurements in LBB of particular vegetation type's height (Szporak et al, 2006). The vegetation types and parameters of Biebrza floodplains are presented in Fig. 5.

3.4 Model calibration and validation

The main role of the model is the reproduction of flood water levels and discharges in the LBB. The model, which is elaborated in a correct manner, should also provide proper reproduction of water flow not exceeding bankful water. For the sake of this, the hydrodynamic model was also calibrated and validated for unsteady state in flood condition and when water level is less than bankful. The values of roughness height coefficients in the riverbed were determined for each cross-section as a result of numerical experiments, calibrating the model for the steady water table elevation, which was measured in the October 2005 (the discharge at BD1 cross-section recorded 22.10 m³/s, while the lateral inflow from Wissa was equal to 2.15 m³/s).

Mean square error (RMSE) for the steady state reached 0.03 m. The calculations for the steady state were performed in order to identify k_s coefficients at the computational cross-sections of the main riverbed for non-vegetation period. The calibrated values of roughness height coefficient for the riverbed varied from 0.1 to 0.7 m. The resultant roughness coefficients were next used for model validation for unsteady state in the period from 1 November 2005 to 30 November 2005. For that period the available data considered constant measurements of water stages by automatic sensors (divers) in the riverbed. The sensors were located at four following places along the river: D1-Pale at 48+648 km, D2-Klimaszewnica at 40+483 km, D3-Kosodka at 24+195 km and D4-Chyliny at 17+843 km (Fig. 6). The sampling rate of water levels was equal to 6 hours. Maximum value of the RMSE is 0.06 m in this case (Tab.2) and takes place at diver D1. Results of the executed simulation confirmed that calibrated values of k_s coefficient can be used in river flow modelling for main channel in non-vegetation period.

The elaborated model was next calibrated for the spring flood period: 1 April 2006 to 30 April 2006, based also on constant measurements of water stages collected by automatic sensors. In the process of model calibration the following parameters were estimated: dimensions of active zones on the floodplain, hydraulic features for low and high vegetation in floodplains, time lag in the $Q_{lateral}$ assessment (equation 20). Because the analyzed flood event took place during the non-vegetation period, k_s coefficients for the riverbed were the same as in the previous simulation.

The dimensions of active zones on the floodplain were determined by flow capacity calculation for each cross-section, which is used in the model. An example of a relationship between the discharge and the width of a cross-section, which takes part in water flow, is shown in Fig.9. Calibrated values of hydraulic parameters of Biebrza floodplain vegetation types are presented in Tab.1. The best correspondence between the observed discharge hydrograph and the one produced by the model at the Burzyn gauge was calculated for a time lag Δt of 24 hours.

Vegetation type	k_s	a_x	a_y	d_p
grass-medows	0.3	-	-	-
mosaic of grass and sedge meadows	0.4	-	-	-
loose tussock sedges	0.4	-	-	-
reed	1.2	-	-	-
glycerietum maximae	0.4	-	-	-
tussock sedges	0.5	-	-	-
phalaridetum arundinaceae	0.6			
willow shrubs	0.4	0.13	0.13	0.03
alder swamp forest	0.8	1.50	1.50	0.18

Table 1. The hydraulic plant characteristics calibrated for various vegetation types in the
LBB (non-vegetation period)

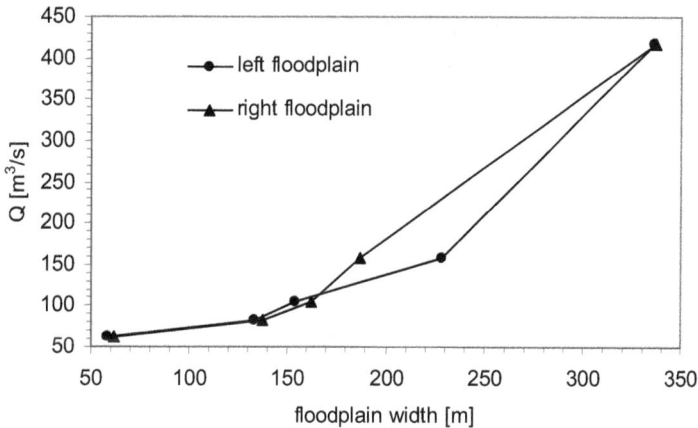

Fig. 9. Relationship between the river discharge and the active part of the floodplain (BD3
cross-section)

Calculation results for the flood in 2006 were compared with the water level observations at
four automatic sensors (Fig. 10-13). The RMSE water stage varies from 0.07 m to 0.14 m
(Tab.2) and indicates that the developed model correctly reproduces the range of variability
of water levels for flood simulation. On the other hand, values of correlation coefficient (R)
(Tab.2, Fig. 12-13), which is a measure of compliance of observed and calculated water stage
hydrographs variation, shows that 1D model is unable to reproduce flood dynamics in a
vast part of floodplain (Fig.10). Two divers D3 and D4 are located in places in the river,
where surrounding floodplain is very wide, and the calculated R coefficients for these
sensors have small values: 0.43 and 0.69 respectively. The R coefficient for the two divers
located in a narrow part of valley is higher than 0.9, and observed and calculated water
stage hydrographs are congruent (Fig. 10-11).

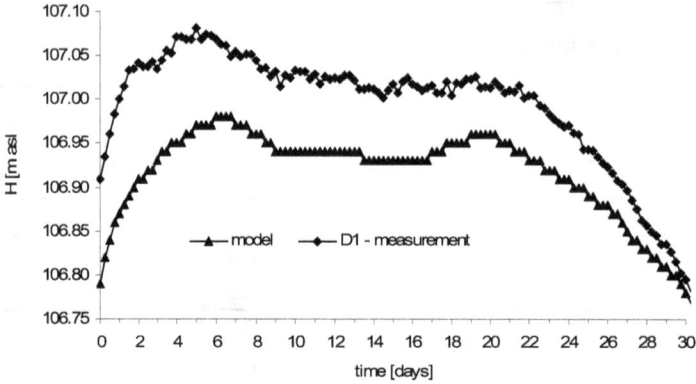

Fig. 10. Observed and calculated water stage at D1-Pale diver (01.04.2006 – 30.04.2006) (H – water level [m asl.])

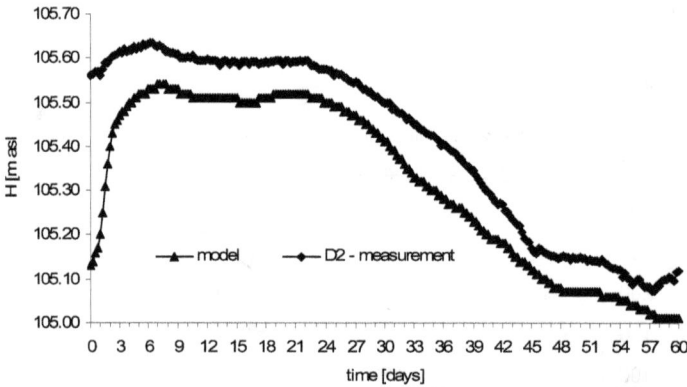

Fig. 11. Observed and calculated water stage at D2-Klimaszewnica diver (01.04.2006 – 30.04.2006) (H – water level [m asl.]

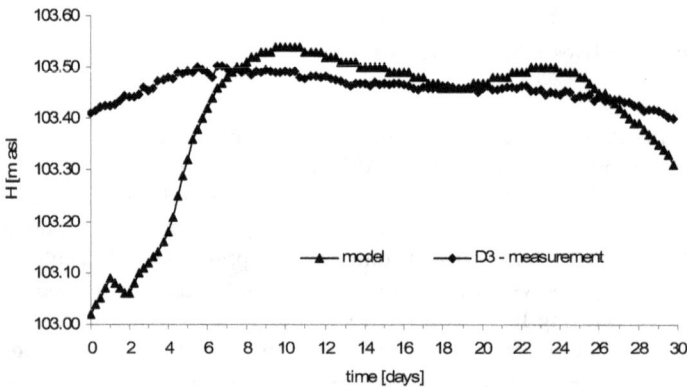

Fig. 12. Observed and calculated water stage at D3-Kosódka diver (01.04.2006 – 30.04.2006) (H – water level [m asl.])

The elaborated model was validated for the flood in 2007 (1 March – 30 April), which was characterized by a significantly higher water level than the flood in 2006. The maximum water level in Burzyn gauge was 101.76 m a.s.l. in 2006 and amounted to 102.28 m a.s.l. in 2007. In this case the model gives similar results to the ones of water stage observations performed at D1-D4 divers (Tab.2).

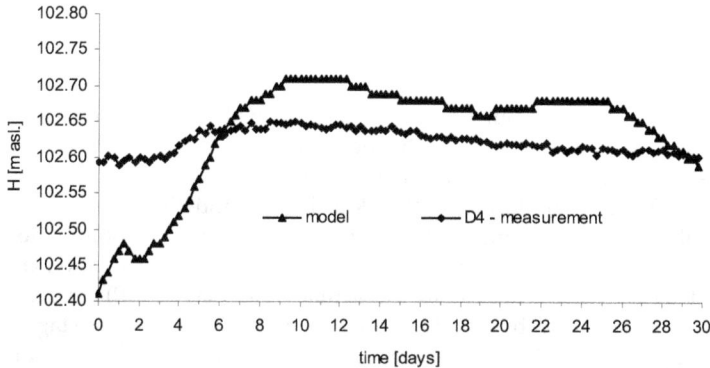

Fig. 13. Observed and calculated water stage at D4-Chyliny diver (01.04.2006 – 30.04.2006) (H – water level [m asl.])

diver's name	km	2005 01.11-30.11		2006 01.04-30.04		2007 01.03-30.04	
		RMSE [m]	R	RMSE [m]	R	RMSE [m]	R
D1 Pale	48+648	0.06	0.90	0.08	0.91	0.07	0.90
D2 Klimaszewnica	40+483	0.05	0.92	0.11	0.96	0.09	0.93
D3 Kosódka	24+195	0.04	0.96	0.14	0.43	0.12	0.46
D4 Chyliny	17+843	0.05	0.94	0.07	0.69	0.07	0.72

Table 2. Variation of the mean square error (RMSE) and correlation coefficient (R) for water stage-divers (model calibration and validation)

3.5 Scenarios determination

Various types of land use were determined for three scenarios and two of them present protective activities carried out nowadays in the National Park, which consist of preservation of open grasslands by mowing, elimination of biomass in selected areas, and mechanical scrub cutting. In the first analyzed scenario (scenario I), wet meadows, sedge plant communities and pastures are areas of extensive agricultural use. In this scenario, these areas are mowed for cattle bedding, used for hay production and as feeding areas. In the second scenario (scenario 2), there is willow scrub and birch tree cutting, in addition to the former scenario. The overgrowing process in many cases seem to be a natural succession so the third scenario (scenario 3) allows the natural succession of willow shrubs and birch forest upon non-forest ecosystems in the valley.

The numerical calculations for the three scenarios, which show changes in the land use, were made for the maximum flood event in the LBRB in 1979. In scenario I, roughness height coefficients for wet meadows, tall sedge plant communities, and pastures were given according to values in Tab.1. In the model scenario II, it was described by $k_s = 0.4$ m for the floodplain area with willow scrubs. In the last simulation (scenario 3), roughness coefficients for the study site were established like willow scrubs (Tab.2) because of natural succession in this area.

3.6 Results and discussion

The results were analyzed in four river cross sections which represent a variability of the hydromorphological shape of the main channel and floodplain in the Lower Biebrza River Basin. The distances from BD17, where the river course ends, to the selected cross sections are 7.62 km, 29.62 km, 33.04 km and 39.96 km. Fig. 14 and Fig. 15 show the calculated hydrographs of water stages and water stage profile of the maximum flood event of 03.04.1979 for the various land use scenarios. The average depth in scenario I and scenario II was decreased by 0.25 m compared to the actual state (scenario 0). In scenario III, the average depth was increased by 0.20 m. Some differences in the water stage at the outlet (near BD17) are related to the downstream boundary condition which is formed by the rating curve, which is the same for all scenarios. Therefore in calculation, the influence of the vegetation structure on the rating curve was excluded which results in lack of influence of a land use type on a calculated water stage at BD17. Tab. 3 shows average water depth in the floodplain, top flood width, and flood duration for the various land use scenarios.

The results for scenario I and scenario II are similar, because of willows scrubs covering a small area in the valley (Fig. 5). The highest changes of flooding time were calculated for P4 localized in the northern part of the Lower Basin, where the width of the river valley is lower than in the middle and southern parts of the research area. Time of flood in scenario I and II, when the resistance of flow is lower, was decreased by about 40 %. When the resistance of flow is higher because of growing vegetation (scenario III), flood duration was increased by 17 % compared to the actual state.

An increasing valley width resulted in decreased flooding for scenarios I and II by 8 to 6 % along the river from P17 to BD14. In scenario III, flood duration was increased by 5, 7, and 9 % for P17, BD6, and BD14 in comparison to the scenario 0. Changes in the average depth do not exceed 10 %. The variation of flood extent is connected with the cross sections geometry. The highest decrease of width occurs at P4 for scenarios I and II and is about 40 %. In these scenarios it is about 10 % at P17 and BD6. The flood extent was increased by several percent in scenario III at the analyzed cross sections. In this scenario birch forest and willow shrubs occurred over large areas of wetland. The flood extent, water depth and flood duration increased which means that the area of plant communities dependent on rich surface water from the river like Phragmition and Magnocaricion will increase at the cost of more valuable plant communities (like sedge-moss meadows). Long lasting inundation also does not support the development of woody vegetation except the alluvial forests. Using the hydrodynamic model it was impossible to predict to what extent changes in flooding characteristic will affect the vegetation, changing in the next turn its structure and location. We can only state here that secondary succession probably does not form climax vegetation on the floodplain.

Unsteady 1D Flow Model of Natural Rivers with Vegetated Floodplain – An Application to Analysis of Influence of
Land Use on Flood Wave Propagation in the Lower Biebrza Basin

163

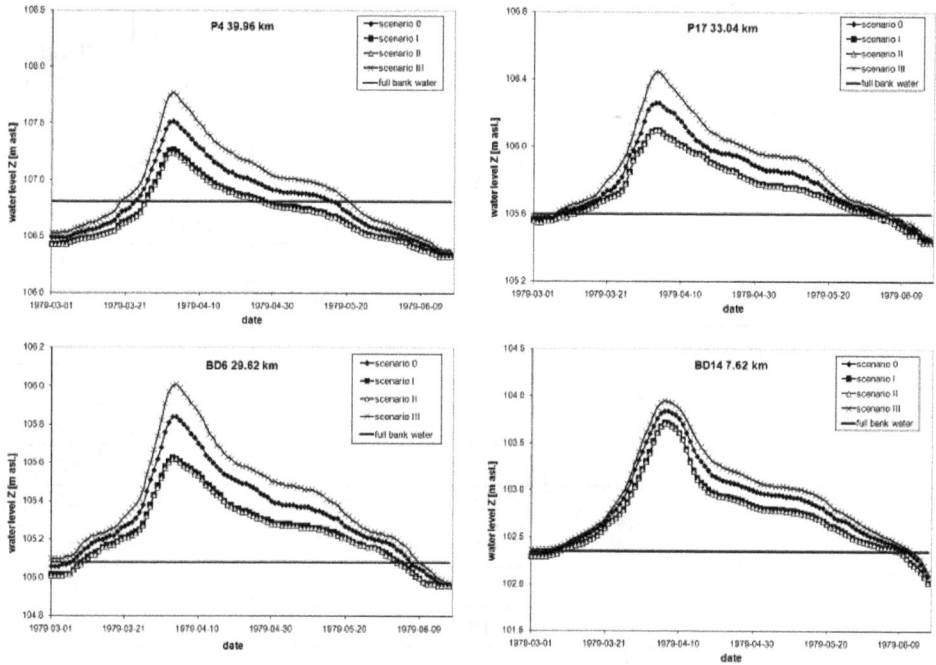

Fig. 14. Calculated stage hydrograph at select cross sections for different land use scenarios

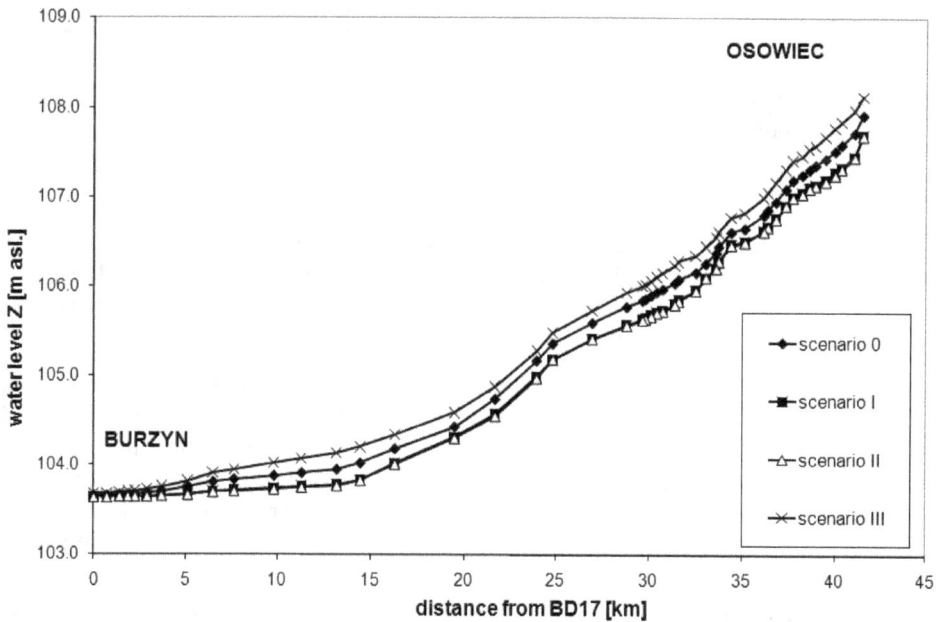

Fig. 15. Calculated water stage profile for different land use scenarios (03.04.1979)

Number	Cross-section name	Distance along the river from BD17 [km]	Scenario	Average depth [m]	Top width [m]	Flood duration [days]
1	P4	39.96	0	0.58	2346	53
			I	0.51	1432	32
			II	0.51	1332	31
			III	0.64	2400	62
2	P17	33.04	0	0.24	3439	86
			I	0.19	3010	79
			II	0.19	3008	79
			III	0.29	3744	90
3	BD6	29.62	0	0.32	5103	92
			I	0.28	4685	86
			II	0.28	4666	85
			III	0.35	5367	100
4	BD14	7.62	0	0.49	6950	97
			I	0.44	5655	91
			II	0.44	5592	91
			III	0.53	7182	104

Table 3. Variability of the average water depth on the floodplain, top flood width and flood duration for different land use scenarios

4. Conclusions

The application of a flow law in the model, in the form of the Darcy-Weisbach formula, made it possible to consider the influence of vegetation on water flow conditions. It allowed for implementation of equations mass and momentum conservation processes between the main riverbed and the floodplains in 1D St Venant.

In order to explore the potential of the model use for the analysis of unsteady flood flows, simulations of water flow were performed for the Lower Biebrza Basin, which is a natural reference valley. The model was calibrated and validated for floods which took place in 2006-2007 based on the data achieved as a result of executed field research. During model calibration process there were generally two difficulties: the determination of inflow from a sub-catchment and delineation of those areas of a very flat and vast valley that take part in water flow.

The solution to the first problem was applied in the model in the form of uniformly distributed lateral inflow estimated as a result of numerical experiments, and, even if it leads to a satisfactory match of flow hydrographs at the closing water gauge, it is a very rough solution. A proper solution should involve the elaboration of a hydrological model for the LBB with the consideration of river water-ground water interaction. The application of flow capacity calculations allowed for delineation of that part of the valley which takes active part in a flood wave propagation. In future this approach should be coupled with a field monitoring.

The measurement technique used for inundation dynamics monitoring, which involves installation of automatic sensors of D-Diver type in the river bed, is considered to be the most reasonable measurement solution in the field of achieving data on water levels in the moment of high water occurrence. That sort of data is indispensable to perform proper identification and validation of a hydrodynamic model. In order to properly investigate the variability of inundation in the Lower Biebrza Valley, that type of sensors should not only be installed at selected points located in the river, but also on the floodplains. The location of measurement points may be chosen on the basis of collected measurement data and GIS analyses against the indications of existing digital elevation model. Such an organized monitoring network should lead to a more complete recognition of spatial and temporal variability of inundation in that area.

The application of aerial photographs, vegetation maps and time consuming field monitoring made it possible to delineate the low and high vegetation communities on the floodplains and to determine the values of roughness of high and mean geometrical parameters for each plant community.

In the case of the Lower Biebrza the biggest differences of water levels occur in those places where the valley is very wide and the simplification of water flow to a one dimensional case results in impossibility to reproduce real dynamics of the floods. The water flow in this part of the valley can be performed by a 2D model.

The elaborated model, which was calibrated and validated for the area of the LBB, allows for recognition the dynamics of inundation process and the range of inundation in that region. It is particularly vital for the implementation of an effective policy of natural values protection in the area of the Biebrza National Park

The obtained results show that vegetation structure on the floodplain influences the flood wave propagation in the LBRB. An intensive land use on the floodplain decreased the resistance of flow, which decreased flood duration, flood extent, and average flood water depth. These flood characteristics are strongly connected with the cross section geometry. The valley in the northern part is narrow, which causes flood duration shortening; for an intensive land use on the floodplain the time was decreased by about 40 % compared to the current state, while in the other part of the Lower Basin - by less than 10 %. The influence of the natural succession on the flood wave is less significant than the influence of intensive agriculture. In this case, flood duration changes by a few percent.

The achieved results were obtained with the use of a one dimensional hydrodynamic model which does not cover water flow across the river. It means that the used model represents simplified flow conditions in the Lower Biebrza River Basin, which is a wide area and for which this aspect is very important. The essential influence also has assigned ineffective flow areas that contain water which is not actively conveyed. Ineffective flow areas are used to describe portions of a cross section in which water ponds, but velocity of that water is close to or equal to zero.

Moreover, the hydrodynamic model is an appropriate tool for assessment of various agricultural practices with regard to the effective management to protect the unique wetland site of the Biebrza National Park.

The developed model makes it possible to find a solution for unsteady flow problems in natural rivers with vegetated floodplains. It may be used as a tool to estimate new water surface level for renaturalized rivers, especially for flood conditions, as well as to ensure suitable conditions for habitat diversity in projects of environmental flood management. It is an appropriate tool for estimation of floodplain vegetation influence on flow conditions, which can be considered a 1D problem.

5. References

Bretschneider H., Schulz A. 1985: Anwendung von Fließformeln bei naturnahem Gewässerausbau. *DVWK - Schriften, Heft 72.*

Chormanski J. 2003: Methodology for the flood extent determination in the Lower Biebrza Basin. *Ph.D. thesis, Warsaw Agricultural University,* 186.

DVWK – MERKBLÄTTER 220, 1991: Hydraulische Berechnung von Fließgewässern, *DK 551.51/54 Fließgewässer, DK 532.543 Hydraulik, DVWK - Merkblätter 220/1991,* Kommissionsvertrieb Verlag Paul Parey, Hamburg und Berlin.

Cunge, J., F. Holly Jr., and A. Verwey, Practical Aspects of Computational Hydraulics, *Pitman Publishing Ltd.,* London, UK, 1980.

Helmio , T. 2002, Unsteady 1D flow model of compound channel with vegetated floodplains, *J. Hydrol.,* 269(1–2), 89–99.

Hooijer A. 1996: Floodplain hydrology: An ecologically oriented study of the Shannon Callows, Ireland. Thesis, Virje Universiteit Amsterdam

Horritt, M.S. and Bates, P.D. 2002. Evaluation of 1-D and 2-D numerical models for predicting river flood inundation. Journal of Hydrology, 268, 87-99.

Indlekofer, H., 1981, Überlagerung von Rauhigkeitseinflüssen beim Abfluß in offenen Gerinnen. *Mitt. Institut für Wasserbau und Wasserwirtschaft,* RWTH Aachen, Heft 37, s. 105-145.

Kaiser, W., 1984, Fließwiderstandsverhalten in Gerinnen mit durchströmten Ufergehölzzonen. *Thesis presented for the degree of Doctor in Applied Sciences* TH Darmstadt.

Kubrak J., Nachlik E. (red.), 2003: Hydrauliczne podstawy obliczania przepustowości koryt rzecznych. *Wydawnictwo SGGW,* 317 stron.

Laks I., Kałuża T., 2007, Analiza przejścia fali wezbraniowej z 1997roku na odcinku od Jeziorska do Obornik w modelu obliczeniowym, uwzględniającym aktywną część przekroju. Nauka Przyr. Technol. 1, 2, #27.

Lindner K., 1982: Der Strömungswiderstand von Pflanzenbeständen. *Mitteilungen aus dem Leichtweiss - Institut für Wasserbau der TU Braunschweig*, H. 75.

Nuding, A., 1991, Fließwiederstandsverhalten in Gerinnen mit Ufergebusch. Entwicklung eines Fließgewässer mit und ohne Geholzufer, unter besonderer Berucksichtigung von Ufergebusch, *Wsserbau-Mitteilugen* Nr.35, Technische Hochschule Darmstadt.

Oswit J. 1991: Roślinność i siedliska zabagnionych dolin rzecznych na tle warunków wodnych. [Vegetation and wetland habitats against the background of water conditions]. *Rocz. Nauk Rol.*, 221: Wyd. Nauk. PWN 229

Pasche E., 1984: Turbulenzmechanismen in naturnahen Fließgewässern und die Möglichkeit ihrer mathematischen Erfassung. *Thesis presented for the degree of Doctor in Applied Sciences*, RWTH, Aachen.

Pasche, E., and G. Rouve, 1985, Overbank flow with vegetatively Roughened flood plains. *Journal of Hydraulic Engineering* 111(9), 1262-1278.

Rickert K., 1988, Hydraulische Berechnung naturnaher Gewässer mit Bewuchs. *DVWK-Fortbildung*, H. 13.

Ritterbach E.1991: Wechselwirkungen zwischen Auenökologie und Fließgewässerhydraulik und Möglichkeiten der integrierenden computergestützten Planung. *Mitteilungen für Wasserbau und Wasserwirtschaft*, Rheinisch - Westfälische Technische Hochschule Aachen.

Rouvé G., DFG Deutsche Forchungsgemeinschaft, 1987: Hydraulische Probleme beim naturnahen Gewässerausbau Ergebnisse aus Schwerpunktprogramm "Anthropogene Einflüsse auf hydrologische Prozesse", Band 2.

Schumacher, F., 1995, Zur Durchflußberechnung gegliederter naturnah gestalteter Fließgewasser. *Mitteilung* Nr. 127, TU Berlin.

Swiatek D., Szporak S., Chormański J.: Hydrodynamic model of the lower Biebrza river flow - a tool for assesing the hydrologic vulnerability of a floodplain to management practice. *Ecohydrology & Hydrobiology* 2008

Swiatek D.2008: Influence of vegetated floodplains on compound channels discharge capacity in 1D modelling. Publs. Inst. Geophys. Pol. Acad. SC.E-10 (406), pp.155-162.

Swiatek D., Chormanski J. 2007: Verification of the numerical river flow model by use of remote sensing. Balkema – Proceedings and Monographs in Engineering, Water and Earth Sciences, *Proceedings of the international conference W3M "Weatlands: Modelling, Monitoring, Management"*. Wierzba, Poland, 22-25 September 2005

Swiatek, D., J.Kubrak, J.Chormanski, 2006, Steady 1 D water surface model of natural rivers with vegetated floodplain: An application to the Lower Biebrza, *Proceedings of the International Conference on Fluvial Hydraulics River Flow*, Vol. 1, p. 545-553.

Szporak S., Koziol A., Miroslaw-Swiatek D., Kubrak J. ,2006, *Determination of plant characteristics used in discharge assessment of the Biebrza River valley. Przegląd Naukowy Inżynierii i Kształtowania Środowiska*, Rocznik XV (33) z 2.

Szymkiewicz, R., 2010, Numerical Modeling in Open Channel Hydraulics, *Springer*, DOI:10.1007/978-90-481-3674-2.

Zalewski M., Janauer G.A., Jolankai G. 1997: A new paradigm for the sustainable use of aquatic resources. *UNESCO IHP Technical Document in Hydrology* 7. IHP - V Projects 2.3/2.4, UNESCO Paris, Ecohydrology

8

Hydrology and Methylmercury Availability in Coastal Plain Streams

Paul Bradley and Celeste Journey

U.S. Geological Survey,
USA

1. Introduction

The primary matrix of occurrence strongly influences the impacts of contaminants on environmental and human health. In groundwater and surface-water settings, water 1) dominates environmental transport and distribution, 2) influences contaminant reactivity, transformation and, by extension, toxicity, and 3) mediates direct and indirect exposure pathways. The role of hydrology in determining contaminant risk in groundwater and surface-water environments varies with contaminant type. Consequently, this chapter focuses on mercury (Hg), a widely distributed environmental pollutant, in order to illustrate the critical role that hydrology plays in determining contaminant risk. A comprehensive review of all of the mechanisms by which hydrology affects Hg risk is beyond the scope of this chapter. Rather, this chapter will discuss a few specific mechanisms that illustrate the critical link between hydrology and Hg risk in the environment.

1.1 Hg in the environment

Mercury occurs naturally in the environment, primarily in subsurface mineral deposits (Grigal, 2002, 2003; Pacyna et al., 2006; Selin, 2009; Swain et al., 2007). Although volcanic activity and volatilization from mineral outcrops can mobilize Hg to the surface atmosphere, the importance of Hg as an environmental contaminant increasingly is attributed to mining activities and subsequent anthropogenic releases to the surface biosphere. Historically, mercury has been mined from mercuriferous (cinnabar) belts in western North America, central Europe, and southern China. A notable example is the Almadén mine, in operation since Roman times (Selin, 2009). Annual releases of Hg to the surface atmosphere are estimated to be 5000-6600 Mg y^{-1} (Driscoll et al., 2007; Pacyna et al., 2006; Selin, 2009; Swain et al., 2007).

Direct use in a variety of applications, as a liquid metal or chemical constituent, has resulted in widespread environmental releases and, in some cases, heavily contaminated sites resulting from Hg point sources (Selin, 2009; Swain et al., 2007). Until recently, Hg was a common constituent of commercial products throughout the world. Recognition of the environmental and human health risks of Hg has prompted regulation and a shift toward Hg-free substitutes in many industrialized countries. However, even today, the extent of regulation and enforcement varies considerably "across jurisdictions and industrial sectors" (Selin, 2009; Swain et al., 2007). Countervailing environmental concerns over energy conservation combined with a lack of cost-effective alternatives has continued the use of energy efficient Hg-vapour, fluorescent lighting. Thus, domestic exposures to harmful levels of Hg vapour resulting from breakage and/or improper disposal of fluorescent lamps remain significant concerns in industrialized countries, as does continued use of legacy, Hg thermometers. Direct exposure to high concentrations of Hg vapour is a particular concern in less industrialized areas where small-scale (artisanal) mining practice typically involves amalgamation of gold with Hg followed by heating to release Hg vapour and concentrate the gold (Selin, 2009; Swain et al., 2007).

Hg also is present in low concentrations in many natural materials, most notably in coal, oil, and minerals. The mining and use of low-Hg materials in large quantities, particularly combustion of fossil fuels, are primary pathways of Hg release and non-point source, environmental Hg contamination. Approximately 60% of the estimated annual global anthropogenic Hg release to the atmosphere is attributed to combustion of coal and other fossil fuels (Swain et al., 2007). Coal-fired power plants, waste incinerators (municipal and medical), chlor-alkalai facilities, and industrial boilers, contribute about 80% of anthropogenic emissions in the USA (Driscoll et al., 2007; Driscoll et al., 1998; EPRI, 1994; Seigneur et al., 2004). These on-going Hg emissions have resulted in regional and global atmospheric Hg reservoirs and widespread deposition to terrestrial and aquatic environments, albeit at generally low environmental concentrations (Selin, 2009; Swain et al., 2007).

1.2 Hg bioaccumulation and environmental risk

Atmospheric Hg deposition represents a substantial environmental threat even at low concentrations, due to the potential transformation to neurotoxic and highly bioaccumulative methylmercury (MeHg)(Bloom, 1992; Brumbaugh et al., 2001; Hall et al., 1997) by microorganisms indigenous to wetlands, lake sediments, and other saturated environments. MeHg bioconcentration factors in the order of 10^4 to 10^7 have been reported in aquatic food webs (Grigal, 2003; Rudd, 1995; Ullrich et al., 2001). Thus, in the USA, Canada and in many other industrialized nations, the primary risk of mercury (Hg) in the environment, including the risk to human health, is due to accumulation of Hg in aquatic biota (Environment Canada, 2011; Mergler et al., 2007; Selin, 2009; Swain et al., 2007; U.S. Environmental Protection Agency, 2009a).

MeHg contamination in fish is the leading cause of fish consumption advisories in the United States (U.S. Environmental Protection Agency, 2009a). A comparable percentage of the lakes (40% of total area) and streams (36% of total river distance) in the United States are Hg impaired (U.S. Environmental Protection Agency, 2009a). In 2008, the United States Environmental Protection Agency listed 3361 fish consumption advisories, affecting 50 states and covering more than 6.8×10^6 ha of lake and 2.1×10^6 km of river (U.S. Environmental Protection Agency, 2009a). Hg-driven fish consumption advisories, likewise, are common throughout Canada (Environment Canada, 2011). For this reason, identification of surface-water environments that are susceptible to bioaccumulation of Hg above accepted human and wildlife adverse impact thresholds and improved understanding of the key geochemical, hydrological, and biological characteristics that contribute to Hg vulnerability in the environment are global health priorities (Benoit et al., 2003; Mergler et al., 2007).

1.3 Purpose

A number of surface-water settings in North America are characterized by elevated levels of Hg bioaccumulation in fish (Bauch et al., 2009; Krabbenhoft et al., 1999; Scudder et al., 2009). Much of the current understanding of the factors contributing to elevated Hg bioaccumulation in aquatic habitats of North America is based on research conducted in the extensive peatland environments of Canada and in the organic-enriched surface waters of the northeastern USA, where Hg bioaccumulation in top predator fish and piscivorous bird species is well documented (Driscoll et al., 2007). Recent studies have demonstrated that Coastal Plain stream environments also are particularly prone to elevated Hg concentrations in fish and other indigenous aquatic communities (Bauch et al., 2009; Bradley et al., 2011; Bradley et al., 2010; Glover et al., 2010; Guentzel, 2009; Scudder et al., 2009), but considerably less is known about the specific ecological interactions contributing to elevated Hg bioaccumulation in this physiographic setting. However, recent research indicates that the elevated Hg risk associated with Coastal Plain streams is inextricably linked to the hydrologic characteristics of the Coastal Plain physiographic region.

The concentration of MeHg in fish tissues can be attributed to interactions between three conceptual components of the aquatic MeHg biocycle: 1) production and accumulation of MeHg, often in near-stream wetland environments, 2) transport of MeHg from source areas to the stream aquatic habitat, and 3) uptake by and trophic transfer in the aquatic foodweb (Figure 1)(Bradley et al., 2009). In this chapter, the general impact of hydrology on microbial production and in situ persistence of MeHg in saturated sediment environments is discussed with specific emphasis on characteristics relevant to the southeast region of the USA. The role of hydrology in the transport of MeHg from the site and matrix of production to the point of entry into the food web in Coastal Plain stream systems is illustrated by recent research in a paired basin study in South Carolina (Bradley et al., 2010; Bradley et al., 2009). Although water quality and quantity also affect the composition, trophic structure, and trophic transfer efficiency of indigenous communities, the role of hydrology in the uptake and accumulation of Hg in Coastal Plain aquatic food webs is beyond the scope of this chapter.

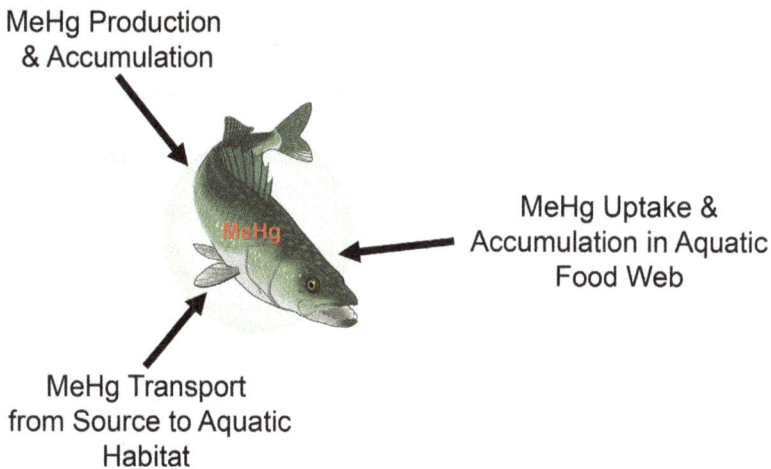

MeHg Production
& Accumulation

MeHg Uptake &
Accumulation in Aquatic
Food Web

MeHg Transport
from Source to Aquatic
Habitat

Fig. 1. Conceptual model of factors affecting Hg bioaccumulation in fish.

2. Hydrologic controls on methylmercury production

Hydrologic processes affect the production and accumulation of MeHg in the environment, in general, and in the Coastal Plain region of the USA, in specific, in a number of ways. In this section, two fundamental roles are presented as examples. The supply of Hg, primarily inorganic Hg, is a prerequisite to MeHg production and accumulation in saturated environments and hydrologic transport is critical to Hg supply to the landscape and to sites of active methylation. Likewise, the geochemical characteristics of saturated wetland sediments are conducive to microbial methylation of Hg and wetland environments are among the most important sources of MeHg in the environment.

2.1 Hydrologic controls on Hg supply to riparian wetlands

Wet deposition is a substantial pathway for transfer of atmospheric inorganic Hg to terrestrial and aquatic environments. Hg is emitted from natural sources primarily in elemental form (Hg(0)) and from anthropogenic sources as Hg(0), divalent Hg (Hg^{2+}), or particulate Hg (Hg(P))(Selin, 2009). The dominant form of Hg in the surface atmosphere is Hg(0), with mean concentrations estimated to be 1.6 ng m^{-3} (Selin, 2009). Delivery of Hg to terrestrial and aquatic surfaces occurs year round via wet and dry deposition on open areas and via throughfall (wash off of foliar Hg deposition during rainfall events). Mean concentrations of Hg^{2+} and Hg(P) in the surface atmosphere are estimated in the range of 1-100 pg m^{-3} (Selin, 2009). However, because Hg^{2+} and Hg(P) are more soluble in water, these are the primary forms of Hg deposited in wet and dry deposition and throughfall. Elevated atmospheric concentrations can occur regionally, downwind of major emission sources such as power facilities. Atmospheric Hg deposition can also occur on a more seasonal basis as litterfall following direct deposition (wet and dry) and/or direct uptake of gaseous Hg by plants (Selin, 2009).

The highest wet deposition rates in the USA occur in the southeast, with elevated rates also occurring in the midwest to northeast of the USA downwind of major North American industrial centers. Elevated Hg wet deposition in the midwest and northeast of the USA correspond to a region of high Hg emissions. The cause of elevated Hg wet deposition in the southeast is less clear, but may be related to scavenging of reactive Hg(0) from the higher altitude global pool by convective storm events during the summertime (Selin, 2009). The Mercury Deposition Network (MDN) in the USA routinely monitors wet deposition only. Dry deposition is not currently monitored systematically on a national scale. Thus, while it is clear that wet deposition processes, particularly throughfall (Grigal, 2002), are substantial sources of Hg to the landscape, the relative importance of the various mechanisms of Hg transfer from the atmosphere to the landscape is a matter of considerable uncertainty. In forested systems, substantially more Hg is deposited as throughfall than as dry or wet deposition in open areas (Grigal, 2002). In the southeastern USA, dry Hg deposition generally is estimated to be substantially lower than wet deposition (Brigham et al., 2009).

Terrestrial systems are an important indirect source of atmospheric Hg to aquatic systems via runoff (Lee et al., 1994; Lorey & Driscoll, 1999). On an area basis, terrestrial systems receive more atmospheric Hg from direct deposition, throughfall or litterfall than do freshwater aquatic systems (Grigal, 2002). Terrestrial landscapes are a significant environmental Hg reservoir, estimated to contain a mass of Hg many times greater than the annual flux from the atmosphere (Gabriel & Williamson, 2004; Grigal, 2002; Mason et al., 1994).

Thus, hydrologic transport from the landscape to down-gradient wetland environments is a long-term control on Hg methylation in the environment (Gabriel & Williamson, 2004). The predominant form of Hg in terrestrial soils is Hg^{2+}. Although Hg^{2+} can bind to OH^-, and Cl^-, it is primarily associated with organic matter, specifically reduced S groups that are present in the environment in sufficient concentration to bind all Hg (Schuster, 1991; Skyllberg, 2008). Wet erosion and overland flow of particulate organic carbon (POC) is considered a major pathway of Hg transport from upslope to wetlands in steep and erosive systems with significant surface runoff (Balogh et al., 2000; Balogh et al., 1997, 1998; Grigal, 2002). This transport pathway appears to predominate in agricultural watersheds and is particularly sensitive to agricultural and land management practices (Balogh et al., 2000; Balogh et al., 1997, 1998). In contrast to agriculture-dominated watersheds, transport in the dissolved phase dominates in forested watersheds (Balogh et al., 1997, 1998; Hurley et al., 1995). Occurrence and down-gradient transport of Hg in groundwater is strongly related to the presence and mobility of dissolved organic carbon (DOC) and is expected to occur primarily in shallow groundwater flow paths (Grigal, 2002). Wet erosion and overland flow of POC are not expected to be primary Hg transport mechanisms in highly permeable low-gradient environments like the Coastal Plain of the USA (Bradley et al., 2010).

2.2 Hydrologic controls on Hg methylation

Most of the attention on environmental Hg pollution is focused on MeHg, a potent neurotoxin with reported bioconcentration factors on the order of 10^4 to 10^7 in aquatic food webs (Rudd, 1995; Ullrich et al., 2001). Previous studies have demonstrated microbial Hg methylation under Fe(III)-reducing and SO_4-reducing conditions, and Fe(III)-reducing and SO_4-reducing microorganisms are widely considered responsible for the bulk of Hg methylation in the environment (Compeau & Bartha, 1985; Gilmour & Henry, 1991; Gilmour et al., 1992; Grigal, 2003; Morel et al., 1998; Ullrich et al., 2001). MeHg production and accumulation are promoted under anaerobic conditions, whereas aerobic conditions support demethylation processes (Ullrich et al., 2001). The quantity and quality of DOC plays an important role in the bioavailability of Hg to methylating microorganisms and in the uptake and bioaccumulation of MeHg in the aquatic food web (Ravichandran, 2004).

Oxygen supply is limited in saturated sediments, because of the low solubility of oxygen in water and limitations on advective resupply in the sediment matrix. Aerobic microbial activity in environments with high bioavailable electron donor (organic carbon) can lead to rapid oxygen depletion and the onset of reducing conditions immediately following saturation. Extended experience in groundwater remediation has demonstrated the onset of substantial anaerobic activity at dissolved oxygen concentrations below 0.5 mg L^{-1} (Barcelona, 1994; Chapelle et al., 1995; Wiedemeier et al., 1998). Such conditions are routinely satisfied in wetland environments, particularly in peatland and organic rich bottomland floodplains, (Grigal, 2002, 2003) and wetlands are recognized areas of Hg methylation and elevated MeHg concentrations (Bradley et al., 2011; Bradley et al., 2010; Brigham et al., 2009; Grigal, 2002, 2003; Hurley et al., 1995; St. Louis et al., 1994a). Positive correlations between fish Hg burdens, dissolved MeHg concentrations, and basin wetland densities (Brigham et al., 2009; Chasar et al., 2009; Glover et al., 2010; Grigal, 2002; Guentzel,

2009; Hurley et al., 1995; St. Louis et al., 1994a) are widely reported, indicating that wetlands are the proximal source of MeHg in stream biota.

3. Hydrology and MeHg availability in Coastal Plain streams

In light of the demonstrated importance of wetlands as areas of substantial MeHg production and accumulation and as primary sources of MeHg to nearby lake and stream aquatic environments, transport of MeHg from wetlands to adjacent lake and stream aquatic habitats is a fundament control on environmental Hg bioaccumulation. Much of the current understanding of the controls on MeHg production, transport to primary aquatic habitats, and subsequent uptake and accumulation in aquatic foodwebs of lakes and streams in North America is based on research conducted in peatland, wetlands, and organic-rich surface-water environments of Canada and the northeastern USA. In contrast, comparatively little is known about the fundamental controls on Hg bioaccumulation in the Coastal Plain region of the southeastern USA, despite the recognized pattern of elevated fish Hg concentrations (Brumbaugh et al., 2001; Glover et al., 2010; Guentzel, 2009; Krabbenhoft et al., 1999; Scudder et al., 2009) in this geographically extensive physiographic region (Fenneman, 1928, 1938; Vigil, 2000). Consequently, the remainder of this chapter will focus on the role of hydrology as a control on MeHg availability in Coastal Plain stream settings. In the following subsections, the pattern of Hg bioaccumulation and the potential contribution of Coastal Plain hydrologic characteristics are discussed.

3.1 Hg bioaccumulation in Coastal Plain streams

In the summer and fall of 1998, the National Water Quality Assessment (NAWQA) and Toxics Substances Hydrology (Toxics) Programs of the U.S. Geological Survey (USGS) conducted a national pilot survey of Hg concentrations in the sediment and water (Krabbenhoft et al., 1999) and in axial muscle tissues of top predator fish (Brumbaugh et al., 2001) from 106 sites in 20 stream basins across the US. Among other findings, the results identified the Edisto River in South Carolina as having among the highest top predator fish Hg concentrations in the nation. Corresponding stream and sediment MeHg and total Hg concentrations also were among the highest reported in the USA, with the MeHg to total Hg ratios in the sediment and water of the Edisto basin being the highest observed in the study (Krabbenhoft et al., 1999). A follow-up assessment by the USGS assessed data from a total of 367 sites. This study included the data from the original 107 sites in the pilot survey, 159 stream sites from a second USGS national survey conducted in 2002 and 2004-5, and 101 stream sites from 4 USGS regional studies (Bauch et al., 2009; Scudder et al., 2009). While the highest Hg concentrations in fish were observed in gold or Hg-mined basins in the western USA, comparable concentrations were observed in unmined basins where atmospheric Hg was considered the primary source of Hg to the aquatic environment. The highest fish Hg concentrations in unmined basins were observed in "black-water" (high DOC) Coastal Plain streams in the eastern and southeastern USA (Bauch et al., 2009; Scudder et al., 2009)(Figure 2). While previous studies had demonstrated that elevated bioaccumulation of Hg is common in the organic-rich surface waters of the industrialized northeastern USA, these results indicated that stream habitats within the Coastal Plain physiographic region of the predominantly forested/agricultural southeastern USA also were among the most Hg vulnerable ecosystems in North America (Bauch et al., 2009; Brumbaugh et al., 2001; Krabbenhoft et al., 1999; Scudder et al., 2009).

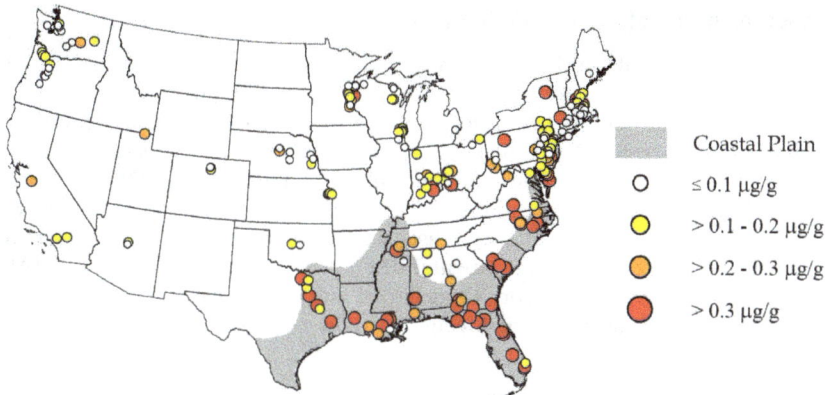

Fig. 2. Spatial distribution of Hg (µg/g wet weight) in piscivorus game fish, 1998-2005 (modified from Scudder et al., 2009).

3.2 Coastal Plain geology and hydrologic implications

The Coastal Plain physiographic region of the southeast USA (Fenneman, 1928, 1938; Vigil, 2000) covers more than 1 million km2, greater than the combined area of France, Germany, and the UK (Hupp, 2000). The Coastal Plain extends from New Jersey to eastern Texas and is primarily the result of alluvial (from adjacent mountain and Piedmont regions) and marine deposits of Late Cretaceous and Holocene age (Fenneman, 1928, 1938; Hupp, 2000; Vigil, 2000). The geomorphology of the modern Coastal Plain is largely due to fluvial processes during the last sea-level low stand (approximately 15,000 years ago) and to subsequent oceanic transgression (Hupp, 2000). Shallow surface sediments in the Coastal Plain primarily consist of deposits of quartz sand, glauconitic sand, silt, and clay. The predominantly coarse-grained sandy character of the Coastal Plain sediments favor efficient vertical recharge and generally low surface runoff (Atkins et al., 1996; Aucott, 1996).

Coastal Plain rivers of the southeastern USA are characteristically low-gradient meandering streams with generally broad floodplains, which are subject to extended and frequent flooding (Hupp, 2000). Coastal Plain stream systems typically exhibit two distinct hydrological seasons, low-flow season typically from June to October and high-flow season when extensive areas of floodplain may become inundated. Coastal Plain streams are often divided into two major types, according to the location of the stream headwaters and the associated geochemical characteristics of the stream water. Alluvial rivers originate in mountain or Piedmont uplands, often exhibit an abrupt reduction in gradient downstream of the Fall Line, and typically carry significant loads of mineral sediment. Alluvial streams are often further characterized as brown-water and red-water systems according to the coloration of the sediment load, with the latter deriving their characteristic red color from iron-oxide coated Piedmont sediment (Hupp, 2000). In contrast, black-water streams arise entirely or almost entirely on the Coastal Plain and typically have low gradients and low sediment loads. Extended leaching of tannins from organic-rich, riparian bottomlands and wetlands generates the characteristic colour and low pH of black-water Coastal Plain streams (Hupp, 2000).

3.3 Conceptual model of Coastal Plain hydrology and MeHg transport

Low topographic gradients and shallow water tables yield low-gradient stream systems with extensive riparian wetlands (Glover et al., 2010; Guentzel, 2009; Hupp, 2000) throughout the Coastal Plain. Characteristically coarse-grained sandy sediments favor efficient hydrologic transport within the shallow groundwater system and between connected groundwater and surface-water systems (Atkins et al., 1996; Aucott, 1996). Coastal Plain sediments generally exhibit efficient vertical recharge and low surface runoff (Atkins et al., 1996; Aucott, 1996), with discharge from the shallow flow system often representing 72-100% of the total groundwater discharge to Coastal Plain streams (Atkins et al., 1996). These characteristics support a conceptual model of Coastal Plain hydrodynamics, which has important implications for MeHg transport between wetland source areas and adjacent stream habitat under flood conditions.

3.3.1 Flood hydrology and MeHg transport in black-water Coastal Plain streams

In Coastal Plain stream reaches, the gradient and the direction of shallow groundwater flow is generally toward the stream channel, under normal to low-flow conditions (Figure 3A). Under these conditions, wetlands and channel margins are the primary areas of hydrologic exchange between groundwater and surface-water compartments. Surface-water connectivity between wetland areas and stream channel habitats often is restricted to small

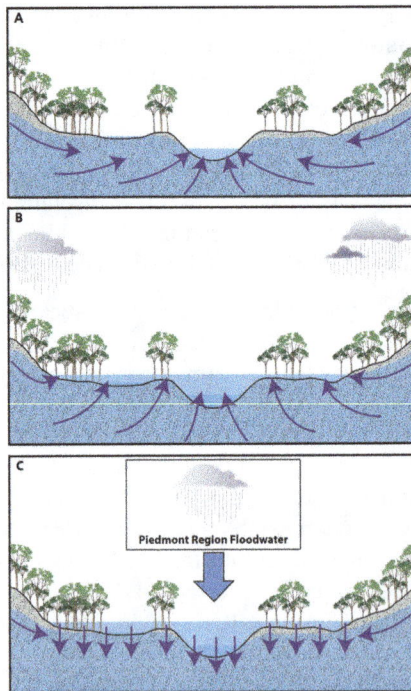

Fig. 3. Conceptual model of flood hydrodynamics in Coastal Plain rivers: (A) low to normal flow conditions, (B) flood conditions driven by Coastal Plain precipitation, and (C) flood conditions caused by downstream transport of Piedmont floodwater.

surface drainages. Rainfall occurring within the Coastal Plain is expected to recharge groundwater with little to no surface runoff, owing to the generally high permeability of the predominantly coarse-grained sandy Coastal Plain sediment. Consequently, high-flow events caused by Coastal Plain rainfall are expected to maintain the general pattern of groundwater flow toward the stream, with flooding predominated by rising groundwater and subsequent discharge across the surface of the riparian floodplain (Figure 3B). Under this scenario, advective transport of pore-water MeHg to the overlying water column and toward the stream channel habitat is enhanced by the increased hydraulic gradient and the expanded surface area for groundwater/surface-water exchange.

3.3.2 Flood hydrology and MeHg transport in alluvial Coastal Plain streams

For alluvial Coastal Plain rivers (i.e. rivers with contributing basins that extend upstream beyond the Coastal Plain), water column MeHg concentrations in Coastal Plain stream reaches are expected to reflect the contribution of upstream mountain and Piedmont drainages, which exhibit comparatively low wetlands coverage (generally less than 2%; (Glover et al., 2010; Guentzel, 2009; NLCD, 2001). The impact of this upstream MeHg signature is expected to be greatest at the upstream margin of the Coastal Plain region and to decrease with distance downstream.

Thus, flood events in Coastal Plain reaches of alluvial stream systems can result from two distinct hydrologic mechanisms, each with important and markedly different implications for MeHg transport to and availability in the adjacent stream aquatic habitat. Flooding events caused by rainfall within the Coastal Plain would be expected to follow the internal groundwater flood mechanism discussed above (Figure 3B) and efficiently transport wetland porewater MeHg toward the stream channel habitat even under flood conditions. In contrast, high-flow events caused by floodwaters from the upstream mountain and Piedmont regions may cause a reversal of the hydraulic gradient and infiltration of mountain and Piedmont floodwater into the shallow subsurface (Figure 3C). Net effects of a flow reversal might include dilution of porewater MeHg concentrations in the shallow subsurface, displacement and downward advection of sediment porewater MeHg, increased MeHg demethylation in the sediment porewater, and decreased transport of wetland sediment MeHg to the stream aquatic habitat. The availability of MeHg in the aquatic habitat of the Coastal Plain portion of such alluvial systems would be expected to reflect the relative importance of these alternative flood mechanisms.

4. Hydrology & MeHg availability: South Carolina Coastal Plain example

The results of the USGS national surveys identified a black-water Coastal Plain stream (Edisto River) in South Carolina as being among the highest in the USA with respect to bioaccumulation of Hg in the tissues of top predator fish. Elevated top predator fish Hg concentrations in South Carolina are not unique to the Edisto, however. Rather, fish Hg concentrations in excess of the criteria for wildlife and human health are common and the substantial variation in Hg bioaccumulation within the state provides an opportunity to better understand the primary controls on Hg bioaccumulation in different environmental settings.

4.1 Hg bioaccumulation in South Carolina Coastal Plain streams

Accumulations of Hg in excess of established guidelines for wildlife and human health are common in game fish as well as in a number of other top-predator and lower trophic level fish species in many streams in South Carolina. The South Carolina Department of Health and Environmental Control (SC DHEC) has established fish consumption advisories for Hg that affect approximately half of the state, primarily within the South Carolina Coastal Plain (DHEC). Figure 4A shows the mean Hg concentrations (µg/g wet weight) observed in *Micropterus salmoides* (largemouth bass) collected by SC DHEC during the period 2001-2007. The orange color indicates 8-digit Hydrologic Unit Code (HUC) basins for which mean largemouth bass Hg concentrations exceeded the 0.3 µg/g wet weight United States Environmental Protection Agency criterion for human health (U.S. Environmental Protection Agency, 2001, 2009b). These data reveal a strong spatial trend of increasing largemouth Hg concentrations along a gradient from Blue Ridge to Piedmont to Coastal Plain physiographic provinces (Bradley et al., 2010; Glover et al., 2010; Guentzel, 2009).

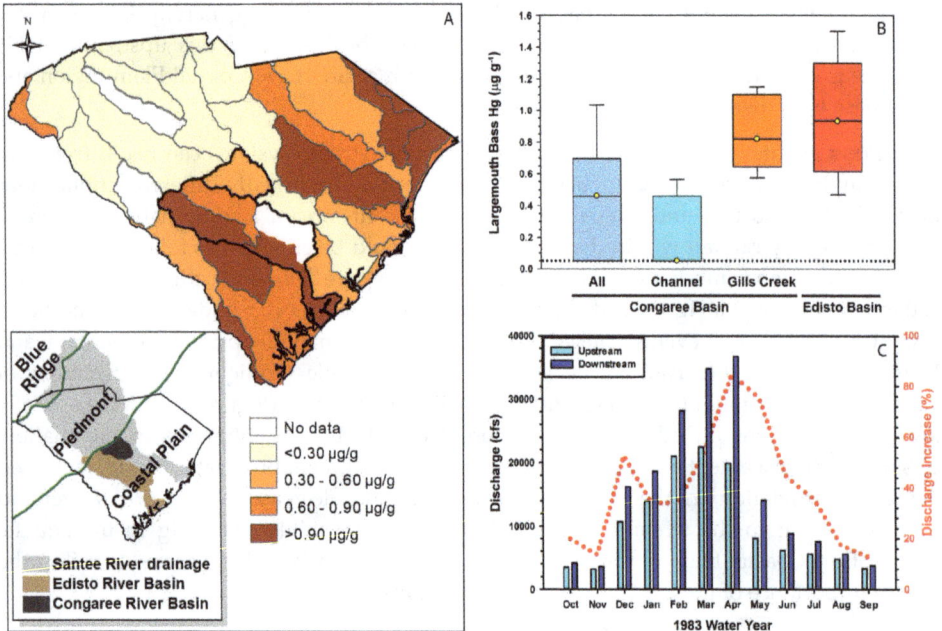

Fig. 4. (A) Mean Micropterus salmoides (largemouth bass) Hg concentrations (µg/g wet weight) in stream basins in South Carolina (SC DHEC). (B) Boxplots (median; interquartile range; 10th and 90th percentiles) of largemouth bass concentrations in Congaree and Edisto (SC DHEC). (C) Increase in discharge between upstream and downstream margin of Congaree in 1983 (USGS).

The pattern of increasing fish Hg burdens from mountains to Coastal Plain in South Carolina corresponds to a pattern of increasing wetland coverages (NLCD, 2001) from Blue Ridge (wetlands coverage: less than 1%) to Piedmont (wetlands coverage: 1-2 %) to Coastal Plain (wetlands coverage: 8-30%) physiographic regions (Guentzel, 2009). Strong correlations between wetlands coverage, dissolved MeHg concentrations, and fish Hg burdens suggest that wetlands coverage is a useful indicator of MeHg source strength (Chasar et al., 2009; Glover et al., 2010; Guentzel, 2009; Scudder et al., 2009). The close correspondence between basin mean fish Hg concentrations and wetlands coverage observed in South Carolina is consistent with this interpretation and with the importance of the MeHg source term as a driver of Hg bioaccumulation in the Coastal Plain region.

4.2 Significant inter-basin variation in Hg bioaccumulation

Substantial variation in median fish Hg concentrations is observed within the Coastal Plain of South Carolina between adjacent stream basins (Figure 4A). The Edisto River and the adjacent Congaree River basins of South Carolina lie within the Coastal Plain physiographic region (Fenneman, 1928, 1938; Vigil, 2000; Figure 4A), which closely corresponds to the Middle Atlantic Coastal Plain and Southeastern Plains Level III Ecoregions (Griffith, 2002). Fish Hg burdens in the Edisto are systematically higher than in the Congaree basin or in the downstream portion of the Santee Basin (Figure 4B). South Carolina Department of Health and Environmental Control (SC DHEC) data for 2001-2007 (Glover et al., 2010; U.S. Environmental Protection Agency, 2010) indicate median concentrations of Hg are at least two times higher in *M. salmoides* from the Edisto River basin than in those collected throughout the Congaree River basin or from the Santee River basin locations downstream from the Congaree River (Figure 4B).

Because MeHg is the primary form of Hg in fish (Bloom, 1992; Rudd, 1995; Wiener & Spry, 1996), the concentration of Hg in fish tissues can be attributed to interactions between three conceptual components of the MeHg biocycle (Bradley et al., 2010; Bradley et al., 2009; Chasar et al., 2009): (1) microbial production and in situ persistence of MeHg; (2) transport of MeHg from the site and matrix of production to the base of the food web; and (3) efficiency of biotic uptake and trophic transfer of MeHg within the food web (Figure 1). An assessment of sediment throughout both basins revealed no statistically significant difference in concentrations of MeHg or net methylation potential in sediments collected from Edisto and Congaree locations (Bradley et al., 2009). Stream channels in both systems were characterized by coarse sands and net methylation potentials at least an order of magnitude lower than in sediments from adjacent wetland and riparian floodplain areas. Likewise, no difference in wetlands coverage is apparent between the Edisto (wetlands coverage: 20.4%) and the Congaree (wetlands coverage: 19.4 %) basins (Bradley et al., 2010; NLCD, 2001). Comparable riparian wetlands coverages (Table 1) and similar ranges of sediment Hg methylation potentials (Bradley et al., 2009) suggest that differences in Hg bioaccumulation between the two systems are not due to systematic differences in MeHg production in adjacent wetland and floodplain sediments.

4.3 Role of hydrology in inter-basin variation in Hg bioaccumulation

The Edisto is a black-water Coastal Plain stream basin, which falls entirely within the Coastal Plain. In contrast, the Congaree River is part of the Santee River drainage, an alluvial Coastal Plain system extending from the Atlantic Ocean to the Blue Ridge region of the Carolinas (Figure 4A). This fundamental difference in hydrology has important implications for the availability of MeHg for uptake/accumulation by the aquatic food web.

4.3.1 Wetland coverage in contributing basin

The Edisto stems from groundwater discharge and precipitation runoff occurring only within the Coastal Plain region and the mean wetlands coverage for the basin is 20.4% (Bradley et al., 2010; NLCD, 2001). In contrast, the contributing area for the Congaree River includes upstream Blue Ridge and Piedmont drainages. The wetlands coverage for the drainage area upstream of the Congaree is approximately 2% (NLCD, 2001), such that the combined wetlands coverage is about 3.5% for the entire drainage area contributing to flow at the downstream margin of the Congaree River (NLCD, 2001). Thus, the generally lower Hg concentrations in largemouth bass from the Congaree basin may reflect hydrologic and geochemical impacts of the Blue Ridge/Piedmont contribution (Bradley et al., 2010).

This hypothesis is supported by the substantially lower largemouth bass Hg concentrations observed in the Blue Ridge/Piedmont-influenced main channel of the Congaree compared to those concentrations in largemouth bass from the Gills Creek drainage (Figure 4B). During 2001-2007, SC DHEC collected largemouth bass (n = 40) from three locations (near the upstream margin, approximate mid-reach, and downstream margin) in the main channel of the Congaree River and the median Hg concentration was below the 0.05 µg/g (wet weight) detection limit. Largemouth bass also were collected (n = 20) from the headwater region of Gills Creek, a small Congaree tributary, which has a wetlands coverage of approximately 9% (NLCD, 2001) and which, like the Edisto River, lies entirely in the Coastal Plain. The median Hg concentration for these bass was 0.82 µg/g (wet weight), comparable to the median Hg concentration for bass from the Edisto River basin. These results indicate that, for Congaree basin black-water tributaries that lie entirely in the Coastal Plain, the hydrologic transport of MeHg from wetlands to the stream aquatic habitat and the extent of Hg bioaccumulation in the food web are comparable to that of the Edisto. However, the aquatic habitat within the Congaree River main channel primarily reflects the upstream Blue Ridge and Piedmont contributing drainage.

4.3.2 Source of water in Congaree basin

Owing to substantially lower wetlands coverages (generally less than 2 %), dissolved MeHg concentrations and the associated availability of MeHg in the aquatic habitat are expected to be low in the Saluda and Broad Rivers and, by extension, in the upstream reaches of the Congaree River (Chasar et al., 2009; Glover et al., 2010; Guentzel, 2009). Low dissolved MeHg concentrations measured in the Saluda, Broad and Congaree Rivers in 2009 (Bradley et al., 2010) are consistent with this expectation, as are the low largemouth bass Hg concentrations (median = 0.05 µg/g wet weight) observed in the Congaree main channel in 2001-2007 (Figure 4B) (U.S. Environmental Protection Agency, 2010).

However, surface-water discharge increases substantially within the Congaree basin, indicating that Coastal Plain water sources also are an important contributor to discharge at the downstream margin of the basin (Figure 4C). During 1982-1983, the USGS collected discharge data at a short-term gage station near the downstream margin of the Congaree River (station 02169740; (NWIS, 2010)). For this period of record, the increase in monthly mean stream discharge between the most upstream Congaree gage (station 02169500; (NWIS, 2010)) and gage 02169740 ranged from about 20% to more than 85%, with the greatest contribution from the Coastal Plain occurring during the hydroperiod (high-flow season). The long-term record at the upstream location (02169500) indicates that discharge in the Congaree River in 1982-1983 was in the normal range, falling within the 50th to 75th percentile range for all observations. Based on these observations, a substantial fraction of the water at the downstream margin of the Congaree River originates within the Congaree basin. The consistently low largemouth bass Hg burdens observed in the downstream portions of the basin despite this substantial Coastal Plain contribution, the 20% wetlands coverage within the basin (NLCD, 2001), and the demonstrated potential for elevated Hg burdens in tributaries that fall entirely in the Coastal Plain (Figures 4A and 4B), suggests that the Blue Ridge/Piedmont-derived component of discharge inhibits MeHg transport from the wetland margins of the Congaree River to the stream aquatic habitat.

4.3.3 Fundamental differences in flood hydrology

Hydrologic connectivity between wetland MeHg source areas and adjacent aquatic habitats is recognized as a significant control on Hg bioaccumulation in aquatic and associated terrestrial communities (Krabbenhoft & Babiarz, 1992; Krabbenhoft et al., 1999; Rypel et al., 2008; St. Louis et al., 1994b; Stoor et al., 2006). In stream reaches dominated by riparian wetlands, flood conditions maximize hydrologic connectivity between the wetland margins and the stream aquatic habitat by maximizing the area for groundwater/surface-water exchange (Poff et al., 1997; Schuster et al., 2008; Ward et al., 2010). However, the direction of water and solute transport during flood conditions is dictated by the hydraulic gradient (Krabbenhoft & Babiarz, 1992). Thus, the observations of Krabbenhoft and others (Krabbenhoft & Babiarz, 1992; Stoor et al., 2006) suggest a mechanism by which the floodwater source may contribute fundamentally to the disparity in Hg bioaccumulation between Coastal Plain rivers like the Edisto and Congaree. The crucial hypothesis is that characteristic coarse-grained sediments favor high hydrologic connectivity throughout the Coastal Plain region, but essential differences in flood hydrodynamics determine the direction of water movement and thus the efficiency of dissolved MeHg transport between wetland MeHg source areas and the adjacent stream aquatic habitat. The validity of this hypothesis was assessed with stream channel and shallow groundwater level data from locations in the Edisto River basin and Congaree River basin(Bradley et al., 2010).

Figure 5A presents groundwater level changes in monitoring well transect ELB near the streamgage (02172305) in McTier Creek within the Edisto basin (Bradley et al., 2010). Observation wells 1-4 were located approximately 1, 3, 21 and 45 m, respectively, from the edge of the stream (Figure 5A; inset). Prior to flooding the gradient was approximately 0.3 m from the well nearest to the channel toward McTier Creek. Approximately 4 h after rainfall began, stream and groundwater levels began rising essentially simultaneously, indicating good hydrologic connectivity between the stream and inland groundwater locations. At the onset, peak, and end of flood conditions the groundwater gradient was upward, indicating discharge of groundwater from the sediment to the overlying water column at all well locations. Similar patterns were observed at other McTier Creek locations during multiple events (Bradley et al., 2010).

The Congaree River is periodically flooded by Blue Ridge/Piedmont-derived red-water (color due to iron-oxide coated piedmont sediment load) as shown in Figure 6B. Figure 6C shows Piedmont red-water from the Broad River drainage inundating the floodplain at the Congaree National Park. Water level data collected within the Congaree National Park illustrate the strong downward hydraulic gradient that characterizes these Piedmont flood events. Prior to flooding, the gradient between well RIC-346 and Cedar Creek at station 02169672 was generally low (Figure 6A). Groundwater levels at RIC-346 were approximately 1.5 m below land surface at the onset of flood conditions and more than 2 m below flood water levels at the peak of flooding, demonstrating a dramatic downward gradient throughout. The rapid rise in groundwater level following the onset of flood conditions indicated vertical infiltration of floodwaters. This pattern was repeated a few days later (Figure 6A). Similar patterns were observed at other Congaree locations during multiple events.

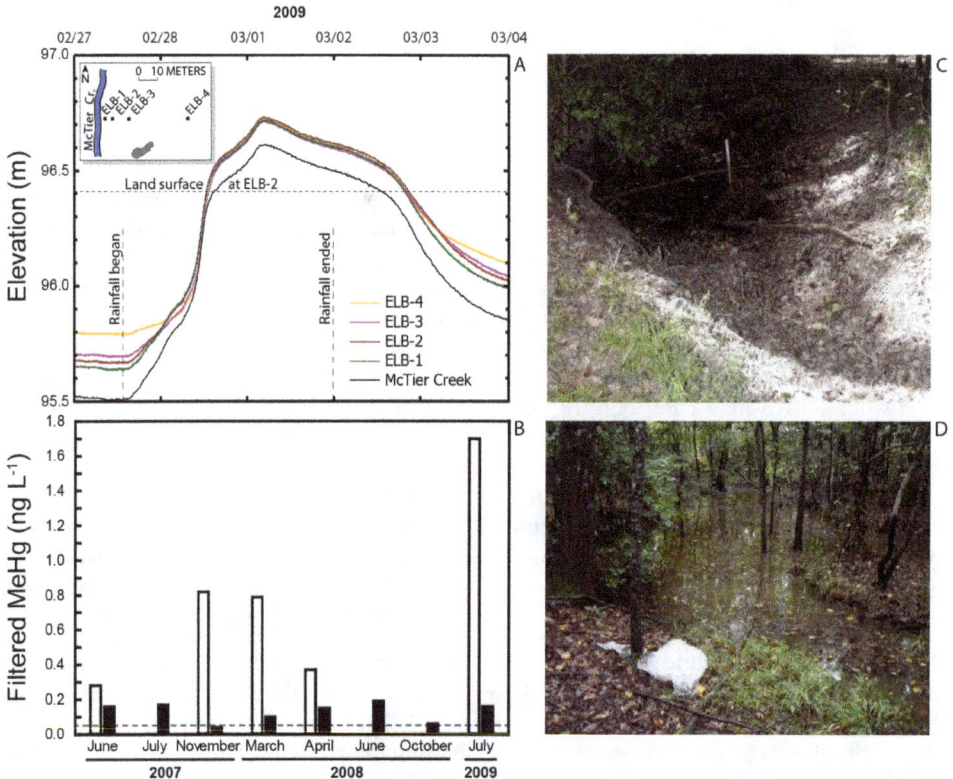

Figure 4. (A) Groundwater and surface-water levels during a flood in McTier Creek. (B) Filtered MeHg concentrations in floodplain depression (white) and in stream (black) at McTier Creek. Floodplain depression; (C) dry and (D) filled by rising groundwater.

4.4 Implications for MeHg availability in South Carolina Coastal Plain streams

The hydrologic pattern observed in the McTier Creek sub-basin of the Edisto River system indicates that in black-water Coastal Plain stream reaches groundwater continues to discharge from the shallow subsurface toward the stream channel aquatic habitat even during flood conditions. This type of hydrologic response favors transport of MeHg from the subsurface source area to the stream channel aquatic habitat. To illustrate, filtered MeHg

Fig. 5. (A) Groundwater and surface-water levels during a flood in Congaree National Park. (B) Confluence of Saluda River (upper left) and Broad River (upper right) forming the Congaree River. The Broad is shown under flood conditions. (C) Red-water Piedmont flood inundating Congaree National Park.

samples were collected from a shallow depression in the McTier Creek floodplain (Figures 5C and 5D). These floodplain depressions are inundated by rising groundwater before flooding creates an overland connection to the stream and provide an opportunity to assess MeHg in discharging groundwater prior to the mixing with stream-channel surface water that occurs during flood conditions. This floodplain depression was assessed eight times during 2007-2009 (Figure 5B). On the five occasions that standing water was present no overland connection to McTier Creek existed. In every instance the dissolved MeHg concentrations observed in the pool were greater than were measured in the adjacent stream channel, with concentrations ranging from approximately two times higher to greater than 10 times higher than in the stream channel (Bradley et al., 2010).

Samples collected from the floodplain depression in November 2007 provided particular insight into the MeHg signature associated with groundwater discharging through the floodplain land surface (Bradley et al., 2010). A localized rainfall event the day before sample collection resulted in rising groundwater and stream water levels. Discharging groundwater partially filled the previously dry depression but did not overtop the floodplain. Dissolved oxygen concentrations, measured in the pool during MeHg sample collection in 2007-2009, ranged from 2-5 mg/L, indicating that redox conditions in the standing water did not favor Hg methylation. Thus, because the pool and the surrounding floodplain were dry less than 12 h prior to sample collection; the 0.8 ng L[-1] dissolved MeHg concentration in the November 2007 riparian depression sample (compared with less than 0.05 ng L[-1] dissolved MeHg concentrations in the stream) reflects the immediate transport of MeHg into the surface-water compartment. These results demonstrate that groundwater flooding in black-water Coastal Plain streams efficiently transfers MeHg from the floodplain sediment porewater to the surface-water compartment.

In contrast, flooding in alluvial Coastal Plain rivers, like the Congaree, that is caused by Piedmont-derived floodwater can decouple wetland MeHg transport to the stream aquatic habitat, preserving low dissolved MeHg concentrations and, consequently, decreasing MeHg availability for biotic uptake and accumulation in the main channel aquatic habitat. For example, dissolved MeHg concentrations observed in the Congaree flood in November 2009 were low, consistent with the comparatively low wetlands coverage of the upstream Blue Ridge/Piedmont drainages (Saluda and Broad River basin wetlands coverages are 2%; (NLCD, 2001) and indicating lower MeHg availability within the Congaree River channel aquatic habitat (Bradley et al., 2010).

4.5 Implications for MeHg bioaccumulation in Coastal Plain streams

These results illustrate that the coarse-grained sediment that characterizes much of the Coastal Plain physiographic region favors efficient exchange of water between streams, wetlands, and shallow groundwater systems. This hydrologic context suggests that black-water Coastal Plain streams, like the Edisto River, are particularly vulnerable to Hg bioaccumulation, because they lie entirely or largely within the Coastal Plain and are primarily subject to groundwater discharge-driven flooding. In contrast, alluvial Coastal Plain stream reaches, like the Congaree River, which experience groundwater-driven flooding as well as external floodwater events, are expected to exhibit reduced MeHg availability in the stream channel aquatic habitat, depending on the relative frequency of the two mechanisms.

The results of this study have regional-scale implications for Hg bioaccumulation, because the Coastal Plain physiographic region extends along the Atlantic and Gulf Coasts of the USA from New Jersey to Texas. The fundamental hydrologic characteristics of the South Carolina Coastal Plain are common in the Coastal Plain physiographic region and a similar relationship between flood dynamics and Hg bioaccumulation is expected throughout the region. The recent USGS national survey of Hg burdens in high trophic level piscivores (Scudder et al., 2009) indicated an elevated incidence of high Hg concentrations in top predator fish from stream reaches along the Atlantic and Gulf Coasts, which correspond closely to the Coastal Plain physiographic region. Thus, the hydrologic characteristics of the Coastal Plain region appear to contribute to an increased vulnerability to Hg bioaccumulation in Coastal Plain rivers.

5. Conclusions

The primary risk of mercury (Hg) in the environment, including the risk to human health, is due to accumulation of Hg in aquatic biota (Mergler et al., 2007) and is inextricably linked to hydrology. Water provides habitat for aquatic biota, limits oxygen supply in saturated soil and sediment, and contributes to the onset of iron- and sulfate-reducing conditions, which support microbial production of toxic alkyl-mercury species (Benoit et al., 2003; Compeau & Bartha, 1985; Fleming et al., 2006). MeHg, in particular, is neurotoxic (Clarkson et al., 2003) and readily accumulated in aquatic foodwebs (Bloom, 1992; Brumbaugh et al., 2001; Hall et al., 1997). MeHg is the primary form of Hg in fish (Bloom, 1992) and wetlands are recognized MeHg source areas (Bradley et al., 2011; Bradley et al., 2009; Brigham et al., 2009; Grigal, 2003; Hall et al., 2008; Rypel et al., 2008; St. Louis et al., 1994b). Hydrologic transport of MeHg from sediment sources in riparian wetlands and floodplains to the stream channel is a fundamental control on the availability of MeHg in the stream aquatic habitat and, thus, on Hg bioaccumulation in the stream foodweb (Bradley et al., 2011; Rypel et al., 2008; Ward et al., 2010). The coarse-grained sandy sediment that characterizes much of the Coastal Plain from New Jersey to Texas in the USA favors efficient transport of MeHg from wetlands to the stream habitat (Atkins et al., 1996; Aucott, 1996; Bradley et al., 2010; Hupp, 2000). The hydrologic characteristics of the Coastal Plain region appear to contribute to an increased vulnerability to Hg bioaccumulation in Coastal Plain rivers (Bradley et al., 2011; Bradley et al., 2010).

6. References

Atkins, J.B., Journey, C.A., & Clarke, J.S. (1996). Estimation of ground-water discharge to streams in the central Savannah River basin of Georgia and South Carolina. *U.S. Geological Survey Water Resources Investigations Report 96-4179*, 36,

Aucott, W.A. (1996). Hydrology of the Southeastern Coastal Plain Aquifer systems in South Carolina and parts of Georgia and North Carolina. *U.S. Geological Survey Professional Paper 1410-E*, 83,

Balogh, S.J., Meyer, M.L., Hansen, N.C., Moncrief, J.F., & Gupta, S.C. (2000). Transport of mercury from a cultivated field during snowmelt. *Journal of Environmental Quality*, 29, no. 3, 871-874,

Balogh, S.J., Meyer, M.L., & Johnson, D.K. (1997). Mercury and suspended sediment loadings in the lower Minnesota River. *Environmental Science and Technology*, 31, no. 1, 198,

— — —. (1998). Transport of mercury in three contrasting river basins. *Environmental Science and Technology*, 32, no. 4, 456-462,

Barcelona, M.J. (1994). Site characterization: What should we measure, where (when?), and why?, 1-9, Available from

Bauch, N.J., Chasar, L.C., Scudder, B.C., Moran, P.W., Hitt, K.J., Brigham, M.E., Lutz, M.A., & Wentz, D.A. (2009). Data on mercury in water, streambed sediment, and fish tissue from selected streams across the United States, 1998-2005. *U.S. Geological Survey, Data Series 307*, 32, Available from http://pubs.usgs.gov/ds/307/

Benoit, J.M., Gilmour, C.C., Heyes, A., Mason, R.P., & Miller, C.L. (2003). Geochemical and Biological Controls over Methylmercury Production and Degradation in Aquatic Ecosystems. *Acs Symposium Series*, 835, 262-297,

Bloom, N.S. (1992). On the chemical form of mercury in edible fish and marine invertebrate tissue. *Canadian Journal of Fisheries and Aquatic Sciences*, 49, 1010-1017,

Bradley, P., Burns, D., Murray, K., Brigham, M., Button, D., Chasar, L., Marvin-
    DiPasquale, M., Lowery, M., & Journey, C.A. (2011). Spatial and seasonal
    variability of dissolved methylmercury in two stream basins in the eastern
    United States. *Environmental Science and Technology*, 45, no. 6, 2048-2055,

Bradley, P., Journey, C., Chapelle, F., Lowery, M., & Conrads, P. (2010). Flood Hydrology
    and Methylmercury Availability in Coastal Plain Rivers. *Environmental Science &
    Technology*, 44, no. 24, 9285-9290, 0013-936X

Bradley, P.M., Chapelle, F.H., & Journey, C.A. (2009). Comparison of methylmercury
    production and accumulation in sediments of the Congaree and Edisto River
    Basins, South Carolina, 2004-06. *U.S. Geological Survey, Scientific Investigations
    Report 2009-5021*, 9, Available from http://pubs.water.usgs.gov/sir2009-5021

Brigham, M.E., Wentz, D.A., Aiken, G.R., & Krabbenhoft, D.P. (2009). Mercury cycling in
    stream ecosystems. 1. Water column chemistry and transport. *Environmental
    Science and Technology*, 43, no. 8, 2720-2725,

Brumbaugh, W.G., Krabbenhoft, D.P., Helsel, D.R., Wiener, J.G., & Echols, K.R. (2001). A
    national pilot study of mercury contamination of aquatic ecosystems along
    multiple   gradients—bioaccumulation   in   fish.   *U.S.   Geological   Survey,
    USGS/BRD/BSR – 2001-0009*, 25, Available from
    http://www.cerc.usgs.gov/pubs/center/pdfdocs/BSR2001-0009.pdf

Canada, E. (2011). Fish Consumption Advisories. *Environment Canada*, Available from
    http://www.ec.gc.ca/mercure-mercury/default.asp?lang=En&n=DCBE5083-1

Chapelle, F.H., McMahon, P.M., Dubrovsky, N.M., Fujii, R.F., Oaksford, E.T., &
    Vroblesky, D.A. (1995). Deducing the distribution of terminal electron-accepting
    processes in hydrologically diverse groundwater systems. *Water Resources
    Research*, 31, 359-371,

Chasar, L.C., Scudder, B.C., Stewart, A.R., Bell, A.H., & Aiken, G.R. (2009). Mercury
    cycling in stream ecosystems. 3. Trophic dynamics and methylmercury
    bioaccumulation. *Environmental Science and Technology*, 43, no. 8, 2733–2739,

Clarkson, T., Magos, L., & Myers, G. (2003). The toxicology of mercury—current
    exposures and clinical manifestations. *New England Journal of Medicine*, 349, 1731-
    1737,

Compeau, G.C., & Bartha, R. (1985). Sulfate-reducing bacteria—principal methylators of
    mercury in anoxic estuarine sediment. *Applied and Environmental Microbiology*, 50,
    no. 2, 498-502,

South Carolina Department of Health and Environmental Control. South Carolina Fish
    Consumption Advisories. In: *Fish Advisories Home*, October 4, 2011, Available
    from http://www.scdhec.gov/environment/water/fish/map.htm

Driscoll, C.T., Han, Y.-J., Chen, C.Y., Evers, D.C., Lambert, K.F., Holsen, T.M., Kamman,
    N., & Munson, R. (2007). Mercury contamination in forest and freshwater
    ecosystems in the northeastern United States. *Bioscience*, 57, no. 1, 17-28,

Driscoll, C.T., Holsapple, J., Schofield, C.L., & Munson, R. (1998). The chemistry and
    transport of mercuury in a small wetland in the Adirondack region of New York,
    USA. *Biogeochemistry*, 40, 9,

EPRI, E.P.R.I. (1994). Mercury Atmospheric Processes: A Synthesis Report, Available from

Fenneman, N.M. (1928). Physiographic dvisions of the United States. *Annals of the Association of American Geographers*, 18, no. 4, 261-353,

— — —. (1938). *Physiography of the eastern United State.* New York: McGraw-Hill Book Co.

Fleming, E.J., Mack, E.E., Green, P.G., & Nelson, D.C. (2006). Mercury methylation from unexpected sources—molybdate-inhibited freshwater sediments and an iron-reducing bacterium. *Applied and Environmental Microbiology*, 72, 457-464,

Gabriel, M.C., & Williamson, D.G. (2004). Principal biogeochemical factors affecting the speciation and transport of mercury through the terrestrial environment. *Environmental Geochemistry and Health*, 26, no. 4, 421-434(14),

Gilmour, C.C., & Henry, E.A. (1991). Mercury methylation in aquatic systems affected by acid deposition. *Environmental Pollution*, 71, 131-169,

Gilmour, C.C., Henry, E.A., & Mitchell, R. (1992). Sulfate stimulation of mercury methylation in freshwater sediments. *Environmental Science and Technology*, 26, no. 11, 2281-2287,

Glover, J., Domino, M., Altman, K., Dillman, J., Castleberry, W., Eidson, J., & Mattocks, M. (2010). Mercury in South Carolina Fishes, USA. *Ecotoxicology*, 19, no. 4, 781-795,

Griffith, G.E.O., J.M.; Comstock, J.A.; Schafale, M.P.; McNab, W.H.; Lenat, D.R.; MacPherson, T.F.; Glover, J.B.; Shelburne, V.B. (2002). Ecoregions of North Carolina and South Carolina. *U.S. Geological Survey*, Available from http://www.epa.gov/wed/pages/ecoregions/ncsc_eco.htm

Grigal, D.F. (2002). Inputs and outputs of mercury from terrestrial watersheds: a review. *Environmental Reviews*, 10, 1-39,

— — —. (2003). Mercury sequestration in forests and peatlands--a review. *Journal of Environmental Quality*, 32, no. 2, 393-405,

Guentzel, J.L. (2009). Wetland influences on mercury transport and bioaccumulation in South Carolina. *Science of the Total Environment*, 407, 1344-1353,

Hall, B., Aiken, G., Krabbenhoft, D., Marvin-DiPasquale, M., & Swarzenski, C. (2008). Wetlands as principal zones of methylmercury production in southern Louisiana and the Gulf of Mexico region. *Environmental Pollution*, 154, no. 1, 124-134,

Hall, B.D., Bodaly, R.A., Fudge, R.J.P., Rudd, J.W.M., & Rosenberg, D.M. (1997). Food as the dominant pathway of methylmercury uptake by fish. *Water, Air, and Soil Pollution*, 100, 13-24,

Hupp, C.R. (2000). Hydrology, geomorphology and vegetation of the Coastal Plain rivers of the south-eastern USA. *Hydrological Processes*, 14, 2991-3010,

Hurley, J.P., Benoit, J.M., Shafer, M.M., Andren, A.W., Sullivan, J.R., Hammond, R., & Webb, D.A. (1995). Influences of watershed characteristics on mercury levels in Wisconsin rivers. *Environmental Science and Technology*, 29, no. 7, 1867-1875,

Krabbenhoft, D.P., & Babiarz, C.L. (1992). The role of groundwater transport in aquatic mercury cycling. *Water Resources Research*, 28, no. 12, 3119-3128,

Krabbenhoft, D.P., Wiener, J.G., Brumbaugh, W.G., Olson, M.L., DeWild, J.F., & Sabin, T.J. (1999). A National pilot study of mercury contamination of aquatic ecosystems along multiple gradients, 147-160, Available from http://toxics.usgs.gov/pubs/wri99-4018/Volume2/sectionB/2301_Krabbenhoft/pdf/2301_Krabbenhoft.pdf

Lee, Y.H., Borg, G.C., Iverfeldt, A., & Hultberg, H. (1994). Fluxes and turnover of methylmercury: Mercury pools in forest soils. In *Mercury pollution: Integration and synthesis*, ed. C. J. Watras & J. W. Huckabee, 329-341. Boca Raton, FL: Lewis Publishers.

Lorey, P., & Driscoll, C.T. (1999). Historical trends of mercury deposition in Adirondack lakes. *Environmental Science and Technology*, 33, no. 5, 718-722,

Mason, R.P., Fitzgerald, W.F., & Morel, F.M.M. (1994). The biogeochemical cycling of elemental mercury: anthropogenic influences. *Geochimica et Cosmochimica Acta*, 58, no. 15, 3191-3198,

Mergler, D., Anderson, H.A., Chan, L.H.M., Mahaffey, K.R., Murray, M., Sakamoto, M., & Stern, A.H. (2007). Methylmercury exposure and health effects in humans--a worldwide concern. *Ambio*, 36, no. 1, 3-11,

Morel, F.M.M., Kraepiel, A.M.L., & Amyot, M. (1998). The chemical cycle and bioaccumulation of mercury. *Annual Review of Ecology and Systematics*, 29, 543-566,

NLCD. (2001). National Land Coverage Data-2001. *Multi-Resolution Land Characteristics Consortium (MRLC)*,

NWIS. (2010). National Water Information System: USGS Water Data for the Nation. 2010,

Pacyna, E.G., Pacyna, J.M., Fudala, J., Strzelecka-Jastrzab, E., Hlawiczka, S., & Panasiuk, D. (2006). Mercury emissions to the atmosphere from antropogenic sources in Europe in 2000 and their scenarios until 2020. *Science of the Total Environment*, 370, 147-156,

Poff, N., Allan, J., Bain, M., Karr, J., Prestegaard, K., Richter, B., Sparks, R., & Stromberg, J. (1997). The natural flow regime: a paradigm for river conservation and restoration. *BioScience*, 47, no. 11, 769-784,

Ravichandran, M. (2004). Interactions between mercury and dissolved organic matter—a review. *Chemosphere*, 55, no. 3, 319-331,

Rudd, J.W.M. (1995). Sources of methyl mercury to freshwater ecosystems—a review. *Water, Air, and Soil Pollution*, 80, 697-713,

Rypel, A.L., Arrington, D.A., & Findlay, R.H. (2008). Mercury in Southeastern U.S. Riverine Fish Populations Linked to Water Body Type. *Environmental Science and Technology*, 42, no. 14, 5118-5124,

Schuster, E. (1991). The behavior of mercury in the soil with special emphasis on complexation and adsorption Processes-A Review of the Literature. *Water, Air, and Soil Pollution*, 56, 667-680,

Schuster, P.F., Shanley, J.B., Marvin-Dipasquale, M., Reddy, M.M., Aiken, G.R., Roth, D.A., Taylor, H.E., Krabbenhoft, D.P., & DeWild, J.F. (2008). Mercury and organic carbon dynamics during runoff episodes from a northeastern USA watershed. *Water, Air, and Soil Pollution*, 187, 89-108,

Scudder, B.C., Chasar, L.C., Wentz, D.A., Bauch, N.J., Brigham, M.E., Moran, P.W., & Krabbenhoft, D.P. (2009). Mercury in fish, bed sediment, and water from streams across the United States, 1998-2005. *U.S. Geological Survey, Scientific Investigations Report 2009-5109*, 74, Available from http://pubs.usgs.gov/sir/2009/5109

Seigneur, C., Vijayaraghavan, K., Lohman, K., Karamchandani, P., & Scott, C. (2004). Global source attribution for mercury deposition in the United States. *Environmental Science and Technology*, 38, 555-569,

Selin, N.E. (2009). Global Biogeochemical Cycling of Mercury: A Review. *Annual Review of Environment and Resources*, 34, no. 1, 43-63,

Skyllberg, U. (2008). Competition among thiols and inorganic sulfides and polysulfides for Hg and MeHg in wetland soils and sediments under suboxic conditions-- illumination of controversies and implications for MeHg net production. *Journal of Geophysical Research*, 113, no. G00C03,

St. Louis, V.L., Rudd, J.W.M., Kelly, C., Beaty, K.G., Bloom, N.S., & Flett, R.J. (1994a). Importance of wetlands as sources of methyl mercury to boreal forest ecosystems. *Canadian Journal of Fisheries and Aquatic Sciences*, 51, 1065-1076,

St. Louis, V.L., Rudd, J.W.M., Kelly, C.A., Beaty, K.G., Bloom, N.S., & Flett, R.J. (1994b). Importance of wetlands as sources of methyl mercury to boreal forest ecosystems. *Canadian Journal of Fisheries and Aquatic Sciences*, 51, 1065-1076,

Stoor, R.W., Hurley, J.P., Babiarz, C.L., & Armstrong, D.E. (2006). Subsurface sources of methyl mercury to Lake Superior from a wetland–forested watershed. *Science of the Total Environment*, 368, 99-110,

Swain, E.B., Jakus, P.M., Rice, G., Lupi, F., Maxson, P.A., Pacyna, J.M., Penn, A., Spiegel, S.J., & Veiga, M.M. (2007). Socioeconomic consequences of mercury use and pollution. *Ambio*, 36, no. 1, 45-61,

U.S. Environmental Protection Agency. (2001). Water quality criterion for the protection of human health — methylmercury. *Office of Science and Technology, Office of Water, USEPA, EPA-823-R-01-001*, Available from http://www.epa.gov/waterscience/criteria/methylmercury/document.html

— — —. (2009a). 2008 Biennial National Listing of Fish Advisories. *U.S. Environmental Protection Agency, EPA-823-F-05-004*, 7, Available from http://www.epa.gov/waterscience/fish/advisories/tech2008.pdf

— — —. (2009b). Guidance for Implementing the January 2001 Methylmercury Water Quality Criterion. *U.S. Environmental Protection Agency, EPA 823-R-09-002*, Available from http://www.epa.gov/waterscience/criteria/methylmercury/pdf/guidance-final.pdf

U.S. Environmental Protection Agency (2010). STORET database access. In, August 09, Available from http://www.epa.gov/storet/

Ullrich, M., Tanton, T.W., & Abdrashitova, S.A. (2001). Mercury in the Aquatic Environment: A Review of Factors Affecting Methylation. *Critical Reviews in Environmental Science and Technology*, 31, no. 3, 241-293,

Vigil, J.F.P., R.J.; Howell, D.G. (2000). A tapestry of time and terrain. *U.S. Geological Survey Geologic Investigations Series 2720*, 1,

Ward, D., Nislow, K., & Folt, C. (2010). Bioaccumulation syndrome: identifying factors that make some stream food webs prone to elevated mercury bioaccumulation. *Annals of the New York Academy of Sciences*, 1195, no. 1, 62-83,

Wiedemeier, T.H., Swanson, M.A., Moutoux, D.E., Gordon, E.K., Wilson, J.T., Wilson, B.H., Kampbell, D.H., Haas, P.E., Miller, R.N., Hansen, J.E., & Chapelle, F.H.

(1998). Technical protocol for evaluating natural attenuation of chlorinated solvents in groundwater. EPA/600/R-98/128. USEPA, Washington DC, USA. 128p.

Wiener, J.G., & Spry, D.J. (1996). Toxicological significance of mercury in freshwater fish. In *Environmental Contaminants in Wildlife: Interpreting Tissue Concentrations*, ed. W. N. Beyer, G. H. Heinz & A. W. Redmon, 297-339. Boca Raton, FL: Lewis Publishers.

9

Contribution of GRACE Satellite Gravimetry in Global and Regional Hydrology, and in Ice Sheets Mass Balance

Frappart Frédéric and Ramillien Guillaume
Université de Toulouse, OMP-GET, CNRS, IRD,
France

1. Introduction

Continental water storage is a key component of global hydrological cycle which play a major role in the Earth's climate system via controls over water, energy and biogeochemical fluxes. In spite of its importance, the total continental water storage is not well-known at regional and global scales because of the lack of *in situ* observations and systematic monitoring of the terrestrial water reservoirs, especially the groundwaters component (Alsdorf & Lettenmaier, 2003). Although some local hydrological monitoring networks exist, the current description of water storage comes primarily from global hydrology models (*e.g.*, WaterGAP Global Hydrology Model - WGHM (Döll et al., 2003) or Global Land Data Assimilation System - GLDAS (Rodell et al., 2004a)), which still suffer from important limits, such as the absence of different compartments of the terrestrial water compartments (*e.g.*, groundwater and floodplains), the errors introduced by external forcings (mostly the precipitation), the lack of reliability of external parameters as the soil type or the vegetation cover, the parameterization. Unfortunately, direct hydrological measurements are fairly limited, while imaging satellite techniques and satellite altimetry only give access to surface water and superficial soil moisture variations, that represent just the more accessible components of total water storage. Since its launch in 2002, the Gravity Recovery and Climate Experiment (GRACE) gravimetry from space mission offers a unique alternative to classical remote sensing technique for measuring changes in total water storage (ice, snow, surface waters, soil moisture, groundwater) over continental areas, representing a new source of information for hydrologists and the global hydrological modelling community.

We propose a review of the major techniques and results in extracting continental hydrology signals from GRACE satellite gravimetry, aiming at studying large-scale processes in hydrology and glaciology and the impact of the climatic variations on the global hydrological cycle. The two first parts describe the principle of the GRACE mission and the nature of the data measured. The third part is devoted to the post-processings applied to the GRACE-based products to improve their quality and accuracy. The last part presents the most significant results obtained for land hydrology and ice sheets mass balance using GRACE.

2. Principle of GRACE space gravimetry mission

GRACE, a joint US-German (NASA/DLR) mission, was launched in March 2002 and placed at low altitude on a quasi-polar orbit (inclination of 89.5°), to map the time variations of the Earth's gravity field.

2.1 Description of the "Low-Low" time-variable gravity measurements

The mission consists of two identical satellites moving along the same orbit at an altitude of ~ 450 km, one following the other with a mean inter-satellite distance of approximately 220 km in a satellite-to-satellite tracking configuration in the low-low mode initially proposed by Wolf (1969). The satellites use K-Band microwave ranging (KBR) system (Tapley et al., 2004a) to monitor continuously the relative motion of the spacecraft (*i.e.*, the range and the range rate), which varies proportionally to the integrated differences of the gravity accelerations felt by the satellites (Schmidt et al., 2008). The KBR system provides measurements of the distance between the two satellites with an accuracy better than 1 micrometre using carrier phase measurements in the K (26 GHz) and Ka (32 GHz) frequencies. Each GRACE satellite contains a 3-axis accelerometer which gives access to the dynamical effects of the non-conservative forces such as solar pressure and atmospheric drag. Once removed the non-gravitational effects from the inter-satellite range, the corrected variations of distance are used to estimate the geopotential with an unprecedented accuracy.

2.2 Time-variations of mass in the Earth system

Models of the Earth's geopotential are classically obtained by inverting GRACE data at monthly timescales and a spatial resolution of a few hundred kilometres (Tapley et al., 2004b). For the very first time, GRACE has enabled the monitoring of the spatiotemporal variations of the mass in the Earth system. The tiny variations of the Earth gravity field are due to the redistribution of fluid mass inside the surface fluid envelopes of the planet (atmosphere, oceans, continental water storage) (Dickey et al., 1999).

2.3 Expected accuracy and resolution for continental hydrology

The main application of GRACE is quantifying the terrestrial hydrological cycle through measurements of vertically-integrated water mass changes inside aquifers, soil, surface reservoirs and snow pack, with a precision of a few millimeters in terms of water height and a spatial resolution of ~ 400 km (Wahr et al., 1998; Rodell & Famiglietti, 1999). Pre-launch studies focused on the ability of GRACE to provide realistic hydrological signals, especially on continents. Wahr et al., (1998) highlighted the need to consider short-wavelength noise and leakage errors, as well as the size of the river basin. This early work largely inspired subsequent studies addressing the accuracy of recovering continental water mass signals from synthetic GRACE geoids (Rodell and Famiglietti 1999, 2001, 2002; Swenson and Wahr, 2002; Ramillien et al., 2004). The results of these simulations suggest that final accuracy would increase with spatial and temporal scales. The comparison of a large a number of modelled outputs of Terrestrial Water Storage (TWS) and the expected GRACE measurements accuracy showed that water storage changes would be detectable at spatial scales greater than 200,000 km^2, at monthly and longer timescales, and with monthly accuracies of roughly 1.5 cm (Rodell & Famiglietti, 1999). Similar conclusions were obtained using a network of hydrological observations of snow, surface water, soil moisture (SM) and groundwater (GW) in Illinois (Rodell & Famiglietti, 2001).

At basin-scale, the accuracy of GRACE measurements is expected to be 0.7 cm equivalent water height (EWH) for a basin with an area of 0.4 x 10^6 km², and about 0.3 cm EWT for a basin with an area of 3.9 x 10^6 km² (Swenson et al., 2003).

Combined with external information, GRACE offers the potentiality to monitor large-scale groundwater changes. Using a dense network of soil moisture data for the High Plains aquifer in the Central USA, Rodell and Famiglietti (2002) demonstrated the feasibility to dectect changes in aquifers since the uncertainties of mean variations of GRACE-derived GW for the ~ 450,000 km² aquifer was approximately 8.7 mm, whereas the amplitudes of the groundwater storage change signals were 20 and 45 mm for annual and 4-year periods respectively.

3. The GRACE data

Three processing centers including the Centre for Space Research (CSR), Austin, Texas, USA, the GeoForschungs Zentrum (GFZ), Potsdam, Germany, and Jet Propulsion Laboratory (JPL), Pasadena, California, USA), and forming the Science Data Center (SDS) are in charge of the processing of the GRACE data and the production of Level-1 and Level-2 GRACE products. They are distributed by the GFZ's Integrated System Data Center (ISDC – http://isdc.gfz-potsdam.de) and the JPL's Physical Oceanography Distributive Active Data Center (PODAAC – http://podaac-wwww.jpl.nasa.gov). Preprocessing of Level-1 GRACE data (*i.e.*, positions and velocities measured by GPS, accelerometer data, and KBR inter-satellite measurements) is routinely made by the SDS, as well as monthly GRACE global gravity solutions (Level-2).

3.1 Level-2 GRACE products

The Level-2 GRACE gravity solutions consist of monthy time-series of Stokes coefficients (*i.e.*, dimensionless spherical harmonics coefficients of the geopotential) developed up to a degree between 50 and 120, that are adjusted for each monthly time-period from raw along-track GRACE measurements. Formal errors associated with the estimated Stokes coefficients are provided to the users. A dynamic approach, based on the Newtonian formulation of the satellite's equation of motion in an inertial frame centered at the Earth's center combined with a dedicated modelling of gravitational and non-conservative forces acting on the spacecraft, is used to compute the monthly GRACE geoids (Schmidt et al., 2008). During the process, atmospheric and ocean barotropic redistribution of masses are removed from the GRACE coefficients using ECMWF and NCEP reanalyses for atmospheric mass variations and ocean tides respectively, and global circulation oceanic models (Bettadpur, 2007; Fletchner 2007). The GRACE coefficients are hence residual values that should represent continental water storage, errors from the correcting models, and noise over land. The monthly solutions differ from one group to another due to differences in the data processing, the choice of the correcting models, and the data selection during the monthly time-period. Four different completely reprocessed releases were distributed yet presenting a continuous improvement in the quality of the GRACE data. Some other research groups also propose alternate processing of the GRACE data at monthly or sub-monthly time-scales using global (harmonics coefficients) and regional approaches. They include the Groupe de Recherche en Géodésie Spatiale (GRGS), Toulouse, France, the Institute of Theoretical

Geodesy (ITG) at the University of Bonn, Germany, the Delft Institute of Earth Observations and Space Systems (DEOS), Delft, Nederlands, the Goddard Space Flight Center (NASA GSFC), Greenbelt, Maryland, USA.

3.2 Conversion into coefficients of water mass anomalies

Variations in continental water storage cause changes of mass distribution in the Earth system, and thus spatiotemporal changes of the geoid, defined as the gravitational equipotential surface that best coincides with the mean sea surface. In global representation, it can be expressed as (Heiskanen & Moritz, 1967):

$$\begin{Bmatrix} \delta C_{nm}(t) \\ \delta S_{nm}(t) \end{Bmatrix} = \frac{1 + k_n'}{(2n+1)} \frac{R_e^2}{M} \iint_S \delta q(\theta, \lambda, t) \begin{Bmatrix} \cos(m\lambda) \\ \sin(m\lambda) \end{Bmatrix} P_{nm}(\cos\theta)\delta S \tag{1}$$

for a given time t, where δq is the surface distribution of mass, M and S are the total mass and the surface of the Earth, respectively, θ and λ are co-latitude and longitude. δC_{nm} and δS_{nm} are the (dimensionless) fully-normalized Stokes coefficients and P_{nm} are associated Legendre functions, n and m are harmonic degree and order respectively. R_e is the mean Earth's radius (~6,371 km). N is the maximum degree of the development, ideally $N \to \infty$. In practice, the Stokes coefficients are estimated from satellite data with a finite degree $N < \infty$ and this maximum value defines the spatial resolution of the geoid ~ $\pi R_e/N$. The load Love number coefficients k_n' account for elastic compensation of Earth's surface in response to mass load variations. A list of values of the first Love number coefficients can be found in Wahr et al. (1998).

By removing harmonic coefficients of a reference "static" gravity field from GRACE solutions, time-variations of the Stokes coefficients ΔC_{nm} and ΔS_{nm} are computed for each monthly or 10-day period Δt.

For a monthly period Δt, the surface water storage anomaly within a region of angular area Ω (i.e., surface of the studied basin divided by R_e^2) can be obtained as the scalar product between time variations of Stokes coefficients $\Delta C_{nm}(t)$ and $\Delta S_{nm}(t)$, and averaging kernel coefficients A_{nm} and B_{nm} corresponding to the geographical region to be extracted (Swenson & Wahr, 2002):

$$\Delta\sigma_\Omega(\Delta t) = \frac{\rho_e R_e}{3\Omega} \sum_{n=1}^{N} \sum_{m=1}^{n} \frac{2n+1}{1+k_n'} \{\Delta C_{nm}(\Delta t)A_{nm} + \Delta S_{nm}(\Delta t)B_{nm}\} \tag{2}$$

where ρ_e is the mean Earth's density (~5,517 kg/m³). In case of a perfect kernel and error free data, A_{nm} and B_{nm} are the harmonic coefficients of the basin function or mask, which is equal to 1 inside the basin and zero outside.

3.3 The problem of aliasing

The monthly or sub-monthly GRACE solutions have a sufficient temporal resolution to monitor the long wavelengths variations of the gravity over land. The phenomenon of short term mass variability with period of hours to day, ocean tides and atmosphere are removed using de-aliasing techniques consisting of removing model outputs of these contributions. If the atmospheric pressure fields from ECMWF allow a reasonable de-aliasing of the higher frequencies caused by non-tidal atmospheric mass changes, errors due to tide models appear in the GRACE solutions, especially the aliasing from diurnal (S1) and semi-diurnal (S2) tides (Han et al., 2004; Ray & Lutcke, 2006). The aliasing and the modelling errors are responsible for the spurious meridional undulations present in the GRACE monthly grids and known as north-south striping. They are associated to spherical harmonics order 15 and its multiples (Swenson & Wahr, 2006a; Schrama et al., 2007; Seo et al., 2008). Thompson et al. (2004) showed that the degree error increased by factors ~ 20 due to atmospheric aliasing, ~ 10 due to the ocean model, and ~ 3 due to continental hydrology model. The S2 aliasing errors have also a strong impact on the C_{20} spherical harmonics coefficient of the geopotential and (Seo et al., 2008; Chen et al., 2009a) and are significant over the Amazon basin but dismish using longer time-series (Schrama et al., 2007; Chen et al., 2009a).

3.4 Accuracy of GRACE measurements: formal, omission and leakage errors

3.4.1 Formal error

The degree amplitude of the GRACE error is defined as:

$$\sigma_N(l) = R_e \sqrt{\sum_{m=0}^{l} \left(\sigma_{C_{lm}}^2 + \sigma_{S_{lm}}^2 \right)} \tag{3}$$

where $\sigma_{C_{lm}}$ and $\sigma_{S_{lm}}$ are the errors on the gravity potential coefficients.

It can be seen as the square-root of the total variance from all terms of a given spatial scale, as the degree l is the measure of the spatial scale of a spherical harmonics (i.e., a half wave-length of $20,000/l$ km). These errors increase at degrees 20 to 30 and become dominent at degrees 40 to 50. As a consequence, GRACE monthly solutions are low-pass filtered at degree 50 or 60 to remove the noise contained in the high frequency domain.

3.4.2 Omission or cut-off frequency error

Error in frequency cut-off represents the loss of energy in the short spatial wavelength due to the low-pass harmonic decomposition of the signals that is stopped at the maximum degree N_1. For the GRACE land water solutions; $N_1=60$, thus the spatial resolution is limited and stopped at ~330 km by construction. This error is simply evaluated by considering the difference of reconstructing the remaining spectrum between two cutting harmonic degrees N_1 and N_2, where $N_2 > N_1$ and N_2 should be large enough compared to N_1 (e.g., $N_2=300$):

$$\sigma_{truncation} = \sum_{n=0}^{N_2} \xi_n - \sum_{n=0}^{N_1} \xi_n = \sum_{n=N_1}^{N_2} \xi_n \tag{4}$$

using the scalar product:

$$\xi_n = \sum_{m=0}^{n} \left(\Delta C_{nm}(\Delta t) A_{nm} + \Delta S_{nm}(\Delta t) B_{nm} \right) \tag{5}$$

These errors are generally lower than 1% of the amplitude of the signal.

3.4.3 Leakage error

Due to the limited space resolution of the GRACE solutions, the estimate of mass variations inside a region of interest such as a drainage basin is affected by the leakage effect in two different ways. First, the mass changes inside a region spread out the whole globe. On the contrary, signals from other regions of the world pollute the estimate inside the considered region of interest. These two opposite effects have to be accounted for to estimate accurate mass variations at the surface of the Earth. Several techniques were proposed to quantify the leakage error. Swenson & Wahr (2002) proposed a geometric formulation of the leakage departure of the shape of the smoothed averaging kernel from that of the exact averaging kernel, Ramillien et al. (2006a) used an « inverse » mask, which is 0 and 1 in and out of the region respectively, developed in spherical harmonics and then truncated at degree 60, Baur et al. (2009) proposed an iterative approach based on forward modelling, Longuevergne et al. (2010) used a method optimizing the basin shape description. The leakage effects are generally estimated using global models as they can not be determined directly from the GRACE observations. Leakage effects are smaller close to the ocean. Over large areas, the signals leaking in and out tend to compensate each other. For the Amazon basin, leakage was found to have a small impact on TWS estimates (Chen et al., 2009b, Xavier et al., 2010). For small basins, such as the High Plain Aquifers (with an area ~ 450,000 km²), an error due to the leakage is of 25 mm in terms of EWH and remains important, compared with annual variations of TWS varying from 100 to 200 mm (Longuevergne et al., 2010).

4. Post-processing techniques

To filter out the spurious north-south stripes present in the GRACE TWS products different techniques based on empirical, statistical, inverse, or regional approaches were applied.

4.1 Empirical methods

4.1.1 Isotropic and non-isotropic filters

The simple isotropic Gaussian filter proposed earlier by Jekeli (1981) was commonly used in pre-launched studies (Wahr et al., 1998), and then applied to real data (Wahr et al., 2004; Tapley et al., 2004b) to demonstrate the abilities of GRACE to retrieve time variations of water mass anomalies over land. The choice of the smoothing radius and the inhability of this type of filter to distinguish between geophysical signals and noise are its major drawbacks. More elaborate filters were then developed to reduce the errors on the GRACE signals. Several averaging kernels based on the separate minimization of measurement and leakage error to signal were proposed, by using Lagrange multipliers and a leakage function defined as the difference in shape between exact and smoothed averaging kernels, with no *a priori* (Swenson & Wahr, 2002), or *a priori* (*i.e.*, the shape of the covariance function assuming it is azimuthally symmetric) information and then then minimizing the sum of variance of the satellite and leakage errors (Swenson & Wahr, 2003; Seo & Wilson, 2005).

To take into account the meridian structures of the noise polluting the GRACE signals, several non-isotropic smoothing kernels were applied, using spectral Legendre coefficients dependent on both degree and order (Han et al., 2005), or Tikhonov-type regularization (Kusche, 2007).

4.1.2 Destriping method

This filtering technique, known as destriping or correlated-error filter, was designed to remove the correlated errors in the Stokes coefficients responsible for the north-south stripes in the GRACE solutions. For a particular order m, the Stokes coefficients are smoothed with a quadratic polynomial in a moving window of width w centered about degree l. The sum only concerns the terms of the same parity as l (Swenson and Wahr, 2006a). This method provides better results at high latitudes than close to the equator as residual errors due to the near-sectorial coefficients ($l \sim m$) are not completely removed (Swenson and Wahr, 2006a; Klees et al., 2008a). Besides, some short-wavelength features are removed, mainly at high latitudes. Monthly CSR, GFZ and JPL destriped and smoothed EWH grids of 1-degree spatial resolution are available over 2002-2010 at: http://grace.jpl.nasa.gov.

4.1.3 Stabilization criteria

The Level-2 GRGS-EIGEN-GL04 models are derived from Level-1 GRACE measurements including KBRR, and from LAGEOS-1/2 SLR data for enhancement of lower harmonic degrees (Lemoine et al., 2007a; Bruinsma et al., 2010). These gravity fields are expressed in terms of normalized spherical harmonic coefficients from degree 2 up to degree 50-60 using an empirical stabilization approach without any smoothing or filtering. This stabilization approach consists in adding empirically determined degree and order dependent coefficients to minimize the time variations of the signal measured by GRACE over ocean and desert without significantly affecting the amplitude of the signal over large drainage basins. Monthly (Release-1) and 10-day (Release-2) TWS grids of 1-degree spatial resolution are available over 2002-2010 at: http://grgs.obs-mip.fr.

4.2 Statistical methods

4.2.1 Wiener filtering

The spatial averaging of monthly GRACE using the Wiener filter data is based on the least-square minimization of the difference between modelled and filtered signals. This filter is isotropic, depending only on the degree power of the signal and noise models. Using hydrology and ocean models outputs for simulating the degree power spectrum of the GRACE signal, the formal error of the GRACE coefficients, and GRACE solutions, Sasgen et al. (2006) were able to determine the optimal coefficients of the Wiener filter, and to demonstrate that this method is robust approach for low-pass filtering the GRACE data which does not need the specification of any averaging radius.

4.2.2 Principal Components Analysis

Principal Components Analysises (PCA) were performed on time-series of Gaussian-filtered EWH grids (Schrama et al., 2007) and Stokes coefficients (Wouters & Schrama, 2007). PCA was able to efficiently separate modes of geophysical signals from modes corresponding to the North-South striping in the EWH grids. The three first modes explain 73.5% of the continental hydrology with a Gaussian prefiltering of 6.25° of radius. The residual modes contain S2 aliasing errors and a semiannual continental hydrology signal contained in the Global Land Data Assimilation Systems (GLDAS) model (Schrama et al., 2007). Directly applied to the Stokes coefficients, PCA also successfully removes the meridian undulations present in the GRACE solutions. The results of this technique are expected to be better as the time-series of the GRACE solutions become longer (Wouters & Schrama, 2007).

4.2.3 Independent Components Analysis

The Independent Component Analysis (ICA) approach is used to extract hydrological signals from the noise in the Level-2 GRACE solutions by considering completely objective constraints, so that the gravity component of the observed signals is forced to be uncorrelated numerically. This blind separation method is based on the assumption of statistical independence of the elementary signals that compose the observations, *i.e.*, geophysical signals and spurious noise, and does not require other *a priori* information (Frappart et al., 2010a). Comparisons at a global scale showed that the ICA-based solutions present less north–south stripes than Gaussian and destriped solutions on the land, and more realistic hydrological structures than the destriped solutions in the tropics. ICA filtering seems to allow the separation of the GIA from the TWS as negative trends were found over the Laurentides and Scandinavia. Unfortunately, this important geophysical parameter does not appear clearly in an independent mode yet. At basin-scale, the ICA-based solutions allowed us to filter out the unrealistic peaks present in the time series of TWS obtained using classical filtering for basins with areas lower than 1 million km² (Frappart et al., 2011a). The major drawback of this approach is that it cannot directly be applied to the GRACE Level-2 raw data, as a first step of (Gaussian) prefiltering is required. Monthly CSR, GFZ and JPL ICA-filtered TWS grids of 1-degree spatial resolution are available over 2002-2010 at: http://grgs.obs-mip.fr.

4.2.4 Kalman smoothing

Daily water masses solutions have been derived from GRACE observations using a Kalman filter approach. They are estimated using a Gauss Markov model, the solution at day $t+1$ slightly differs from solution at day t from the noise prediction (first-order Markov process) estimated using a Kalman filter. A *priori* information on the hydrological patterns from the WGHM model was introduced in the covariance matrix to compensate the small number of available observations over a daily time-span. This approach enables to produce temporarily enhanced solutions without loss in spatial resolution and resolution (Kurtenbach et al., 2009). Daily TWS Stokes coefficients are available over 2002-2010 at: http://www.igg.uni-bonn.de/apmg.

4.3 Inverse methods

4.3.1 Iterative least-squares approach

This method is based on the matrix formalism of the generalized least-squares criteria (Tarantola, 1987) and consists in estimating separately the spherical harmonics coefficients, in terms of EWH, of four different fluid reservoirs (atmosphere, oceans, soil water, and snow) from the monthly GRACE geoids. For each water reservoir, the anomaly of mass is estimated after the convergence of an inverse approach combining the GRACE observations with stochastic properties of the unknown hydrological signal (Ramillien et al., 2004; 2005).

4.3.2 Optimal filter

This method takes into account the statistical information present in each GRACE monthly solution. The noise and full signal variance-covariance matrix is used to tailor the filter to the error characteristics of a particular monthly solution. The resulting filter was found to be both anisotropic and non-symmetric to accommodate noise of an arbitrary shape as the north-south stripes present in the GRACE monthly solutions. It is optimal as it minimizes the difference between the signal and the filtered GRACE estimate in the least-squares sense. This filter was found to perform better than any isotropic, or anisotropic and symmetric filters preserving better the signal amplitudes better (Klees et al., 2008a). Monthly GRACE-derived TWS grids of 1-degree spatial resolution computed using the optimal filter are available over 2002-2010 at: http://lr.tudelft.nl/en/organisation/departments-and-chairs/remote-sensing/physical-and-space-geodesy/data-and-models/dmt-1/

4.4 Regional approaches

Since the begining of the low-low satellite gravimetry missions dedicated to Earth's gravity field measurement (*i.e.*, CHAMP, GRACE, GOCE), determining locally the time variations of the surface water storage has been of great interest for region where very localized strong mass variations occur, such as flood and glaciers fields. Moreover, it represents an alternative to the problem of numerical singularity at poles by choosing a suitable geometry of surface tiles. So far, rectangular surface elements have been considered but other geometries have successfully tested (Eicker, 2008; Ramillien et al., 2011).

4.4.1 Mascons

In this local approach, the mass of water in surface 4-by-4 degree blocks has been explicitly solved using the GRACE inter-satellite KBR Rate (KBRR) data for continental hydrology and collected over the region of interest. This size of the surface elements has been chosen as the limit of *a priori* spatial resolution of the GRACE data (*i.e.*, 400 km). This method uses inherently empirical spatial and temporal constraints among the coefficients that are dependent on geographical locations. The local representation of gravity minimizes the leakage error from other areas due to aliasing or mis-modelling (Rowlands et al., 2005; Lemoine et al., 2007b). It has revealed more information about interannual variability in the subregions than any Stokes coefficients-based approach. Global and mascons methods based on spherical harmonics have provided similar results for the mass trend in the drainage basins, especially when no spatial constraints are taken into account (Rowlands et al., 2010). But, inherent problem of spectral truncation such as leakage are not avoided while using spherical harmonics. 10-day and 4° Mascons solutions and corresponding GLDAS model time series for each continental surface element are available at: http://grace.sgt-inc.com/V2/Global.html.

4.4.2 Regional method

Another type of regional approach has been recently proposed by adjusting the surface mass density from the Level-1 GRACE data, in particular the accurate satellite-to-satellite velocity variations (Ramillien et al., 2011). Once these observations are corrected from known accelerations (atmosphere and ocean mass changes, tides, static gravity field...), along track residuals of KBR-rate are used to compute kinetic energy (or equivently potential anomaly) variations between the twin GRACE satellites, corresponding mainly to continental water redistributions. The linear system of equations to solve between observations (*i.e.*, potential anomalies) and parameters (*i.e.*, surface elements) is built according to the first Newton's law of attraction of masses. Unlike the mascons solutions, spherical harmonics are not used as a basis of regional orthogonal functions, consequently this regional method does not suffer from the drawbacks related to any spectral truncation. Singular Value Decomposition (SVD) and L-curve analysis are considered to compute regularized 10-day 2-by-2 degree solutions of the ill-posed of gravimetry. Lately, another version of the inversion scheme takes spatial correlations versus the geographical distance between surface elements into account (Ramillien et al., in press). Time series of regional solutions over South America have been produced and compared to global solutions and mascons, they reveal very comparable amplitudes of seasonal cycle over the Amazon basin, as well as sudden flooding events.

5. Applications to continental hydrology: results and discussion

5.1 Application in continental hydrology and validation

Once sufficiently long series of GRACE solutions were made available by the former official providers (CSR, JPL, GFZ), various regional studies for validating these products were made. Examples of GRACE-based TWS maps are given in Figure 1 showing the ability of the gravimetry from space mission to restitue both seasonal and interannual variability.

Comparisons with global hydrology model outputs and surface measurements revealed acceptable agreements between GRACE-derived and model changes of continental water storage versus time, especially at seasonal time scale. GRACE provides directly the term of TWS variation of the net balance equation. In general, GRACE-based estimates of TWS compare favourably with those based on land surface models as well as atmospheric and terrestrial water balances (Rodell et al., 2004a; Ramillien et al., 2005; Syed et al., 2005; Klees et al., 2008b). The structures seen on global maps of seasonal amplitudes are comparable to those described by the WaterGap model developed by Döll et al. (2003) and GLDAS (Rodell et al., 2004a). Over the entire Northern Hemisphere, GLDAS water storage simulations with a resolution of 1,300 km and accuracy of 9 mm in terms of EWH were found to have a spatial correlation of 0.65 with GRACE data, suggesting that gravity field changes may be related to TWS variations (Andersen & Hinderer, 2005). However, TWS variations tend to be slightly over-estimated by GRACE (Schmidt et al., 2006; Syed et al., 2008). GRACE-based TWS changes were also validated by accurate observations of superconducting gravimeters during the 2003 heat wave that occured in Central Europe (Andersen et al., 2005). They were used to detect the exceptional drought (Chen et al., 2009b) and flooding (Chen et al., 2010) affected the Amazon basin in 2005 and 2009 respectively (Figure 2a). Correlations of 0.7-0.8 between GRACE-based TWS and *in situ* measurements of water level along the Amazon River were found by Xavier et al. (2010). Combined with other satellite techniques such as imagery, GRACE geoid data were used to study the mechanisms of seasonal flooding in large inundation areas like the Mekong delta (Frappart et al., 2006a) and large river basins as the Amazon River (Frappart et al., 2008; Papa et al., 2008).

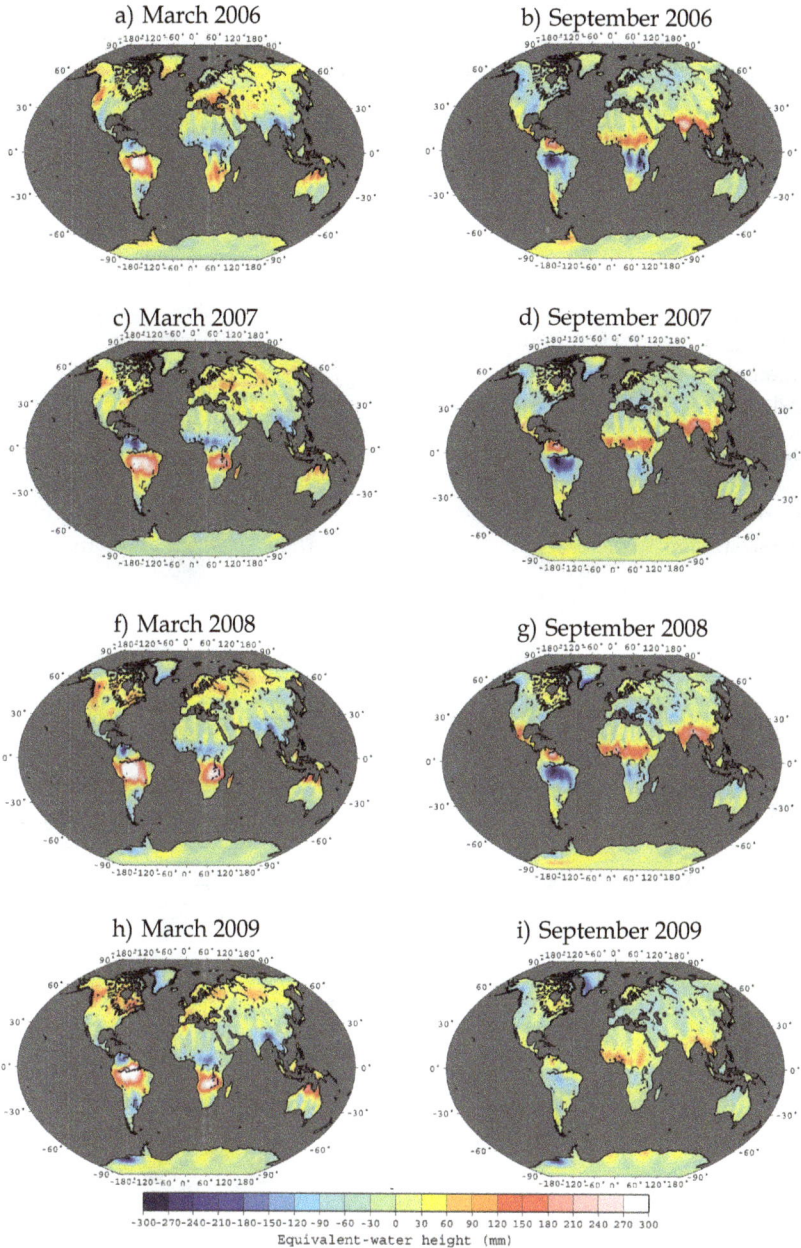

Fig. 1. Examples of GRACE-derived TWS maps processed using a Gaussian filtering of radius 400 km and ICA (Level-2 Release 4 GFZ solutions) for a) March 2005, b) September 2005, c) March 2006, d) September 2006, e) March 2007, f) September 2007, g) March 2008, h) September 2008, i) March 2009, j) September 2009.

5.2 Isolating groundwater changes

Variations in the groundwater storage can be extracted from *TWS* measured by GRACE using external information on the other hydrological reservoirs. The time variations in *TWS* are the sum of the contributions of the different reservoirs present in a drainage basin:

$$\Delta TWS = \Delta SW + \Delta SN + \Delta TSS \tag{6}$$

with:

$$\Delta TSS = \Delta RZ + \Delta GW + \Delta P \tag{7}$$

where *SW* represents the total surface water storage including lakes, reservoirs, in-channel and floodplains water, *SN* is the snow storage, *TSS* is the total soil storage including *RZ* the water contained in the root zone of the soil (generally representing a depth of 1 or 2 m), *GW* the groundwater storage in the aquifers, and *P* the permafrost storage at boreal latitudes.

Post-launch studies of GRACE-based groundwater remote sensing have clearly demonstrated that when combined with ancillary measurements of surface waters and soil moisture, either modeled or observed, GRACE is capable of monitoring changes in groundwater storage with reasonable accuracy. Important seasonal correlations of 0.8-0.9

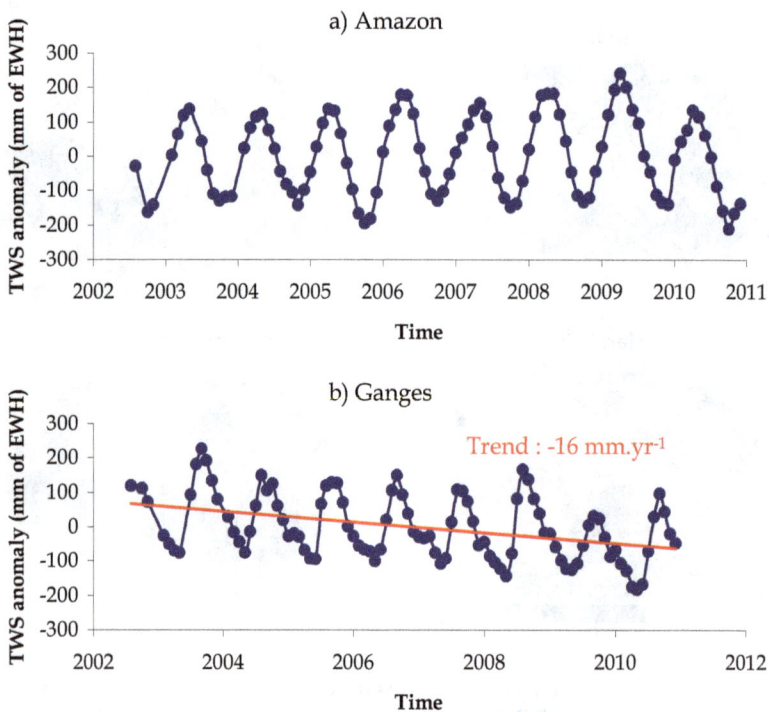

Fig. 2. Time series of TWS anomalies (mm of EWH) from Level-2 Release 4 GFZ GRACE solutions (Gaussian filtering of radius 400 km and ICA) for a) Amazon, b) Ganges.

were found by confronting GRACE to recordings of well network in Illinois (Yeh et al., 2006), Oklahoma (Swenson et al., 2008), the High Plain aquifer (Strassberg et al., 2007) and the Mississippi basin (Rodell et al. 2007). This method was successfully applied to monitor large-scale climatic and anthropogenic impacts on water resources, as the detection of the recent severe drought in the Murray-Darling basin in Southern Australia (Leblanc et al., 2009) or the depletion of the aquifers in India (Rodell et al., 2009) (Figure 2b), but also to estimate aquifer storage parameters as the specific yield or storativity (Sun et al., 2010). For large drainage basins covered with extensive floodplains, changes in water stored in the aquifer is isolated from the TWS measured by GRACE by removing contributions of both the surface reservoir, derived from satellite imagery and radar altimetry, and the root zone reservoir simulated by hydrological models (Frappart et al., 2010b; 2011b). In the Negro River basin, the groundwater anomalies show a realistic spatial pattern compared with the hydrogeological map of the basin (DNPM, 1983), and similar temporal variations to local in situ groundwater observations (Figure 3).

5.3 Estimates of snow mass changes at high latitudes

After a separation of water and snow contributions from observed gravity field using an inverse iterative approach based on the least-squares criteria, Frappart et al. (2006b) used satellite microwave data and global hydrological model outputs to validate GRACE-derived snow mass changes at high-latitudes. In particular, rms errors of 10-20 mm were found for Yenisey, Ob, McKenzie and Yukon basins. Comparison at basin scales between GRACE-derived snow mass variations, GRACE-based TWS variations and river discharge temporal evolution shows that the snow component has a more significant impact on river discharge at high latitudes than TWS, and that snow and streamflow present similar interannual variabilities (Frappart et al., 2011c).

5.4 Estimates of regional geo-hydrological parameters

5.4.1 River discharges

GRACE-based estimates of river discharges are based on the resolution of the water balance equation at basin-scale:

$$\frac{\partial S}{\partial t} = P - ET - R \tag{8}$$

where S is land water storage (TWS), P and ET the basin-wide totals of precipitation and evapotranspiration, and R the total basin discharge, or the net surface and groundwater outflow.

As ET presents large uncertainties, the quantity P-ET is eliminated using the water balance of the atmospheric branch of the water cycle:

$$\frac{\partial W}{\partial t} = ET - P - div\vec{Q} \tag{9}$$

where W is the vertically-integrated precipitable water, $div\vec{Q}$ the divergence of the vertically-integrated average atmospheric moisture flux vector, computed using the following equations:

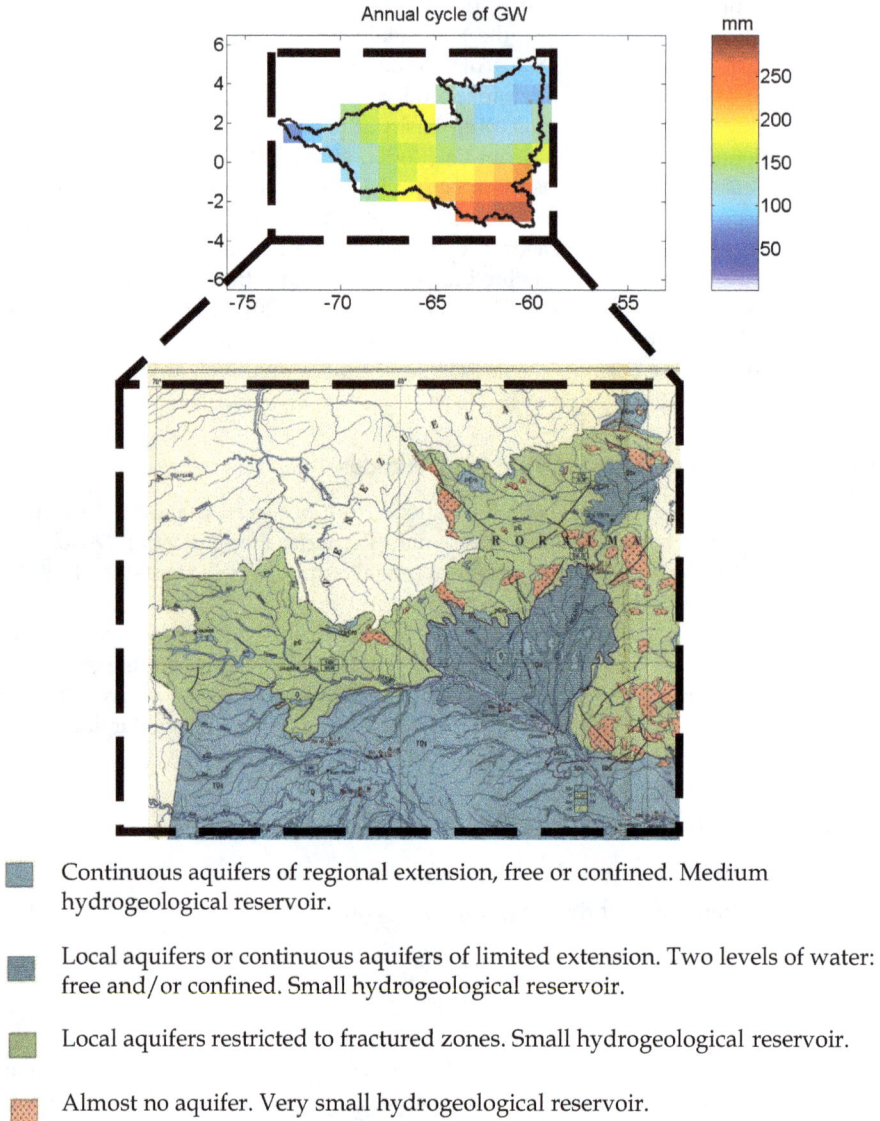

Continuous aquifers of regional extension, free or confined. Medium hydrogeological reservoir.

Local aquifers or continuous aquifers of limited extension. Two levels of water: free and/or confined. Small hydrogeological reservoir.

Local aquifers restricted to fractured zones. Small hydrogeological reservoir.

Almost no aquifer. Very small hydrogeological reservoir.

Fig. 3. When combined with other data, GRACE-derived TWS anomalies provide valuable information on changes affecting the aquifers. Above a comparison between a) Map of annual amplitude of GW in the Negro River Basin. b) Hydrogeological map of Brazil from DNPM (1983), available at: http://eusoils.jrc.ec.europa.eu/esdb_archive/EuDASM/EUDASM.htm, in the framework of the European Commission — Joint Research Centre through the European Digital Archive of Soil Maps (EuDASM) (Panagos et al., in press). Adapted from Frappart et al. (2011b).

$$W = \int_{P_T}^{P_s} \frac{q}{g} dp \tag{10}$$

$$\vec{Q} = \int_{P_T}^{P_s} \frac{q}{g} \vec{V} dp \tag{11}$$

where p_s and p_T are the pressure at the surface and the top of the atmosphere respectively, q is the specific humidity, g is gravitational acceleration and $\vec{V}$ is the horizontal wind vector.

This approach was firstly applied to the Amazon and the Mississippi River basins (Syed et al., 2005), and extended to obtain a global volume of freshwater discharge estimated here is 30,354 ± 1,212 km³/y (Syed et al., 2009).

5.4.2 Vertical water fluxes

GRACE-based TWS can be also used to estimate changes in vertical water fluxes solving the water balance equation (8). Changes of regional evapotranspiration (ET) rate over Mississippi basin were estimated by combining 600-km low-pass filtered GRACE TWS data with observed precipitation and streamflow in the water balance equation (Rodell et al., 2004b). Similarly, Ramillien et al. (2006b) solved the water balance equation for ET rate. Time variations of ET were evaluated over 16 large drainage basins solving the water balance equation and using precipitation from GPCC and runoff from the WGHM model (Döll et al., 2003), and they revealed that GRACE-based ET variations were comparable to the ET value simulated by four differend global hydrology models. Swenson and Wahr (2006b) estimated the difference precipitation minus evapotranspiration using the water balance framework and comparing to surface parameters variations from global analysis.

5.5 Contribution of TWS to sea level variations

Seasonal and interannual changes in land water storage derived from GRACE were used to estimate the contribution of land hydrology to the sea level variations for 27 of the largest drainage basins in the world. Estimation of 2002-2006 sea level contribution of GRACE-derived TWS of the largest basins was made using the GRGS GRACE solutions, and corresponds to a water loss of ~ 0.5 mm per year ESL (Ramillien et al., 2008). Wouters et al. (2011) recently estimated the global mean eustatic cycle, i.e., the total amount of water exchanged between continents and ocean, to be 9.4 ± 0.6 mm equivalent water level over the period 2003-2010. In the Pan Arctic regions, snow volume derived from the Special Sensor Microwave/Imager (SSM/I) radiometric measurements over 1989-2006 and GRACE over 2002-2007 is a key driver of the sea level seasonal cycle, but snow volume trend indicates a negligible and not statistically significant contribution to sea level variations (Biancamaria et al., 2011).

5.6 Optimization and assimilation into hydrological models

The GRACE data have been widely confronted to global hydrological ouputs for comparison and validation, and then to improve the description and parameterization of the hydrological processes in the models. The GRACE geoid data were used as a proxy to test and improve the efficiency of surface waters schemes in the ORCHIDEE land surface hydrology model (Ngo-Duc et al., 2007). Unfortunately, a lack of consistency in the seasonal cycle and a time shift between GRACE TWS and global hydrology models may persist. The comparison between six GRACE-derived TWS and 9 land surface models forced with the same forcings over West Africa permits to identify the processes needing improvement in the land surface models, especially the correct simulation of slow water reservoirs as well as evapotranspiration during the dry season for accurate soil moisture modeling over West Africa (Grippa et al., 2011).

Güntner (2008) pointed out the importance of integrating the GRACE data into hydrological models for improving the reliability of their prediction through advanced methods of multi-objective calibration and data assimilation. A multicalibration approach to constrain model predictions by both measured river discharge and TWS anomalies from GRACE was apllied to the WGHM model, improving simulation results with regard to both objectives. Using only monthly TWS variations, the RMSE was reduced of about 25 mm for the Amazon, 6 mm for the Mississippi and 1 mm for the Congo river basins (Werth et al., 2009). GRACE-derived monthly TWS anomalies were also assimilated into one of the GLDAS models using a Kalman smoother approach. Compared with open-loop simulations, assimilated ones exhibited better performance thanks to improvements in the surface and groundwater estimates (Zaitchik et al., 2008).

5.7 Detecting long-term variations over ice sheets

GRACE-derived mass balance estimates of Antarctica (Velicogna & Wahr, 2006a), Greenland (Velicogna & Wahr, 2006b; Chen et al., 2006) or both ice sheets (Ramillien et al., 2006a) have been determined from the Level-2 solutions of the GRACE project. Velicogna and Wahr (2006b) found a lost rate of -82 ± 28 Gt/y for Greenland with an acceleration of melting in Spring 2004. Mass changes of this ice sheet resolved by drainage basin were estimated using the NASA mascons solutions as -101 ± 16 Gt/y after correction for post-glacial rebound (Luthcke et al., 2006). This latter result is consistent with the one found previously by Ramillien et al. (2006a) using 10-day GRGS GRACE solutions. Velicogna & Wahr (2006a) estimated an extreme decrease of ice mass of 152 ± 80 Gt/y for Antarctica. The problem is that this main value represents the Post-Glacial Rebound (PGR) correction itself that the authors removed from the GRACE data using the IJ05 (Ivins and James, 2005) and ICE-5G (Peltier, 2004). If GRACE data are not corrected from PGR, the mass balance of Antarctica appears close to zero, with large uncertainties due to North-South striping and leakage effects. This result shows it is not clear whether this continent actually looses or gains ice mass during the recent period (Ramillien et al., 2006a). PGR phenomena are not well-modelled, especially over the whole Antarctica where available long-term observational constraints remain rare. Consequently, removing PGR that still cannot be modelled accurately represent a source of important errors. Velicogna & Wahr (2002) have shown earlier that the detection of PGR remains possible by combination of different satellite techniques, using 5 years of simulated GRACE and Laser GLAS data over the whole Antarctica. However, an effective extraction of PGR signals using GRACE products still needs to be demonstrated.

As mentioned by Baur et al. (2009), even if the PGR correction is less than 10 Gt/y over Greenland, mass balance for this continental area using GRACE solutions can vary strongly from simple to twice, according to which provider (CSR, JPL, GFZ) is considered, and which processing is applied. Velicogna and Wahr (2005) found a constant decrease of -84 ±-28 Gt/y for the period 2002-2004. Later, Wouters et al. (2008) adjusted a rate of -171 Gt/y for 2003-2008, with an acceleration during summer. Interestingly, Ramillien et al. (2006a) and Luthcke et al. (2006) found comparable loss rates of about 110 Gt/y considering 10-day GRGS and mascons solutions, respectively, for 2003-2005. Recent re-evaluation of Greenland ice mass loss provides -65 Gt/y using ICA-400 solutions and around 100 Gt/y using classical Gaussian filtered solution for 2003-2010 (Bergmann et al., 2012). It is clear that an interannual variations of Greenland ice sheet exists, as an acceleration of the ice melting in Spring 2004 and 2007-2009 (Velicogna & Wahr, 2006b; Velicogna, 2009). A recent study pointed a clear de-acceleration of the ice melting over Greenland for the very last years (Bergmann et al., 2012), which was confirmed by a recent analysis of Rignot et al. (2011).

Level-2 GRACE solutions have also been used to quantify ice mass loss of coastal glaciers in Western Antarctica, Patagonia (Chen et al., 2007) and in the Gulf of Alaska (Luthcke et al., 2008) as the response of the recent global warming. Chen et al. (2007) found a depletion of mass of -27.9 ± 11 Gt/y for the Patagonian glaciers between April 2002 and December 2007. For glaciers in Alaska, negative trends of -84 Gt/y during 2003-2007 and -102 Gt/y for 2003-2006 were deduced from mascon solutions. Once again, trend estimates depend strongly upon either using a model of PGR, or estimating it together with the current mass loss. Another complication is the coarse footprint of GRACE which mixes signals from neighbouring basins (Horwath and Dietrich, 2009).

6. Conclusion

Many studies have demonstrated the possibility of satellite gravimetry to detect and monitor spatial versus time, at a precision of only tens of mm of EWH and spatial resolution of 400 km. Accuracy of the results still depends upon the level of noise in the GRACE data and the post-processing used. In addition to measurement errors and loss of signals by low-pass filtering, uncertainty on model-predicted GIA represents a major source of error in the mass balance estimates of the ice sheets melting, in particular for Antarctica. Besides, a resolution of 400 km still represents an important limitation for small-scale studies, especially in field hydrology, which often requires surface resolution of tens of kilometers. Improvements in pre- and post-processing techniques should increase the quality of the GRACE products, such as the "regional" or local approaches. Steady improvement of precision and resolution of GRACE products remains encouraging and it opens a wider class of applications than previously possible.

Even if the dying GRACE mission may not be able to provide data of the same accuracy to be exploitable in continental hydrology in 2012, other follow-on projects have already been proposed by a motivated community to ensure the continuity of such a satellite system. Satellite gravimetry remains the only remote sensing technique to map directly large scale mass variations. Any Follow-On GRACE mission in preparation by NASA and GFZ would be of first interest for studying the impacts on surface redistribution forced by climate change.

7. Acknowledgment

This review on the applications of GRACE to land water hydrology and ice sheets mass balance was funded by Université de Toulouse.

8. References

Alsdorf, D.E. & Lettenmaier, D.P. (2003). Tracking fresh water from space. *Science*, Vol.301, No.5639, (September 2003), pp. 1491–1494, ISSN 0036-8075.

Andersen, O. B. & Hinderer, J. (2005). Global inter-annual gravity changes from GRACE: Early results. *Geophysical Research Letters*, Vol.32, No.1, (January 2005), pp. L01402, ISSN 0094-8276.

Andersen, O.B.; Seneviratne, S.I.; Hinderer, J. & Viterbo P. (2005). GRACE-derived terrestrial water storage depletion associated with the European heat wave. *Geophysical Research Letters*, Vol.32, No.18, (September 2005), pp. L18405, ISSN 0094-8276.

Baur, O.; Kuhn, M. & Featherstone, W.E. (2009). GRACE-derived ice mass variations over Greenland by accounting for leakage effects. *Journal of Geophysical Research*, Vol.114, No.B13, (June 2009), pp. B06407, ISSN 0148-0227.

Bergmann, I.; Ramillien, G. & Frappart, F. (2012). Climate-driven interrannual variations of the mass balance of Greenland, *Global and Planetary Change*, Vol.82-83, (February 2012), pp. 1-11, ISSN 0921-8181

Bettadpur, S. (2007). *CSR Level-2 processing standards document for level-2 product release 04, GRACE*. The GRACE project, Center for Space Research, University of Texas at Austin, pp. 327–742.

Biancamaria, S.; Cazenave, A.; Mognard, N.M.; Llovel, W. & Frappart, F. (2011). Satellite-based high latitudes snow volume trend, variability and contribution to sea level over 1989-2006. *Global and Planetary Change*, Vol.75, No.3-4, (February 2011), pp. 99-107, ISSN 0921-8181.

Bruinsma, S.; Lemoine, J.M.; Biancale, R. & Valès, N. (2010). CNES/GRGS 10-day gravity field models (release 2) and their evaluation. *Advances in Space Research*, Vol.45, No.4, (February 2010), pp. 587–601, ISSN 0273-1177.

Chen, J.L.; Wilson, C.R.; Blankenship, D.D. & Tapley, B. D. (2006). Antarctic mass rates from GRACE. *Geophysical Research Letters*, Vol.33, No.11, (June 2006), pp. L11502, ISSN 0094-8276.

Chen, J.L.; Wilson, C.R.; Tapley, B.D.; Blankenship, D.D. & Ivins, E.R. (2007). Patagonia Icefield melting observed by Gravity Recovery and Climate Experiment (GRACE). *Geophysical Research Letters*, Vol.34, No.22, (November 2007), pp. L22501, ISSN 0094-8276.

Chen, J.L.; Wilson, C.R. & Seo, K.W. (2009a). S2 tide aliasing in GRACE time-variable gravity solutions. *Journal of Geodesy*, Vol.83, No.7, (July 2009), pp. 679-687, ISSN 0949-7714.

Chen, J.L.; Wilson, C.R.; Tapley B.D.; Zang, Z.L. & Nyu, G.Y. (2009b). 2005 drought event in the Amazon River basin as measured by GRACE and estimated by climate models. *Journal of Geophysical Research*, Vol.114, No.B5, (May 2009), pp. B05404, ISSN 0148-0227.

Chen, J.L.; Wilson, C.R. & Tapley B.D. (2010). The 2009 exceptional Amazon flood and interannual terrestrial water storage change observed by GRACE. *Water Resources Research*, Vol.46, No.12, (December 2010), pp. W12526, ISSN 0043-1397.

Dickey, J.O.; Bentley, C.R.; Bilham, R.; Carton, J.A.; Eanes J.A.; Herring, T.A.; Kaula, W.M.; Lagerloef, G.S.E.; Rojstaczern S.; Smith, W.H.F.; van den Dool, H.; Wahr J.M. & Zuber, M.T. (1999). Gravity and the hydrosphere: new frontier. *Journal des Sciences Hydrologiques – Hydrological Sciences Journal*, Vol.44, No.3, (June 1999), pp. 407-415, ISSN 0262-6667.

Departamento Nacional da Produção Mineral/DNPM (1983). Mapa hidrogeológico do Brasil, escala 1:5,000,000.

Döll, P.; Kaspar, F. & Lehner, B. (2003). A global hydrological model for deriving water availability indicators: Model tuning and validation. *Journal of Hydrology*, Vol.270, No.1-2, (January 2003), pp.105-134, ISSN 0022-1694.

Eicker, A. (2008). *Gravity Field Refinement by Radial Basis Functions from In-situ Satellite Data*, Dissertation Univ. Bonn D 98, pp. 137, Univ. Bonn, Bonn.

Fletchner, F. (2007). *GFZ Level-2 processing standards document for level-2 product release 04, GRACE*. Department 1: Geodesy and Remote Sensing, GeoForschungZentrum, Potsdam, pp. 327-742.

Frappart, F.; Do Minh K.; L'Hermitte, J.; Cazenave, A.; Ramillien, G.; Le Toan, T. & Mognard-Campbell, N. (2006a). Water volume change in the lower Mekong basin from satellite altimetry and imagery data. *Geophysical Journal International*, Vol.167, No.2, (November 2006), pp. 570-584, ISSN 0956-540X.

Frappart, F.; Ramillien, G.; Biancamaria, S.; Mognard, N. M. & Cazenave, A. (2006b). Evolution of high-latitude snow mass derived from the GRACE gravimetry mission (2002–2004). *Geophysical Research Letters*, Vol.33, No.2, (January 2006), pp. L02501, ISSN 0094-8276.

Frappart, F.; Papa, F.; Famiglietti, J. S.; Prigent, C.; Rossow, W. B. & Seyler, F. (2008). Interannual variations of river water storage from a multiple satellite approach: A case study for the Rio Negro River basin. *Journal of Geophysical Research*, Vol.113, No.D12, (November 2008), pp. D21104, ISSN 0148-0227.

Frappart, F.; Ramillien, G.; Maisongrande, P. & Bonnet, M.P. (2010a). Denoising satellite gravity signals by independent component analysis. *IEEE Geoscience and Remote Sensing Letters*, Vol.7, No.3, (July 2010), pp. 421−425, ISSN 1545-598X.

Frappart, F.; Papa, F.; Güntner, A.; Werth, S.; Ramillien, G.; Prigent, C.; Rossow, W.B. & Bonnet, M-P. (2010b). Interrannual variations of the terrestrial water storage in the Lower Ob' basin from a multisatellite approach. *Hydrology and Earth System Sciences*, Vol.14, No.12, (December 2010), pp. 2443-2453, ISSN 1812-5606.

Frappart, F.; Ramillien, G.; Leblanc, M.; Tweed, S.O.; Bonnet, M.P. & Maisongrande, P. (2011a). An independent component analysis filtering approach for estimating continental hydrology in the GRACE gravity data. *Remote Sensing of Environment*, Vol.115, No.1, (January 2011), pp. 187–204, ISSN 0034-4257.

Frappart, F.; Papa, F.; Güntner, A.; Werth, S.; Santos da Silva, J.; Seyler, F.; Prigent, C.; Rossow, W.B.; Calmant, S. & Bonnet, M.P. (2011b). Satellite-based estimates of groundwater storage variations in large drainage basins with extensive floodplains, *Remote Sensing of Environment*, Vol.115, No.6, (June 2011), pp. 1588-1594, ISSN 0034-4257.

Frappart, F.; Ramillien, G. & Famiglietti, J. S. (2011c). Water balance of the Arctic drainage system using GRACE gravimetry products. *International Journal of Remote Sensing*, Vol.32, No.2, (January 2011), pp. 431-453, ISSN 0143-1161.

Grippa, M.; Kergoat, L.; Frappart, F.; Araud, Q.; Boone, A.; de Rosnay, P.; Lemoine, J-M.; Gascoin, S.; Balsamo, G.; Ottlé, C.; Decharme, B.; Saux-Picart, S. & Ramillien, G. (2011). Land water storage over West Africa estimated by GRACE and land surface models. *Water Resources Research*, Vol.47, No.5, (May 2011), pp. W05549, ISSN 0043-1397.

Güntner, A. (2008). Improvement of global hydrological models using GRACE data. *Surveys in Geophysics*, Vol.29, No.4-5, (October 2008), pp. 375–397, ISSN 1573-0956.

Han, S.-C.; Jekeli, C. & Shum, C. (2004). Time-variable aliasing effects of ocean tides, atmosphere, and continental water mass on monthly mean GRACE gravity field. *Journal of Geophysical Research*, Vol.109, No.B04, (April 2004), pp. B04403, ISSN 0148-0227.

Han, S.-C.; Shum, C.; Jekeli, C.; Kuo, C.; Wilson, C. & Seo, K.W. (2005). Non-isotropic Filtering of GRACE Temporal Gravity for Geophysical Signal Enhancement. *Geophysical Journal International*, Vol.163, No.1, (September 2005), pp. 18-25, ISSN 0956-540X.

Heiskanen, W.H. & Moritz H. (1967). *Physical Geodesy*. W.H. Freeman and Co Ltd, ISBN 978-071-6702-23-37, San Francisco, USA.

Horwath, M. & Dietrich, R. (2009). Signal and error in mass change inferences from GRACE: the case of Antarctica. *Geophysical Journal International*, Vol.177, No.4, (June 2009), pp. 849–864, ISSN 0956-540X.

Ivins, E.R. & James, T.S. (2005). Antarctic glacial isostatic adjustment: A new assessment. *Antarctic Science*, Vol.17, No.4, (December 2005), pp. 541-553, ISSN 0954-1020.

Jekeli C. (1981). *Alternative methods to smooth the Earth's gravity field*. Report 327, Department of Geodesy Sciences and Survey, Ohio State University, Columbus, USA.

Klees, R.; Revtova, E.A.; Gunter, B. C.; Ditmar, P.; Oudman, E.; Winsemius, H. C. & Savenije, H.H.G. (2008a). The design of an optimal filter for monthly GRACE gravity models. *Geophysical Journal International*, Vol.175, No.2, (November 2008), pp. 417–432, ISSN 0956-540X.

Klees, R.; Liu, X.; Wittwer, T.; Gunter, B. C.; Revtova, E.A.; Tenzer, R.; Ditmar, P.; Winsemius, H. C. & Savenije, H.H.G. (2008b). A comparison of global and regional GRACE models for land hydrology. *Surveys in Geophysics*, Vol.29, No.4-5, (October 2008), pp. 335–359, ISSN 1573-0956.

Kurtenbach, E.; Mayer-Gürr, T. & Eicker, A. (2009). Deriving daily snapshots of the Earth's gravity field from GRACE L1B data using Kalman filtering. *Geophysical Research Letters*, Vol.36, No.17, (September 2009), pp. L17102, ISSN 0094-8276.

Kusche, J. (2007). Approximate decorrelation and non-isotropic smoothing of timevariable GRACE-type gravity field models. *Journal of Geodesy*, Vol.81, No.11, (November 2007), pp. 733–749, ISSN 0949-7714.

Leblanc, M. J.; Tregoning, P.; Ramillien, G.; Tweed, S.O. & Fakes, A. (2009). Basin-scale integrated observations of the early 21st century multiyear drought in southeast Australia, *Water Resources Research*, Vol.45, No.4, (April 2009), pp. W04408, ISSN 0043-1397.

Lemoine, J.M.; Bruinsma, S.; Loyer, S.; Biancale, R.; Marty, J.C.; Perosanz, F. & Balmino, G. (2007a). Temporal gravity field models inferred from GRACE data. *Advances in Space Research*, Vol.39, No.10, (May 2007), pp. 1620–1629, ISSN 0273-1177.

Lemoine, F.G.; Luthcke, S.B., Rowlands, D.D.; Chinn, D.S.; Klosko, S.M. & Cox, C.M. (2007b). The use of mascons to resolve time-variable gravity GRACE, In: *Dynamic Planet: Monitoring and Understanding A Dynamic Planet with Geodetic and Oceanographic Tools*, P. Tregoning & C. Rizos, (Eds.), 231-236, Springer, ISBN 978-3-540-49349-5, Berlin, Germany.

Longuevergne, L.; Scanlon, B.R. & Wilson, C.R. (2010). GRACE hydrological estimates for small basins : evaluating approaches on the High Plains aquifer, USA. *Water Resources*, Vol.46, No.11, (November 2010), pp. W11517, ISSN 0043-1397.

Luthcke, S.B.; Zwally, H.J.; Abdalati, W.; Rowlands, D.D.; Ray, R.D.; Nerem, R.S.; Lemoine, F.G.; McCarthy, J.J. & Chinn, D.S. (2006). Recent Greenland Ice Mass Loss by Drainage System from Satellite Gravity Observations. *Science*, Vol.314, No.5803, (November 2006), pp. 1286–1289, ISSN 0036-8075.

Luthcke, S.B.; Arendt, A.A.; Rowlands, D.D.; McCarthy, J.J. & Larsen, C.F. (2008). Recent glacier mass changes in the Gulf of Alaska region from GRACE mascon solutions. *Journal of Glaciology*, Vol.54, No.188, (December 2008), pp.767-777, ISSN 1727-5652.

Ngo-Duc, T.; Laval, K.; Ramillien, G.; Polcher, J. & Cazenave, A. (2007). Validation of the land water storage simulated by Organising Carbon and Hydrology in Dynamic Ecosystems (ORCHIDEE) with Gravity Recovery and Climate Experiment (GRACE) data. *Water Resources Research*, Vol.43, No.4, (April 2007), pp. W04427, ISSN 0043-1397.

Panagos, P.; Jones, A.; Bosco, C. & Senthil Kumar P.S. (in press). European digital archive on soil maps (EuDASM): preserving important soil data for public free access. *International Journal of Digital Earth*, ISSN 1753-8947.

Papa, F.; Güntner, A.; Frappart, F.; Prigent, C., & Rossow, W.B. (2008). Variations of surface water extent and water storage in large river basins: A comparison of different global data sources. *Geophysical Research Letters*, Vol.35, No.11, (June 2008), pp.L11401, ISSN 0094-8276.

Peltier, W.R. (2004). Global Glacial Isostasy and the Surface of the Ice-Age Earth: The ICE-5G (VM2) Model and GRACE. *Annual Review of Earth and Planetary Sciences*, Vol.32, (May 2004), pp. 111-149, ISSN 0084-6597.

Ramillien, G.; Cazenave, A. & Brunau, O. (2004). Global time-variations of hydrological signals from GRACE satellite gravimetry. *Geophysical Journal International*, Vol.158, No.3, (September 2004), pp. 813– 826, ISSN 0956-540X.

Ramillien, G.; Frappart, F.; Cazenave, A. & Güntner, A. (2005). Time variations of land water storage from the inversion of 2-years of GRACE geoids. *Earth and Planetary Science Letters*, Vol.235, No.1-2, (June 2005), pp. 283–301, ISSN 0012-821X.

Ramillien, G.; Lombard, A.; Cazenave, A.; Ivins, E.R.; Llubes, M.; Remy, F. & Biancale, R. (2006a). Interannual variations of the mass balance of the Antarctica and Greenland ice sheets from GRACE. *Global and Planetary Change*, Vol.53, No.3, (September 2006), pp. 198–208, ISSN 0921-8181.

Ramillien, G.; Frappart, F.; Güntner, A.; Ngo-Duc, T.; Cazenave, A. & Laval, K. (2006b). Time-variations of the regional evapotranspiration rate from Gravimetry Recovery And Climate Experiment (GRACE) satellite gravimetry. *Water Resources Research*, Vol.42, No.10, (October 2006), pp. W10403, ISSN 0043-1397.

Ramillien, G.; Bouhours, S.; Lombard, A.; Cazenave, A.; Flechtner, F. & Schmidt, R. (2008). Land water storage contribution to sea level from GRACE geoid data over 2003-2006. *Global and Planetary Change*, Vol.60, No.3-4, (February 2008), pp. 381-392, ISSN 0921-8181.

Ramillien, G.; Biancale, R.; Gratton, S.; Vasseur, X. & Bourgogne, S. (2011). GRACE-derived surface water mass anomalies by energy integral approach: application to continental hydrology. *Journal of Geodesy*, Vol.85, No.6, (June 2011), pp. 313-328, ISSN 0949-7714.

Ramillien, G.; Seoane, L.; Frappart, F.; Biancale, R.; Gratton, S.; Vasseur, X. & Bourgogne, S. (in press). Constrained regional recovery of continental water mass time-variations from GRACE-based geopotential anomalies. *Surveys in Geophysics* ", ISSN 1573-0956.

Ray, R.D. & Lutcke, S.B. (2006). Tide model errors and GRACE gravimetry: towards a more realistic assessment. *Geophysical Journal International*, Vol.167, No.8, (December 2006), pp.1055-1059, ISSN 0956-540X.

Rignot, E.; Velicogna, I.; van den Broeke, M.R.; Monaghan, A. & Lenaerts, J. (2011). Acceleration of the contribution of the Greenland and Antarctic ice sheets to sea level rise. *Geophysical Research Letters*, Vol.38, No.5, (March 2011), pp. L05503, ISSN 0094-8276.

Rodell, M. & Famiglietti, J.S. (1999). Detectability of variations in continental water storage from satellite observations of the time dependent gravity field. *Water Resources Research*, Vol.35, No.9, (September 1999), pp. 2705– 2723, ISSN 0043-1397.

Rodell, M. & Famiglietti, J.S. (2001). An analysis of terrestrial water storage variations in Illinois with implications for GRACE. *Water Resources Research*, Vol.37, No.5, (May 2001), pp. 1327– 1339, ISSN 0043-1397.

Rodell, M. & Famiglietti, J.S. (2002). The potential for satellite-based monitoring of groundwater storage changes using GRACE: The High Plains aquifer, central US. *Journal of Hydrology*, Vol.263, No.1-4, (June 2002), pp. 245–256, ISSN 0022 1694.

Rodell, M.; Houser, P.R.; Jambor, U.; Gottschalck, J.; Mitchell, K.; Meng, C.-J.; Arsenault, K.; Cosgrove, B.; Radakovich, J.; Bosilovich, M.; Entin, J.K.; Walker, J.P.; Lohmann, D. & Toll, D. (2004a). The Global Land Data Assimilation System. *Bulletin of the American Meteorological Society*, Vol.85, No.3, (March 2004), pp. 381–394, ISSN 1520-0477.

Rodell, M.; Famiglietti, J.S.; Chen, J.; Seneviratne, S.I.; Viterbo, P.; Holl, S. & Wilson, C.R. (2004b). Basin scale estimates of evapotranspiration using GRACE and other observations. *Geophysical Research Letters*, Vol.31, No. 20, (October 2004), pp. L20504, ISSN 0094-8276.

Rodell, M.; Chen, J.; Kato, H.; Famiglietti, J.S.; Nigro, J. & Wilson C.R. (2007). Estimating groundwater changes in the Mississippi River basin using GRACE. Hydrogeology Journal, Vol.15, No.1, (February 2007), pp. 159-166, ISSN 1431-2174.

Roddell, M.; Velicogna, I. & Famiglietti, J. (2009). Satellite-based estimates of groundwater depletion in India. *Nature*, Vol.460, No.7257, (August 2009), 999-1003, ISSN 0028-0836.

Rowlands, D.D.; Luthcke, S.B.; Klosko, S.M.; Lemoine, F.G.R.; Chinn, D.S.; McCarthy, J.J.; Cox, C.M. & Andersen, O.B. (2005). Resolving mass flux at high spatial and temporal resolution using GRACE intersatellite measurements. *Geophysical Research Letters*, Vol.32, No.4, (February 2005), pp. L04310, ISSN 0094-8276.

Rowlands, D.D.; Luthcke, S.B.; McCarthy, J.J.; Klosko, S.M.; Chinn, D.S.; Lemoine, F.G.; Boy, J.P. & Sabaka, T.J. (2010). Global mass flux solutions from GRACE: a comparison of parameter estimation strategies – Mass concentrations versus Stokes coefficients. *Journal of Geophysical Research*, Vol.115, No.B1, (January 2010), pp. B01403, ISSN 0148-0227.

Sasgen, I.; Martinec, Z. & Fleming, K. (2006). Wiener optimal filtering of GRACE data. *Studia Geophysica et Geodaetica*, Vol.50, No.4, (October 2006), 499–508, ISSN 1573-1628.

Schmidt,R.; Schwintzer, P.; Flechtner, F.; Reigber, C.; Güntner, A.; Döll, P.; Ramillien, G.; Cazenave, A.; Petrovic, S.; Jochmann, H. & Wünsch, J. (2006). GRACE observations of changes in continental water storage. *Global and Planetary Change*, Vol.50, No.1-2, (February 2006), pp. 112–126, ISSN 0921-8181.

Schmidt, R.; Flechtner, F.; Meyer, U.; Neumayer, K. -H.; Dahle, Ch.; Koenig, R. & Kusche, J. (2008). Hydrological signals observed by the GRACE satellites. *Surveys in Geophysics*, Vol.29, No.4-5, (October 2008), pp. 319–334, ISSN 1573-0956.

Schrama, E. J. O.; Wouters, B. & Lavallée, D.A. (2007). Signal and noise in Gravity Recovery and Climate Experiment (GRACE) observed surface mass variations. *Journal of Geophysical Research*, Vol.112, No.B8, (August 2007), B08407, ISSN 0148-0227.

Seo, K.W. & Wilson, C.R. (2005). Simulated estimation of hydrological loads from GRACE. *Journal of Geodesy*, Vol.78, No.7–8, (March 2005), pp. 442–456, ISSN 0949-7714.

Seo, K.W.; Wilson, C.R.; Chen, J. & Waliser D. (2008). GRACE's spatial errors. *Geophysical Journal International*, Vol.172, No.3, (January 2008), pp.41-48, ISSN 0956-540X.

Strassberg, G.; Scanlon, B.R. & Rodell, M. (2007). Comparison of seasonal terrestrial water storage variations from GRACE with groundwater-level measurements from the

High Plains Aquifer (USA). *Geophysical Research Letters*, Vol.34, No. 14, (July 2007), pp. L14402, ISSN 0094-8276.

Sun, A.Y.; Green, R.; Rodell, M. & Swenson S. (2010). Inferring aquifer storage parameters using satellite and in situ measurements: Estimation under uncertainty. *Geophysical Research Letters*, Vol.37, No. 10, (May 2010), pp. L10401, ISSN 0094-8276.

Swenson, S. & Wahr, J. (2002). Methods for inferring regional surface-mass anomalies from Gravity Recovery and Climate Experiment (GRACE) measurements of time-variable gravity. *Journal of Geophysical Research*, Vol.107, No.B9, (September 2002), pp. 2193, ISSN 0148-0227.

Swenson, S. & Wahr, J. (2003). Monitoring changes in continental water storage with GRACE. *Space Science Reviews*, Vol.108, No.1–2, (January 2003), pp. 345–354, ISSN 0038-6308.

Swenson, S.; Wahr, J. & Milly, P. C. D. (2003). Estimated accuracies of regional water storage variations inferred from the Gravity Recovery and Climate Experiment (GRACE). *Water Resources Research*, Vol.39, No.8, (August 2003), pp. 1223, ISSN 0043-1397.

Swenson, S. & Wahr, J. (2006a). Post-processing removal of correlated errors in GRACE data. *Geophysical Research Letters*, Vol.33, No.8, (April 2006), pp. L08402, ISSN 0094-8276.

Swenson, S. & Wahr, J. (2006b). Estimating Large-Scale Precipitation Minus Evapotranspiration from GRACE Satellite Gravity Measurements. *Journal of Hydrometeorology*, Vol.7, No.2, (April 2006), pp. 252-270, ISSN 1525-755X.

Swenson, S., Famiglietti, J.; Basara, J. & Wahr, J. (2008). Estimating profile soil moisture and groundwater variations using GRACE and Oklahoma Mesonet soil moisture data. *Water Resources Research*, Vol.44, No.1, (January 2008), pp. W01413, ISSN 0043-1397.

Syed, T.H.; Famiglietti, J.S.; Chen, J.; Rodell, M.; Seneviratne, S.I.; Viterbo, P. & Wilson, C.R. (2005). Total Basin Discharge for the Amazon and Mississippi River Basins from GRACE and a Land-Atmosphere Water Balance. *Geophysical Research Letters*, Vol.32, No.24, (December 2005), pp. L24404, ISSN 0094-8276.

Syed, T.H.; Famiglietti, J.S.; Rodell, M.; Chen, J. & Wilson, C.R. (2008). Analysis of terrestrial water storage changes from GRACE and GLDAS. *Water Resources Research*, Vol.44, No.2, (February 2008), pp. W02433, ISSN 0043-1397.

Syed, T.H.; Famiglietti, J.S. & Chambers, D.P. (2009). GRACE-Based Estimates of Terrestrial Freshwater Discharge from Basin to Continental Scales. *Journal of Hydrometeorology*, Vol.10, No.1, (February 2009), pp. 22–40, ISSN 1525-755X.

Tapley, B.D.; Bettadpur, S.; Watkins, M. & Reigber, C. (2004a). The Gravity Recovery and Climate Experiment : Mission overview and Early results. *Geophysical Research Letters*, Vol.31, No.9, (May 2004), pp. L09607, ISSN 0094-8276.

Tapley, B. D.; Bettadpur, S.; Ries, J.C.; Thompson, P.F. & Watkins, M.M. (2004b). GRACE measurements of mass variability in the Earth system. *Science*, Vol.305, No.5683, (July 2004), pp. 503– 505, ISSN 0036-8075.

Tarantola, A. (1987). *Inverse Problem Theory: Methods for data fitting and model parameter estimation*, Elsevier Science Publishers B.V., ISBN 0-444-42765-1, Amsterdam, The Netherlands.

Thompson, P.F.; Bettadpur, S.V. & Tapley, B.D. (2004). Impact of short period, non-tidal, temporal mass variability on GRACE gravity estimates. *Geophysical Research Letters*, Vol.31, No.6, (March 2004), pp. L06619, ISSN 0094-8276.

Velicogna, I. & Wahr, J. (2002). A method for separating Antarctic postglacial rebound and ice mass balance using future ICESat Geoscience Laser Altimeter System, Gravity

Recovery and Climate Experiment, and GPS satellite data. *Journal of Geophysical Research*, Vol.107, No.B10, (October 2002), pp. ETG 20-1, ISSN 0148-0227.

Velicogna, I. & Wahr, J. (2005). Greenland mass balance from GRACE. *Geophysical Research Letters*, Vol.32, No.18, (September 2005), pp. L18505, ISSN 0094-8276.

Velicogna, I. & Wahr, J. (2006a). Measurements of Time-Variable Gravity Show Mass Loss in Antarctica. *Science*, Vol.311, No.5768, (March 2006), pp. 1754–1756, ISSN 0036-8075.

Velicogna, I. & Wahr, J. (2006b). Acceleration of Greenland Ice Mass Loss in Spring 2004. *Nature*, Vol.443, No.7109, (September 2006), pp. 329–331, ISSN 0028-0836.

Velicogna, I. (2009). Increasing rates of ice mass loss from the Greenland and Antarctic ice sheets revealed by GRACE. *Geophysical Research Letters*, Vol.36, No.19, (October 2009), pp. L19503, ISSN 0094-8276.

Wahr, J.; Molenaar, M. & Bryan F. (1998). Time variability of the Earth's gravity field: hydrological and oceanic effects and their possible detection using GRACE. *Journal of Geophysical Research*, Vol.103, No.B12, (December 1998), pp. 30205– 30229, ISSN 0148-0227.

Wahr, J.; Swenson, S.; Zlotnicki, V. & Velicogna, I. (2004). Time variable gravity from GRACE: First results. *Geophysical Research Letters*, Vol.31, No.11, (June 2004), pp. L11501, ISSN 0094-8276.

Werth, S.; Güntner, A.; Schmidt, R. & Kusche, J. (2009). Evaluation of GRACE filter tools from a hydrological perspective. *Geophysical Journal International*, Vol.179, No.3, (December 2009), pp. 1499-1515, ISSN 0956-540X.

Wolf, M. (1969). Direct measurement of the Earth's gravitational potential using a satellite pair. *Journal of Geophysical Research*, Vol.74, No.22, (November 1969), pp. 5295-5300, ISSN 0148-0227.

Wouters, B. & Schrama, E.O.J. (2007). Improved accuracy of GRACE gravity solutions through empirical orthogonal function filtering of spherical harmonics, *Geophysical Research Letters*, Vol.34, No.23, (December 2007), pp. L23711, ISSN 0094-8276.

Wouters, B.; Chambers, D. & Schrama, E.J.O. (2008). GRACE observes small-scale mass loss in Greenland. *Geophysical Research Letters*, Vol.35, No.20, (October 2008), pp. L20501, ISSN 0094-8276.

Wouters, B.; Riva, R. E. M.; Lavallée, D. A. & Bamber J.L. (2011). Seasonal variations in sea level induced by continental water mass: First results from GRACE. *Geophysical Research Letters*, Vol.38, No.3, (February 2011), pp. L03303, ISSN 0094-8276.

Xavier, L.; Becker, M.; Cazenave, A.; Longuevergne, L.; Llovel, W. & Rotunno Filho, O. C. (2010). Interannual variability in water storage over 2003–2008 in the Amazon Basin from GRACE space gravimetry, in situ river level and precipitation data. *Remote Sensing of Environment*, Vol.114, No.8, (August 2010), pp. 1629–1637, ISSN 0034-4257.

Yeh, P.J.F.; Swenson, S.C.; Famiglietti, J.S. & Rodell, M. (2006). Remote sensing of groundwater storage changes in Illinois using the Gravity Recovery and Climate Experiment (GRACE). *Water Resources Research*, Vol.42, No.12, (December 2006), pp. W12203, ISSN 0043-1397.

Zaitchik, B.F.; Rodell, M. & Reichle, R.H. (2008). Assimilation of GRACE Terrestrial Water Storage Data into a Land Surface Model: Results for the Mississippi River Basin. *Journal of Hydrometeorology*, Vol.9, No.3, (June 2008), pp. 535-548, ISSN 1525-755X.

Part 2

Groundwater Modeling

10

Integration of Groundwater Flow Modeling and GIS

Arshad Ashraf[1] and Zulfiqar Ahmad[2]
[1]National Agricultural Research Center, Islamabad
[2]Department of Earth Sciences, Quaid-i-Azam University, Islamabad
Pakistan

1. Introduction

The development of a sufficient understanding on which to base decisions or make predictions often requires consideration of a multitude of data of different types and with varying levels of uncertainty. The data for the development of numerical groundwater flow model includes time-constant parameters and time-variant parameters. The time-constant parameters were mainly extracted from thematic data layers generated from GIS and image processing of remote sensing data. Integrated approaches in GIS play a rapidly increasing role in the field of hydrology and water resources development. It provides suitable alternatives for efficient management of large and complex databases developed in different model environments. Remote sensing (RS) technology is capable of providing base for quantitative analysis of an environmental process with some degree of accuracy. It provides an economic and efficient tool for landcover mapping and has its advantages in planning and management of water resources (Ashraf and Ahmad 2008 & Ahmad et al., 2011). One of the greatest advantages of using remote sensing data for hydrological investigations and monitoring is its ability to generate information in spatial and temporal domain. Though, the presence of groundwater cannot be directly ascertained from RS surveys, however, satellite data (GRACE- Gravity Recovery and Climate Experiment and GOCE-Gravity field and steady-state Ocean Circulation Explorer) provides quick and useful baseline information on the parameters that control the occurrence and movement of groundwater such as geomorphology, direction of groundwater flows, lineaments, soils, landcover/landuse and hydrology etc.

Reliable investigation of waterlogging problem can be extremely useful in chalking out suitable water management strategies by reclaiming existing waterlogged areas. The problems of water logging and salinity mostly exist in the irrigated areas like in Indus plains of South Asia. This twin menace generally results from over irrigation, seepage losses through canal and distributory system, poor water management practices and inadequate provision of drainage system. Analysis of high watertable in waterlogged areas and drainage of irrigated areas have not been paid adequate attention in the planning process, partly due to lack of requisite data and partly due to resource crunch in the country. In order to develop suitable water management strategies and controlling the extent of waterlogging in the area, reliable investigation and clear apprehension of the groundwater flow system bear great significance. It is important not only to facilitate the reconstruction of the ecological environment but also to accommodate the sustainable development of the water resources and economy of this region.

According to Choubey (1996), a rapid and accurate assessment of the extent of waterlogged areas could be made using remotely sensed data. Remote sensing (RS) and geographical information system (GIS) offers convenient solutions to map the extent and severity of waterlogging and salinity, particularly in large areas (IDNP 2002). Arora and Goyal (2003) highlighted the use of GIS in development of conceptual groundwater model. Using the logical conditions and analytical functions of GIS domain, the recharge/discharge zones can be delineated effectively (Ashraf et al., 2005). Such zones often provide clues of causative factors of the waterlogging problem. It should be realized that by just using last century's schemes no longer solves challenges related to today's groundwater situation (Zaisheng et al., 2006). In recent years, groundwater numerical simulation models like Feflow which is based on finite element method have been widely applied to groundwater dynamics simulation due to its appealing features such as visualization, interaction and diversified solving methods (Yang and Radulescu, 2006; Russo and Civita, 2009; Peleg and Gvirtzman, 2010; Dafny et al., 2010).

In the present study, remote sensing and numerical groundwater flow modeling were integrated in GIS environment to identify and analyze waterlogged areas and associated groundwater behavior spatially and temporally. A rapid and accurate assessment of the extent of waterlogged areas could be made using remotely sensed and groundwater data. The GIS is used for spatial database development, integration with a remote sensing (RS) and numerical groundwater flow modeling capabilities. Remote sensing technique in conjunction with (GIS) provided a quick inventory of waterlogged area and its monitoring. Finite element groundwater flow model (Feflow 5.1) was developed to configure groundwater equipotential surface, hydraulic head gradient and estimation of the groundwater budget of the aquifer. Steady-state simulations were carried out to describe three-dimensional flow field (head distribution) over the entire model domain. The calibrated heads were used as initial conditions in the transient-state modeling. For transient and predictive modeling, management strategies for pre-stress period and post stress periods were developed on the basis of present water use and future requirements. During transient simulations, model was rerun for period of about 15 years i.e. from 2006 to 2020 to simulate groundwater flow behavior in the model domain. Scenarios of impact of extreme climate events (drought/flood condition) and variable groundwater pumpage on groundwater levels were studied. The integration of GIS with groundwater modeling and Remote sensing (RS) provided efficient way of analyzing and monitoring resource status and land conditions of waterlogged areas. The approach would help in providing an effective decision support tool for evaluating better management options to organize schemes of future monitoring of groundwater resource and waterlogging risks on local and regional basis.

1.1 GIS Functions and methods

The effort to perform analysis of management scenarios will be substantially reduced by an easily accessible database, a convenient interface between database and groundwater models, visualization and utilities for model inputs and results (Pillmann and Jaeschke,

1990). GIS is an application-oriented spatial information system with a variety of powerful functions to handle for decision support of problems related to spatial dimension. All of the data in a GIS are georeferenced, that is, linked to a specific location on the surface of the Earth through a system of coordinates. Geographical information attaches a variety of qualities and characteristics to geographical locations. These qualities may be physical parameters such as ground elevation, soil moisture or groundwater levels, or classifications according to the type of vegetation and landcover, ownership of land, zoning, and so on. Such occurrences as accidents, floods, or landslides may also be included. A general term 'attributes' is often used to refer to the qualities or characteristics of places and is considered as one of the two basic elements of geographical information, along with locations. GIS technology integrates common database operations such as query and statistical analysis with the unique visualization and benefits of geographic analysis offered by maps. GIS is the most appropriate tool for spatial data input and attribute data handling. It is a computer-based system that provides the following four sets of capabilities to handle geo-referenced data (Aronoff, 1989):

- Data input
- Data management (data storage and retrieval)
- Data manipulation and analysis
- Data output

All the geographically related information that is available can be input and prepared in GIS such that user can display the specific information of interest or combine data contained within the system to generate further information which might answer or help resolve a specific problem. The hardware and software functions of a GIS include compilation, storage, retrieval, updating and changing, manipulation and integration, analysis and presentation. All of these actions and operations are applied by a GIS to the geographical data that form its database. The data representation phase deals with putting all information together into a format that communicates the result of data analysis in the best possible way. Geographic Information System can be represented with geometric information such as location, shape and distribution, and attribute information such as characteristics and nature. Any spatial features of the earth's surface are represented in GIS by the *Polygons* (features which occupy a certain area, e.g. lakes, landuse, geological units etc.); *Lines* (linear features, e.g. drainage, contours, boundaries etc); *Points* (points define the discrete locations of geographic features like cities/towns, boreholes, wells, spot heights etc. and by *Attribute data* (properties of the spatial entities). GIS stores information about the real world as a collection of thematic layers with characteristics of common coordinate system. The spatial entities of the earth's features can be represented in digital form by two data models: vector or raster models.

1.1.1 Vector data model

The method of representing geographic features by the basic graphical elements of points, lines and polygon is said to be the *vector method* or *vector data model*. *The* related vector data are always organized by *themes*, which are also referred to as *layers* or *coverages i.e. administrative* boundaries, infrastructure, soil, vegetation cover, landuse, hydrology, land parcel and others. In a vector model the position of each spatial feature i.e. a point (or node), line (or arc) and area (or polygon), is defined by the series of x and y coordinates. The interrelationship between these features is called a topological relationship. Any change in a point, line or area will influence other factors through the topological relationship. A *label* is required so that user can load the appropriate attribute record for a given geographic feature.

1.1.2 Raster data model

The method of representing geographic features by pixels is called the *raster method* or *raster data model*. The object space is divided into a group of regularly spaced grids or pixels to which the attributes are assigned. Pixels (a term derived for a picture element) are the basic units for which information is explicitly recorded. A raster pixel represents the generalized characteristics of an area of specific size on or near the surface of the earth. The actual ground size depicted by a pixel is dependent on the resolution of the data, which may range from less than a square meter to several square kilometers. Raster data are organized by themes, which are also referred to as layers for example; a raster geographic database may contain the following themes: bed rock geology, vegetation cover, landuse, topography, hydrology, rainfall, temperature. Image data utilizes techniques very similar to raster data, however typically lacks the internal formats required for analysis and modeling of the data. The raster data model has advantages of being simple in structure, provision of easy and efficient overlaying and compatibility with RS image data while the vector data model has compact data structure, provision of efficient network analysis and projection transformation, and can generate accurate map output. The usage of the two models depends on the objectives and planning needs of the GIS project.

The most popular form of primary raster data capture is remote sensing - a technique used to derive information about the physical, chemical, and biological properties of objects without direct physical contact. Information is derived from measurements of the amount of electromagnetic radiation reflected, emitted, or scattered from objects. A variety of sensors, operating throughout the electromagnetic spectrum from visible to microwave wavelengths, are commonly employed to obtain measurements (Lillesand and Kiefer, 2004). Remote sensing provides information of natural indicators and landuse features that can be related with the characteristics of sub-surface aquifer system (recharge and discharge sources) and presence of soil moisture/shallow groundwater condition.

By integrating remote sensing with GIS, an even greater potential for environmental applications is achieved (Ehlers et al., 1989; Shelton and Estes 1980). The RS data has its advantages of spatial, spectral and temporal availability of data covering large and inaccessible areas within short time and thus proves to be an effective tool in assessing, monitoring surface and groundwater resources. The conventional techniques have the limitation to study these parameters together because of the non-availability of data, integration tools and modeling techniques. RS data provides an economic and efficient tool for landuse mapping and has its advantages in the development and management of water resources for optimum use.

1.2 Numerical groundwater flow methods

The range of numerical methods is quite large, obviously being of use to most fields of engineering and science in general. There are two broad categories of numerical methods i.e.

gridded or discretized methods and non-gridded or mesh-free methods. In the common finite difference method and finite element method (FEM) the domain is completely gridded (form a grid or mesh of small elements). The analytic element method (AEM) and the boundary integral equation method (BIEM - sometimes also called BEM, or Boundary Element Method) are only discretized at boundaries or along flow elements (line sinks, area sources, etc.). Gridded Methods like finite difference and finite element methods solve the groundwater flow equation by breaking the problem area (domain) into many small elements (squares, rectangles, blocks, triangles, tetrahedra, etc.) and solving the flow equation for each element (all material properties are assumed constant or possibly linearly variable within an element), then linking together all the elements using conservation of mass across the boundaries between the elements (http://en.wikipedia.org/wiki/Hydrogeology). This results in a system which overall approximates the groundwater flow equation i.e. approximating the Laplace equations for flow in a porous media and the partial differential equations governing solute transport.

Finite differences are a way of representing continuous differential operators using discrete intervals (Δx and Δt). MODFLOW is a well-known example of a general finite difference groundwater flow model. It is one of the most widely used and tested software program developed by U.S. Geological Survey for simulating groundwater flow. It is a three-dimensional modular finite difference model that uses variable grid spacing to model groundwater flow. Many commercial products have grown up around it, providing graphical user interfaces to its input file based interface, and typically incorporating pre- and post-processing of user data. Many other models have been developed to work with MODFLOW input and output, making linked models which simulate several hydrologic processes possible (flow and transport models, surface water and groundwater models and chemical reaction models), because of the simple, well documented nature of MODFLOW. The model has been adopted by the GMS (Groundwater Modeling System) and 'Visual MODFLOW' Graphical User Interfaces. GMS has the ability to import borehole data and create a 3D visualization of the geology, the ability to import GIS files directly into the conceptual model using the 'map' module. Visual MODFLOW is also an user friendly software that has ability to generate 3D visualization graphics and import GIS data.

Finite Element programs are more flexible in design (triangular elements vs. the block elements most finite difference models use) and there are some programs available (SUTRA, a 2D or 3D density-dependent flow model by the USGS; Hydrus, a commercial unsaturated flow model; FEFLOW, a commercial modeling environment for subsurface flow, solute and heat transport processes; and COMSOL Multiphysics (FEMLAB) a commercial general modeling environment). Regional model are used for large scale systems e.g. the P.R.A.M.S. (Perth Regional Aquifer Modeling System) model of the Swan coastal plain aquifer system. Packages need to be selected on the basis of suitability for the intended modeling purpose (e.g. flow modeling, transport modeling, salt water interface modeling etc). Once the model is chosen and the preparation of the input dataset is achieved, one can then proceed for modeling the groundwater behavior that represent the user's requirement. The modeling process comprises of steady-state simulation and transient simulation of the prolific groundwater system, and predictive simulation of groundwater flow/solute transport to study different scenarios.

1.3 GIS and groundwater modeling

As the geographical location of every item of information stored in a GIS is known, GIS technology makes it possible to relate the quality of groundwater at a site with the health of its inhabitants, to predict how the vegetation in an area will change as the irrigation facilities increases, or to compare development proposals with restrictions on land use. This ability to overlay gives GIS unique power to make decisions about places and to predict the outcomes of those decisions. The ability to combine and integrate data is the backbone of GIS. *Spatial modeling* technique in GIS can answer locational and quantitative questions involving, *e.g.,* the particulars of a given location, the distribution of selected phenomena, the changes that have occurred since a previous analysis, the impact of a specific event in the form of *"Where?* or *How much?* and/or *What if ?"* scenarios-making in a more realistic way. Some of the advantages of GIS application in groundwater modeling include the following:

- GIS provides decision support for groundwater management i.e. groundwater pumping for domestic, industrial and agricultural supplies and other actions influencing the regional water cycle: infrastructure, tunnels, waste dump-sites, sewerage systems etc.
- GIS saves much time of collecting large number of geographical data required for groundwater modeling for both pre-processing and post-processing stages and to improve the model results.
- GIS can handle large datasets through integration with Database Management System (DBMS) component which provides foundation for all analysis techniques.
- With GIS, complex maps can be created and edited much faster that would not be possible by hand, and because the data is stored digitally, the maps are produced with the same level of accuracy each time.
- GIS can utilize a satellite image to extract useful information i.e. landuse, surface hydrology, soil properties etc. which serves as source data for model conceptualization.
- GIS has capability to integrate with many hydrological models and techniques, and transform spatial data according to the modeling requirements.

Application of GIS technology to hydrological modeling requires careful planning and extensive data manipulation work. In general, the following three major steps are required:

- Development of spatial database
- Extraction of model layers
- Linkage to computer models

GIS provide spatial data handling and graphical environment to analyze and visualize spatial data related to numerical groundwater flow modeling. The frequently applied GIS functions, which support groundwater modeling during its various steps, are summarized in Table 1. In fact the modeling steps are interlinked to each other and can be varied according to the planning requirements.

In preprocessing, the data is transferred from the GIS to the model while in postprocessing, the data is transferred from the model to the GIS. Developing conceptual models, choosing a computer code, and developing model designs are the preliminary steps in a groundwater modeling project after first establishing the purpose of the project (Anderson & Woessner, 1992). These steps define the numerical models representing a given field situation,

Phase	GIS functions	Modeling Steps
Pre processing *Data collection,* *Conceptualization*	Data input, Digitization, Data conversion (import/export), Coordinate transformation, Map retrieval	Collection of required GW data
	Conversion of vector and raster layers, Data integration, Image processing, buffering, Surface generation, Linking of spatial and attribute data	Model conceptualization
Model design	Map calculations, Neighborhood operations, Interpolation, Theissen polygons, buffering, Surface generation	Boundary delineation, Mesh generation, 3-dimensional layering of the aquifer
Calibration *&Verification*	Data layers integration	Parameter zonation, Recharge estimation, Water balance
	Overlay analysis	Steady-state and Transient-state simulations
	Statistical analysis	Parameters estimation
Post processing *Predictions, Data* *presentation*	Data retrieval	Prediction, Scenarios
	Data visualization, Presentation of simulated results	Map composition

Table 1. GIS functions involved in different phases of groundwater flow modeling

including: (a) the most important sources and sinks of water in the field system and how they are to be simulated; (b) the available data on the geohydrologic system; (c) the system geometry (generally the number and type of model layers and the areal extent of these layers); (d) the spatial and temporal structure of the hydraulic properties (generally using zones of constant value or deterministic or stochastic interpolation methods); and (e) boundary condition location and type. The groundwater models output may include: water pressure (head) distribution; flow rates, flow directions; plume movement and particle tracking; water chemistry changes and budgeting. Integration of GIS and hydrologic models follows one of the two approaches; a) To develop hydrologic models that operate within a GIS framework, b) To develop GIS techniques that partially define the parameters of existing hydrologic models (Jain et. al., 1997). In most of groundwater modeling softwares such as Feflow, Modflow, GMS there is an interface that links vector data through compatible GIS formats i.e. *.shp, .lin, .dxf* etc. and raster data formats; *.tif, .bmp, .img* etc.

The GIS data assists in identifying the spatial variability of hydraulic conductivity and recharge, the values of which are estimated under steady-state condition. The effect and sensitivity of different parameters values are tested during the modeling process through calibration technique using packages like PEST and UCODE. Calibration optimizes the input parameter values against the historical water data. The simulation output of the models can be integrated with any thematic data in GIS for verification, impact and/or characterization analysis. There are inbuilt presentation tools in the models that can be used to create labeled contour maps of input data and simulate results. One can fill colors to model cells containing different values and report-quality graphics may be saved to a wide variety of file formats i.e. *surfer*, *dxf*, *hpgl* and *bmp*. The presentation tools can even create and display two dimensional animation sequences using the simulation results (calculated heads, drawdowns or concentration).

2. Spatial analysis of waterlogged areas

The study area lies between longitudinal range 73°-74° 5'E and latitudinal range 32°- 32° 45'N in Indus basin, Pakistan (Fig. 1). It covers an area of about 3,417 sq. km within Rivers Jhelum and Chenab flowing in the northwest and southeast, and Upper Jhelum Canal (UJC) and Lower Jhelum Canal (LJC) in the northeast and southwest. The area is well connected with other main cities of the country through metalled roads and rail links. There lie three schemes of Salinity Control and Reclamation Project SCARP-II of Water and Power Development Authority (WAPDA) i.e. Sohawa, Phalia and Busal in the area. The relief increases northward with elevation ranges between 200 and 238 m above mean sea level. The climate of the area is mainly semi arid. May, June and July are the hottest months and the mean maximum and minimum temperature during this period are about 39.5°C and 25.4°C respectively (Qureshi et al. 2003). The coldest months are December, January and February and the maximum and minimum temperatures during this period are about 21.5°C and 5.1°C respectively. The rains are erratic and are received in two rainy seasons i.e. about two third of the total rain is received during monsoon season (mid June to mid September) while remaining one third, during the period from January to March.

Major landforms of the area include piedmont plain and basin, flood plains, scalloped interfluves and channel levee remnants (Fig. 1). Waterlogging is a problem in the central parts of the scalloped interfluves and in some parts of the channel-levee remnants. The meander flood plain has nearly level surface and contains local depressions which causes drainage problem. The Piedmont basin contains clayey soils characterized by swelling and shrinking. During the rainy season runoff water from the adjoining land collects in the area, making these soils waterlogged (Soil Survey Report 1967). The soils of the major part of study area are formed from alluvial sediments brought by the rivers from the Himalayan mountain ranges in the north. The alluvial complex forms a unified, highly permeable aquifer, in which water is generally unconfined. The groundwater flows in general from northeast to southwest of study area with the hydraulic gradient ranging from 2.14 to 7.15 ft/mile (PPSGDP 2000). Drainage pattern is mainly dendritic in nature. The waterlogging has affected the main irrigated area of the Indus basin (Fig. 2). The main factors involved in causing the problem of waterlogging and salinity in the area include the following:

- Micro relief in the land surface resulting in appearance of waterlogging in patches, which act as sinks

- Shallow watertable resulting from lenses of clay material present at shallow depth creating perched watertable conditions
- High seepage rate from canal system resulting from non-lining of irrigation channels
- Lack of water management practices/absence of complementary drainage system
- High precipitation areas
- Lower evaporation than recharge resulting in appearance of swamps

Major contribution to the recharge of the groundwater is from seepage of the rivers, canal irrigation network including main and link canals, distributaries, minors and watercourses/fields, precipitation and return flows of the groundwater use, besides subsurface inflows from nearby zones. The presence of large canal network in the area provides the main source of recharge to the groundwater. Aquifer discharge sources are pumpage from public and private tubewells (water wells), evapotranspiration, outflows to the rivers and drains. The groundwater is mainly abstracted to augment where or when the canal supplies are not adequate especially in the Rabi (wet) season.

The wastelands comprising waterlogged areas, swamps and marshy land along Jhelum and Chenab riverbeds demarcated through RS analysis were related with associated factors like landforms, canal system, topography, climate and groundwater etc.

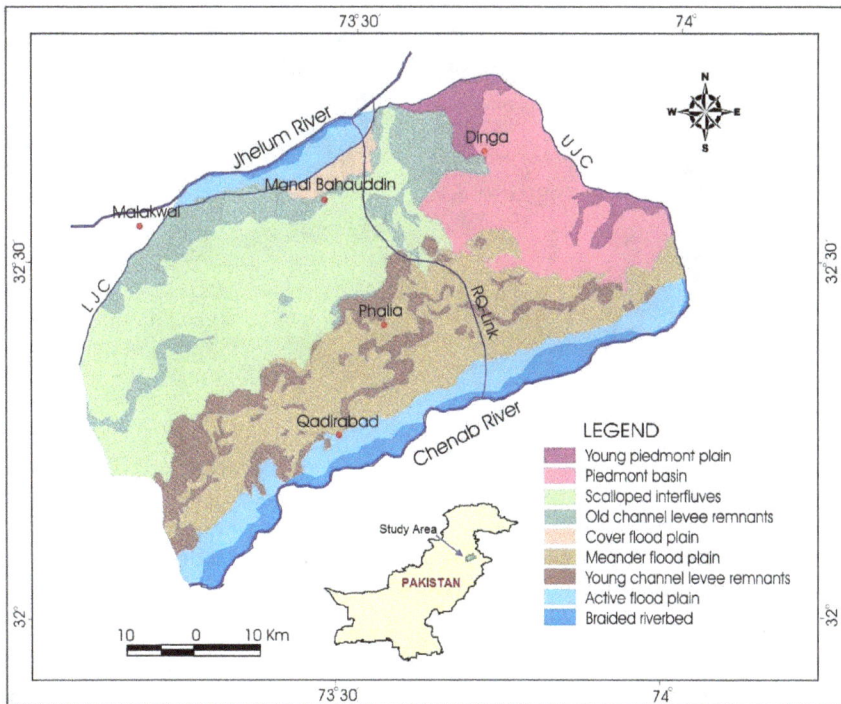

Fig. 1. Location and landforms extent in the study area

The following thematic layers were developed through on-screen digitizing in ILWIS 3.1 GIS software:

- Geomorphology
- Soils
- Infrastructure
- Surface hydrology
- Cities and towns
- Administrative units

The attribute database of each layer was developed and linked to its respective data map for GIS analysis. The spatial functions of GIS were used to derive the following thematic maps for spatial analysis;

- Surface elevation
- Slope
- Buffer zones
- Theissen Polygons
- Watertable depth
- Recharge zones
- Landcover/Landuse

Fig. 2. Effect of waterlogging on irrigated area

The landcover/landuse map was developed through image processing of remote sensing data using ERDAS imagine 8.5 software. These maps including the landcover map have been registered together in GIS environment for integration and analysis (Fig. 3). The ancillary data consists of topographic maps of scale 1:50,000, geomorphology & soil maps, hydrogeology and groundwater levels of observation wells and climatic data.

The landcover map in raster form was converted into vector form and a layer of wasteland polygons was developed which comprises of 585 polygons. Major waterlogged areas were selected by eliminating smaller polygons having an area less than 0.02 km² leaving total of 361 polygons which covered an aggregate area of about 71 sq. km. The polygon data was analyzed with different thematic data through spatial modeling in GIS. Most of the waterlogged areas (about 28%) were found in the active flood plain followed by about 17% and 13% in the scalloped interfluves and young channel levee remnants. The low-lying

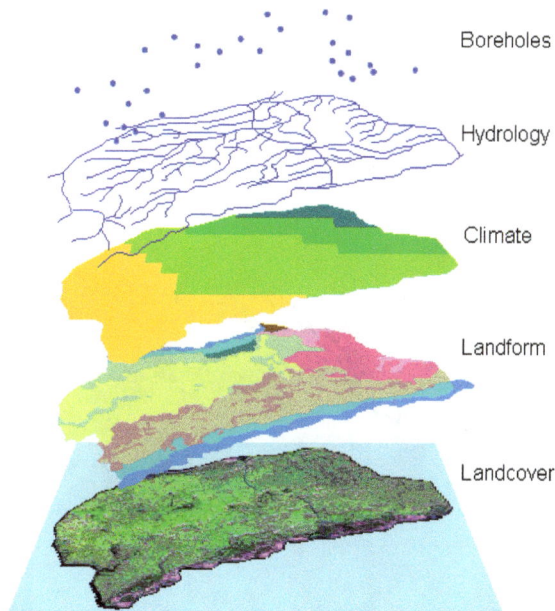

Fig. 3. GIS has capability to integrate different types of spatial data

areas i.e. parts of the piedmont basin and central part of the scalloped interfluves have especially become victims of waterlogging and salinity (Soil Survey Report 1967). The meander flood plain constitutes about 12% of waterlogged areas. About 11% of waterlogged areas occur in the Piedmont basin which is actually the lowest end of the piedmont plain where, due to absence of slope, basin conditions have developed. The braided riverbed constitutes about 9 km^2 of waterlogged areas mostly comprising marshy areas stretched along the Chenab River and to some extent along the Jhelum River. The buffer zones (polygons of 500m interval) were created around the main canal system to analyze the influence of seepage on the waterlogged areas. About 21 sq. km of waterlogged area lie within the buffer zone of 0-500m while only 4.4 sq. km within zone of 1000-1500m indicating a decrease in coverage with the increase in distance from the canal network (Fig. 4). In fact the canal irrigation is a major source of recharge to the groundwater in this area. The watertable beneath the waterlogged areas varies temporally and spatially. During the year 1990, about 4% waterlogged areas lies in zone <1.5m depth whereas about 28% in 1.5-3.0m depth. The watertable zone of 3-6m depth stretched over about 65% area, comprising 63% of the waterlogging polygons. This shows a local effect of the topography and lithological conditions in developing of the waterlogged areas. Analysis of the waterlogged areas with mean annual rainfall indicated maximum of 154 polygons under rainfall range of 500- 600 mm followed by 127 polygons in less than 500mm rainfall range.

The development of a sufficient understanding on which to base decisions or make predictions often requires consideration of a multitude of data of different types and with varying levels of uncertainty (Middlemis 2000). The data for the development of numerical groundwater flow model included the following parameters:

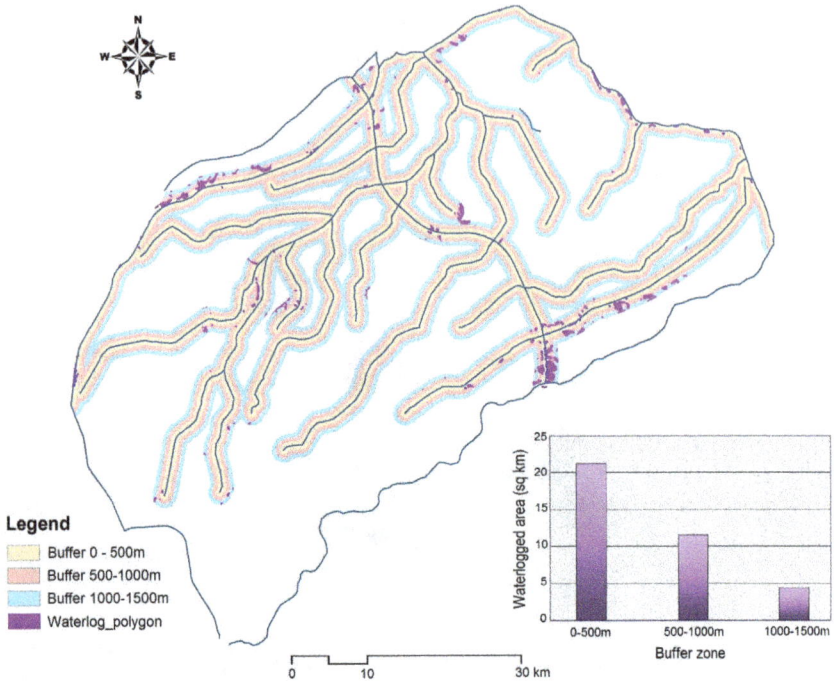

Fig. 4. Spatial analysis of the waterlogged areas using buffering technique of GIS

a.   Time Constant Parameters

- Aquifer geometry (areal and vertical distribution of subsurface strata, aquifer thickness etc.). It also requires river network, landcover, soil, surface and subsurface hydrological data input.
- Hydraulic parameters

b.   Time Variant Parameters

- Hydro-meteorological data
- Water level monitoring data
- Number, distribution and pumpage from irrigation/drainage tubewells
- River and canal flows and hydraulic features

The time-constant parameters were mainly extracted from thematic data layers generated through GIS and image processing of remote sensing data. The remote sensing data of LANDSAT TM & ETM sensors was used of periods 1990 and 2001. The general approach involves processing of RS image data for analysis of different landcovers and features present in the study area and integration of results in GIS system along with results generated through numerical groundwater flow modeling. The image data was geometrically rectified and analyzed through visual and digital interpretation for the study of landcover/landuse types. The former interpretation was used for qualitative analysis while the latter for quantitative analysis of the landcover/landuse in the RS image data.

Initially, unsupervised classification method was used to analyze the statistical patterns of various clusters in the image. Later, supervised method of classification was used which is based on background knowledge and experience of the interpreter, available ground information and ancillary data. The surface hydrology including canal irrigation network, streams and drains were digitized partially from RS image data and topographic maps. The statistics of this hydrological network is shown in Table 2. Main canals i.e. UJC and LJC extends upto 62 km and the distributaries mainly of UJ Canal extends upto 398 km in the study area. There are several large streams (about 96 km in length) that originate from the Himalayan Mountains in the north and drain into the Chenab River in the Southeast. The analytical functions of ILWIS, Arcview and ArcGIS softwares were mainly used for digitization, analysis and visualization of the spatial data in the present study.

Channel_id	Type	Length (km)
1	River	126.6
2	Link canal	115.9
3	Main canal	62.0
4	Branch	63.3
5	Distributary	397.9
6	Minor	127.0
7	Sub minor	9.8
8	Drain	109.0
9	Stream	96.2
	Total	1107.5

Table 2. Statistical analysis of hydrological network in the study area

Through visual interpretation of the image, distribution of landcover and features like surface drainage and irrigation network, water bodies, waterlogged areas and soil moisture condition were analyzed. Landforms and structures that can reveal subsurface condition of groundwater were also studied. The false color composite (FCC) of Landsat bands 7, 4, 2 (Red, Green, Blue) showed waterlogged areas in true colors i.e. dark bluish-green spots. The dark appearance of these areas is due to low reflectivity of high moisture contents of standing water consisting of evergreen natural vegetation grown in waterlogged areas. Waterlogged areas are differentiated from fresh water in bands 4 and 5 due to presence of natural vegetation in it. Moist soil can be differentiated from the swamps and waterlogged area in bands 3, 5 and 7, which are indicating higher reflection of moist soil in these bands. The image classification of the RS data indicated about 70% of the land under various types of crops followed by 8% grassland and 6% forest cover. About 10% land is bare soil and 4% is waterlogged and swampy area.

2.1 Development of numerical groundwater flow model

Numerical modeling employs approximate methods to solve the partial differential equation (PDE) which describes the flow in porous medium. The emphasis here is not on obtaining an exact solution but on obtaining reasonably approximate solution (Thangarajan,

2004). Feflow is a fully integrated modeling environment with a full-featured graphical interface and powerful numeric engines that allow the user to perform any flow or contaminant transport modeling. The components ensure an efficient process for building the finite element model, running the simulation, and visualizing the results. The conceptual model of the study area was developed for carrying out finite element groundwater flow modeling. Based on the hydrogeological conceptual model of the study area, the mathematical model was established as follows:

$$\frac{\partial}{\partial x}\left(Kh\frac{\partial H}{\partial x}\right) + \frac{\partial}{\partial y}\left(Kh\frac{\partial H}{\partial y}\right) + \frac{\partial}{\partial z}\left(Kh\frac{\partial H}{\partial z}\right) + \varepsilon = \mu\frac{\partial H}{\partial t} \tag{1}$$

Initial conditions:

$$H(x,y,z,t)\big|_{t=0} = H_0(x,y,z) \ \ (x,y,z) \in D \tag{2}$$

Boundary conditions:

$$K\frac{\partial H}{\partial n}\bigg|_{\Gamma_2} = q(x,y,z,t) \ \ (x,y,z) \in \Gamma_2 \ \ t \geq 0 \tag{3}$$

Where K is permeability coefficient (or hydraulic conductivity coefficient) (m/d) (Due to the lack of experimental data, the aquifer is taken as isotropic); h is the distance from the impervious bed of the aquifer to the free water surface, i.e. aquifer thickness (m); H is water head (or water table) (m); ε is the inflow and outflow factors, i.e. the volume of water that vertically flows into or out of the aquifer in unit time and unit area with inflow being positive and outflow being negative (m³/d/m²); μ is storativity; t is time (d); H_0 is initial value of water head (m); D is study area enclosed by Γ_2, and Γ_2 is the second kind boundary; n is the direction of outer normal line of the second kind boundary; and q is the volume of water that laterally flows into or out of the aquifer in unit time and unit area on the second kind boundary (m/d).

The finite elements may have both different spatial dimensions and shapes. The order of the underlying interpolation scheme may typically be linear, quadratic or cubic. Continuity may be prescribed not only for the variable themselves but also for their derivatives. The procedures to be followed by UNESCO (1999) are given below;

i.   Discretization of the flow domain into a set of elements, where each element is defined by a number of nodes, for instance 3 or 6-node triangles, 4- 8- or 9-node quadrilaterals.
ii. Expression of field parameters such as piezometric head, hydraulic conductivity etc., in the following form:

$$h(x,y,z) = \sum_{j=1}^{N} h_i\psi_j \tag{4}$$

where 'h' is piezometric head, (x, y, z) are Cartesian coordinates, 'N' is the number of nodal points in the discretized element grid and 'Ψ' is the interpolation function or otherwise called shape function.

iii. Formulation of the groundwater flow equation (partial differential equation or PDE) in integral form.

iv. Element-wise integration of the integral form of the groundwater equation.

v. Assembly of the algebraic matrix equations that result from the integration step into global system of linear equation of the form

$$[M]\{\frac{dh}{dt}\}+\{K(h)\}[h] = [f] \tag{5}$$

Where *[M]* denotes a matrix.

vi. Time integration.

vii. Solution of the global system of linear equations.

The hydrogeological system of the site was modeled as multi-layered using Finite element Model - Feflow. A model grid consisting of superelement mesh of five elements was drawn over the model area using the background information of landuse, landforms and drainage/canal network of the area (Fig. 5). The superelement mesh represents the basic structure of the study domain. The Finite element mesh was generated from the superelement mesh using triangulation option of 6-noded prism for 3-dimensional model. The 3-D mesh consists total of 5,343 elements and 3,928 nodes (Fig. 6). The model layers were developed from point data using Akima's bivariate interpolation method. It is difficult to introduce precise distribution of recharge over the area due to complex distribution of various types of soils in the area. However, equal distribution of the recharges from various sources, on macro level, generally gives the results within practical limits (Thangarajan, 2004). A simple water balance model of an area can be carried out on the fact that there exists a balance between the quantity of water entering into the area, amount store and water leaving the same area during certain period of time. For any hydrologic system, general mass conservation equation can be written as:

$$I - O = \Delta S \tag{6}$$

Where I = Total inflows, O = Total outflows and ΔS = change in groundwater storage.

Since flow in saturated zone is simulated using groundwater flow model, the net recharge to the groundwater reservoir was computed by assuming aggregate water balance for the unsaturated zone and the land surface. The net recharge of the model area was thus estimated as below:

$$Qnet = Qmc + Qdm + Qwf + Qriv + Qrf + Qrtw - (Qet - Qtw - Qdr) \tag{7}$$

Where $Qnet$ = net recharge to the aquifer, Qmc = Recharge from main and link canals, Qdm = recharge from distributary/minors, Qwf = recharge from watercourses and irrigated fields, $Qriv$ = recharge/discharge from rivers, Qrf = recharge from rainfall, $Qrtw$ = recharge from return flow of tubewell pumpage, Qet=discharge by evapotranspiration, Qtw = discharge by tubewells, Qdr = discharge from drains. The recharge from inflows from adjacent areas was considered implicitly in the groundwater modeling (Sarwar, 1999). The losses of the canal system and other recharge sources were maintained according to the specified limits of the irrigation department and WAPDA (PPSGDP 2000 & WAPDA, 1993).

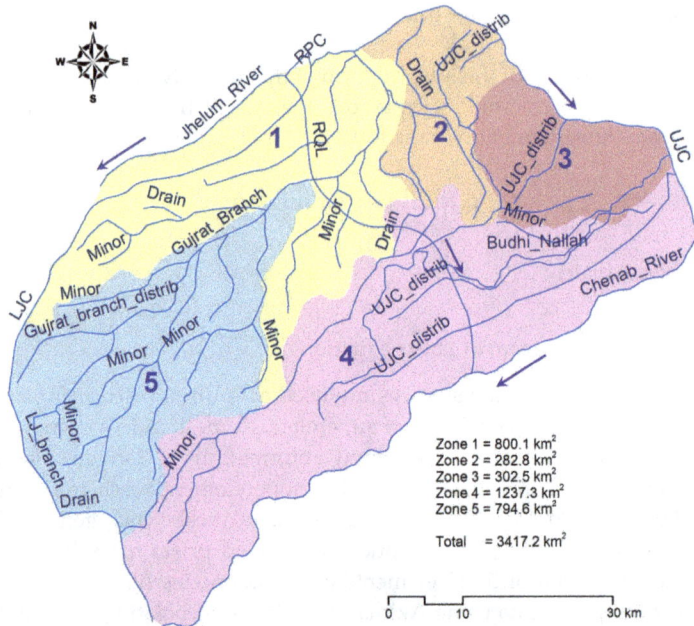

Fig. 5. Delineation of superelements and recharge zones for groundwater flow modeling in Feflow

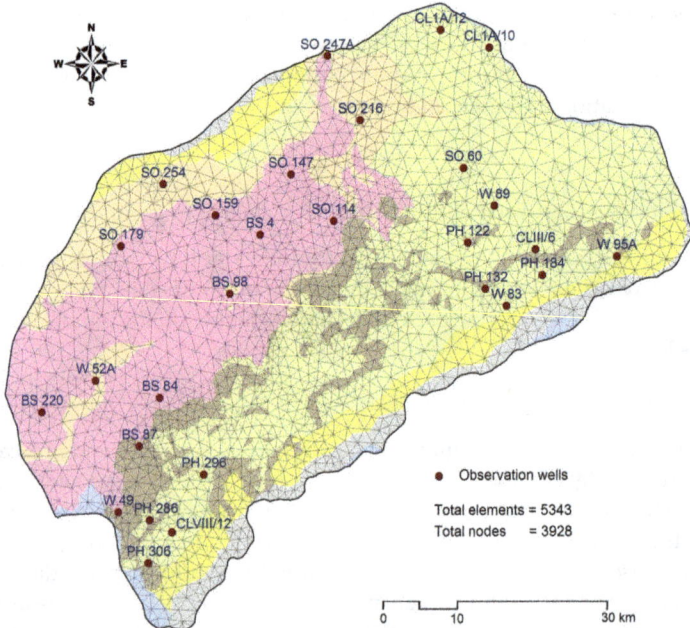

Fig. 6. Finite element mesh and point data of observation wells drapped over the soil map

2.2 Model calibration

The groundwater levels of June 1985 (pre-monsoon period) of 28 observation wells were used as initial condition for executing steady-state simulation. Steady-state simulations were carried out to describe three-dimensional flow field (head distribution) over the entire model domain. Automatic parameter estimation (PEST) method was applied for calibration of the steady-state model (Doherty 1995). The hydraulic conductivity values taken from test holes data were used to develop conductivity zones using Theissen polygons for model calibration. In order to estimate the recharge of the model domain, the model area was divided into five recharge zones (shown in Fig. 5) on the basis of hydrological setup, geomorphology and land capability of the area. The recharge of the zones was characterized by variable infiltration rates of different soils and varying pumpage of groundwater. Initially, steady-state calibration was performed which was fully implicit. The hydraulic conductivity and the recharge values estimated previously were used as initial conditions in the steady-state calibration. The values were adjusted during the calibration runs until the calculated head values became close to the observed heads. Similarly, the specific yield zones were developed using its field data for use in transient state-calibration. The model was rerun for six-month period i.e. April-September 1985, for transient-state calibration. The mean residuals of observed and calculated heads in steady-state and transient-state calibrations are 0.06m and 0.002m with variances of 1.46m and 1.86m, respectively. The calibration results indicated a reasonable agreement between the calculated and observed heads (Fig.7). The sensitivity of the model results was evaluated to variations in hydrologic parameters and modeling assumptions. The errors are expressed as equation 8.

$$RMSE = \sqrt{\frac{1}{n}\sum_{i=1}^{n}(H_i - H_i')^2} \tag{8}$$

Where $RMSE$ is root mean square error (m); n is total number of measurements; H_i is simulated value of groundwater table at the end of ith month (m); and H_i' is observed value of groundwater table at the end of ith month (m). The flow diagram of the methodology followed in the present study is shown in Fig. 8.

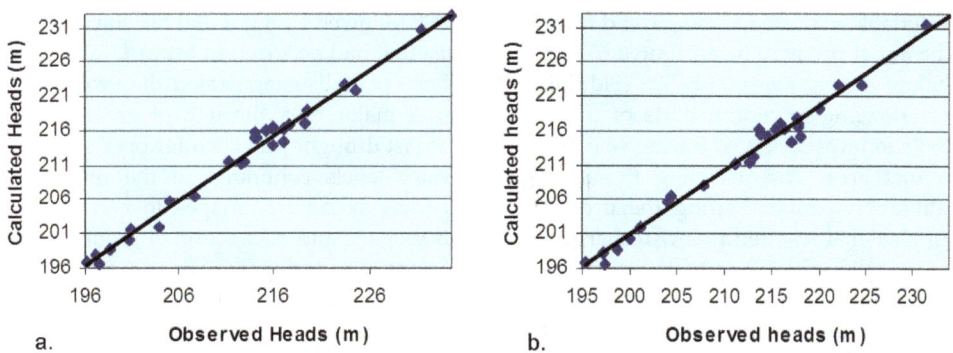

Fig. 7. a. Steady-state calibration b. Transient-state calibration

Stage

Purpose and Objectives

Field Survey — Image georectification — Data collection — Develop conceptual model

Boundary demarcation — Data Input (Digitization, Keyboard entry) — Numerical formulation

Visual Interpretation — Digital Interpretation — Data editing, polygoninzation — Selection of code for Numerical Groundwater flow Modeling

Image classification — Database Development — Model Study Plan for Design and Data input

No — Attribute data linking

Accuracy OK? — Calibrate Model — Revise calibration and/or conceptual model and/or collect more data

Yes — Development of criteria for selection of sensitive areas — No

Modification of Classes — Verification Model OK?

Data Integration in GIS — Yes

Quantitative Analysis — Overlay Analysis — Prediction scenarios

Landcover Map — Statistical Analysis — Output Data

Output Results & Report on Waterlogged Areas

Conceptualization

Calibration

Prediction

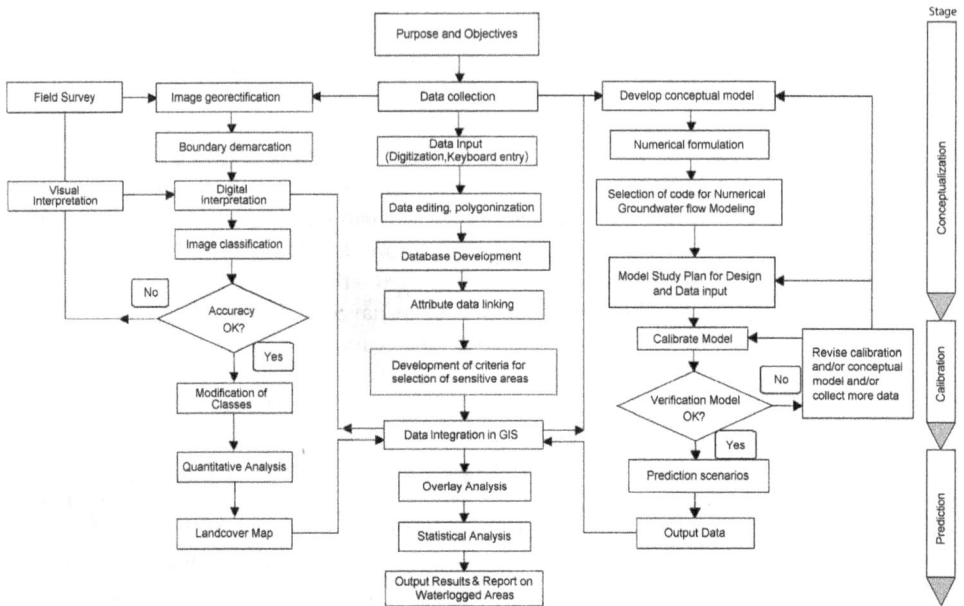

WRRI, NARC & QAU, Islamabad

Fig. 8. Flow diagram of integrating Groundwater flow modeling and GIS

2.3 Model simulation for waterlogging analysis

Strategy of management for pre-stress period and post stress period was developed on the basis of availability of observed data until 2005 and projected hypothetical data for selected periods from 2006 to 2020. The steady-state calibrated model was run for pre-stress period of variable time steps until 2005 (Fig. 9). During the calibrated period of 1985-2005, the watertable had shown an average decline of 0.96m at the rate of 0.05 m/year. The water budget of the steady-state model shows total flux-in of the model area of about 4.6036E+05 m³/d including boundary inflows of 1.6931E+05 m³/d and areal recharge of 2.9105E+05 m³/d. The total flux-out from the model area is estimated as -4.6028E+05 m³/d. The watertable indicated a rising trend from 1988 to 1999 followed by a gradual decline onward. The initial rise may be attributed to record rainfalls that had occurred in Muzaffarabad and Jhelum during period 1997-98 (Siddiqi, 1999). Those rainfalls exaggerated the problem of waterlogging in central parts of the study area. A major breakthrough of groundwater depletion was observed during year 1999 when the last drought prevailed for over 3-4 years in that area. The declining trend of groundwater levels continued in the remaining calibration period. During the drought period, there became shortages in canal water supplies that resulted in low recharge from canals seepage and abstraction of groundwater for irrigation use. The situation had affected the extent of waterlogging which was reduced significantly in different parts of the area. The analysis of velocity variations of groundwater flow in three layers of the model indicated high velocity (>0.12 m/day) in the northwestern part near Rasul Barrage in the first layer. The velocity range 0.04-0.08 m/day was dominant in the central and southeastern parts of the model area. The patches of low velocity zone

(<0.02 m/day) were found in the northern, northeastern and western parts which extend downward in other layers also. In the second layer, velocity zone 0.02-0.04 m/day dominated in most of the central parts. In the third layer, the velocity of less than 0.02 m/day was dominated in most of the northeastern parts of the model area. The regional groundwater flow component in the southern part is indicating existence of a potential aquifer zone in this layer of the model area.

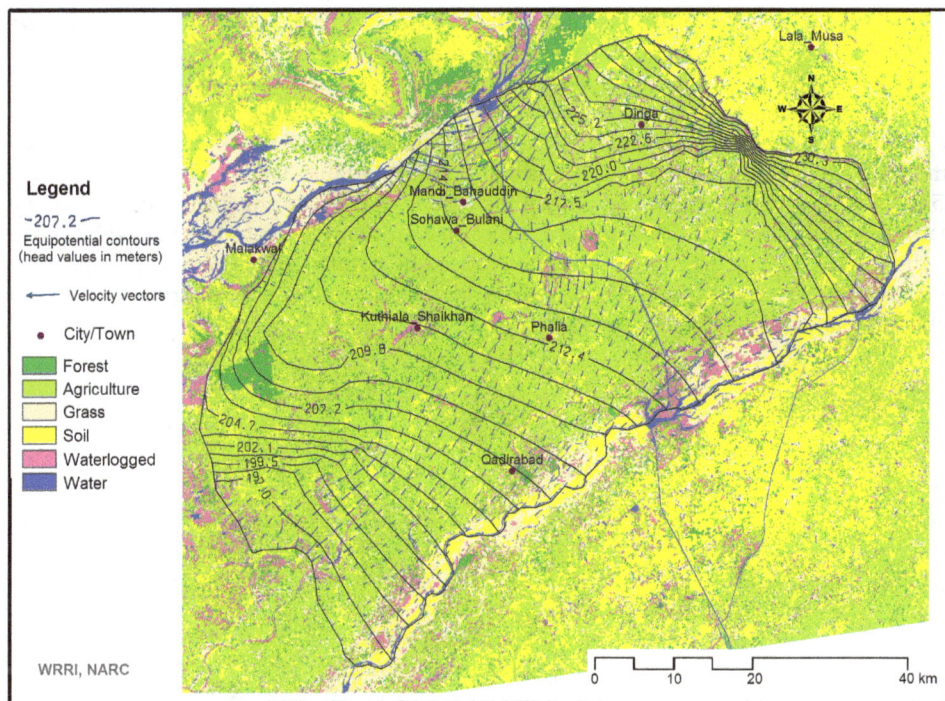

Fig. 9. Integration of spatial data layers in GIS for analysis and output visualization

The calibrated model was rerun to predict the future changes in piezometric heads from period 2006 to 2020. Time series records of previously observed data of precipitation, annual recharge and withdrawals from tubewells were examined which formed basis to generate projected data for groundwater flow simulation. The predictive period of 15 years i.e. 2006-2020, was chosen to perceive long-term impact of droughts/floods on regional groundwater system. Based on the annual incremental increase/decrease in recharge and/or groundwater pumpage, recharge for various stress periods was adjusted for calibration of groundwater flow model. The predictive model showed an average decline of 0.81m per year in watertable. The waterlogging indicated initially an increase in coverage from year 1990 to 2000 and than a gradual decrease from year 2000 to 2020 (Fig. 10). During pre-stress and post-stress periods, variation in head values ranges between 196 and 234 meters above sea level (masl). In upper reaches of the model domain, fluctuation in heads is low due to presence of less extensive alluvial deposits in piedmont plain. Quantitative analysis of the groundwater aquifer was performed for the base year 2005 using geo-processing techniques of GIS software. Figure 11 shows variation in watertable depth during 2005 in different tehsils (sub-districts) of Mandi Bahauddin and Gujrat districts. Results indicated maximum head drop of about 21m in the Phalia tehsil, while minimum head drop of 14m in the Gujrat tehsil. High variation in groundwater velocity was observed at several locations in the model area (Fig. 12). Maximum range of velocity from 0.006 to 0.09 m/day is found in Mandi Bahauddin and minimum range from 0.003 to 0.035 m/day in the Kharian tehsil. Overall there was relatively low variation in heads observed in the Gujrat district. The comparison in coverage of watertable depth of base year 2005 and predictive year 2020 indicated a noticeable reduction in waterlogged areas i.e. a decrease in waterlogged area (<1.5m watertable depth) from 226.9 km² in the base year 2005 to 74.8 km² in the year 2020 (Table 3). The changes are significant in Malakwal and Phalia tehsils. Similarly, the coverage of 1.5-3.0m watertable depth range indicated a decrease from 1299.4 km² in 2005 to 1104.8 km² in the year 2020. The coverage of less than 1.5m watertable depth and to some extent of 1.5m - 3.0m watertable depth had shown increase in extents during 1990-2000 period and than gradual decline from 2000 onward indicating shrinking of waterlogging extent in the predictive year 2020. This condition may be caused by natural factors i.e. low precipitation, reduce river flows, and/or human factors like over exploitation of groundwater for agriculture and domestic use in future.

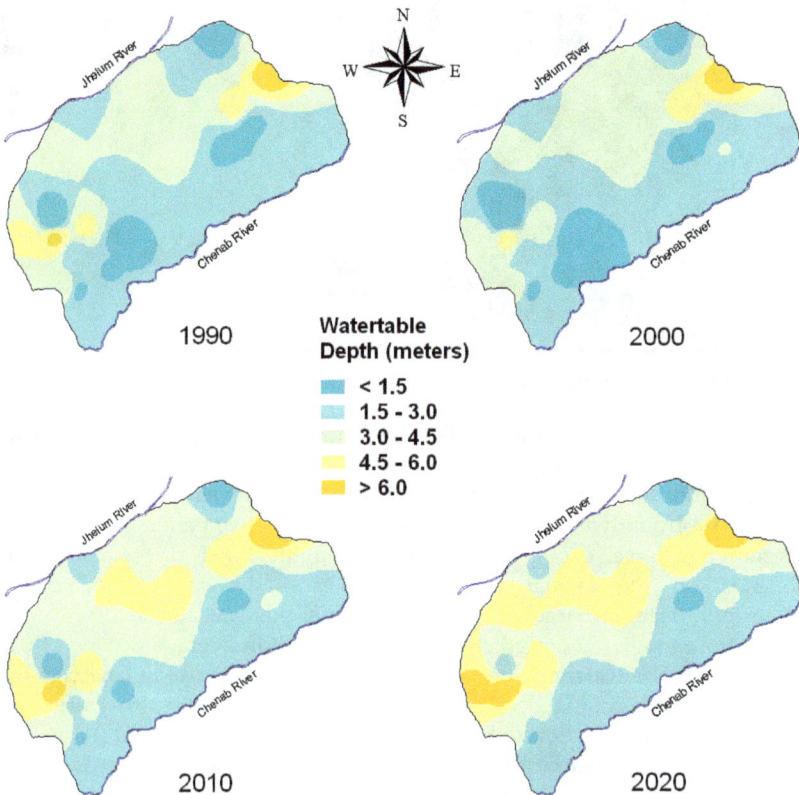

Fig. 10. Development of watertable depth maps for change analysis during 1990-2020

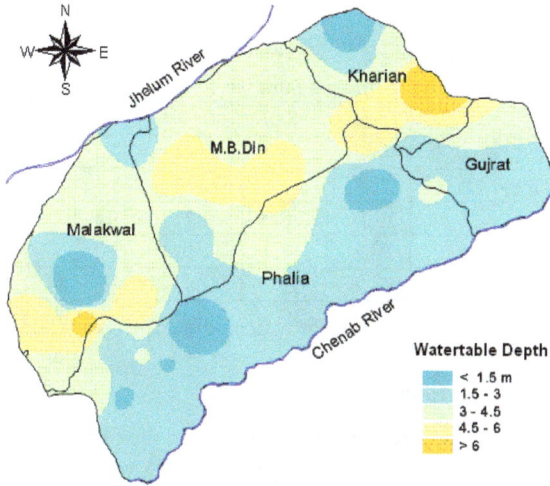

Fig. 11. Analysis of the groundwater behavior in different tehsils during base year 2005

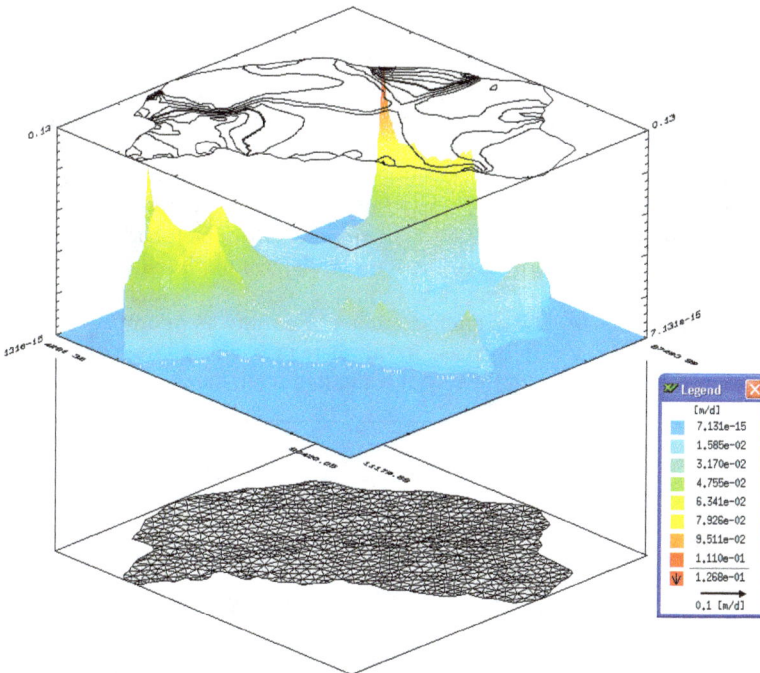

Fig. 12. 3-D view of velocity variations during 2005

Tehsil	2005 Area (Km²)					2020 Area (Km²)				
	< 1.5m	1.5-3.0m	3.0-4.5m	4.5-6.0m	6.0-12.0m	< 1.5m	1.5-3.0m	3.0-4.5m	4.5-6.0m	6.0-12.0m
M.B.Din	0.0	119.3	482.2	206.3	0.0	0.0	25.5	473.8	308.5	0.0
Malakwal	60.5	114.4	344.9	128.2	10.4	0.0	39.3	222.8	308.6	87.7
Phalia	125.1	781.2	200.6	51.5	1.0	39.7	775.4	262.9	75.8	5.6
Gujrat	0.0	223.2	74.0	12.0	0.0	0.0	201.4	93.8	14.0	0.0
Kharian	41.3	61.2	98.2	79.2	48.2	34.1	63.2	97.7	81.0	52.0
Total	226.9	1299.4	1199.8	477.2	59.7	73.8	1104.8	1151.0	788.0	145.4

Table 3. Change in area coverage of different watertable depths for 2005 and 2020

2.4 Scenarios of extreme conditions

In order to observe impact of natural and human induced factors on groundwater levels and ultimately on waterlogging behavior of the study area, various scenarios were developed. In the first scenario, severe drought condition was assumed to prevail for five-year period 2006-2010. The rainfall data of the severe drought that occurred during 1999-2002 was used as reference in developing the scenario. The hypothetical simulation indicated a mean decline of 0.7m in watertable depth from that of the base year 2005 (Table 4). In the second scenario, extreme wet condition was assumed to prevail for a period of five years from 2011 to 2015. It showed a mean decline of only 0.42m in watertable from that of the base year. High rainfall years will have adverse effects especially in the river flood plains and depression areas having poor drainage system. This condition may exaggerate the problem of waterlogging in the central and southern parts of the area. For effective control of water logging and minimizing the risk of this menace, management strategies and measures like installation of tubewells for vertical drainage; construction of subsurface-drains and tile-drains; planning and designing of future canals on proper lines; lining of water channels and optimizing water-use need to be adopted. Scenarios of groundwater pumpage from deep tubewells were developed to see impact of groundwater withdrawals on watertable depth. The increase in wells from 33 to 66 numbers (double) had shown decline in watertable from depth 0.14m to 0.42m (nearly 3 folds). Similarly 60% increase in rate of pumpage (from 5000m³/day to 8000m³/day) indicated increase in watertable depth to 0.25m for case 4 and 0.71m for case 6. The overall situation of groundwater pumpage indicates linear response of watertable depth with increase in numbers of wells and their rates of discharge. This strategy may help in mitigating the risk of waterlogging in the study area.

Scenario	Description	Simulation Period	Net change in Watertable depth (m)
1	Severe drought prevails for 5-year period	2006-2010	0.70
2	Wet condition prevails for 5-year period	2011-2015	0.42
3	Pumpage from 33 TWs for 3-year period	2006-2008	0.14
4	60% increase in pumpage from 33 TWs for 3-year	2006-2008	0.25
5	Pumpage from 66 TWs for 3-year period	2006-2008	0.41
6	60% increase in pumpage from 66 TWs for 3-year	2006-2008	0.71

Table 4. Watertable behavior in different scenarios (WT depth in 2005 =2.95m)

3. Conclusions

Rapidly developing computer technology has continued to improve modeling methods in hydrology and water resource management. GIS application has provided help in an accurate and manageable way of estimating model input parameters, integration of disparate data layers, conceptualizing of model recharge and discharge sources and visualization of the model output. GIS-based modeling, as a side benefit, also provides an updated database that can be used for non-modeling activities such as water resource planning and facilities management. The characteristics and causative factors of waterlogging/salinity can be investigated not only through remote sensing data that provide a rapid and accurate assessment of the land and water resources but also through numerical groundwater modeling which provides insight of the controlling groundwater behavior. The RS data of high spectral and spatial resolutions would be helpful in reliable assessment of landcover/landuse and identification of potential recharge/discharge areas that could ultimately enhance the quality of groundwater modeling results. The developed model would thus provides a decision support tool for evaluating better management options for sustainable development of land, surface and groundwater resources on micro as well as on macro levels in future.

4. Acknowledgments

We thank Dr. Alan Fryar Associate professor and Director of graduate study, Department of Earth and Environmental Sciences, University of Kentucky, USA and Dr. Gulraiz Akhtar, Department of Earth Sciences of Quiad-i-Azam University, Islamabad for their valuable input for execution of this work. The data support by Water and Power Development Authority (WAPDA), Pakistan Meteorological Department (PMD) and National engineering services Pakistan (NESPAK) for this research work is also highly appreciated.

5. References

Ahmad, Z., Arshad, A, Fryar, A and Akhter, G. (2011) Composite use of numerical groundwater flow modeling and geoinformatics techniques for monitoring Indus Basin aquifer, Pakistan, Environmental Monitoring and Assessment, 173(1-4):447-457.

Anderson, M. P. and Woessner W. W. (1992) Applied groundwater modeling simulation of flow and advective transport. Academic Press.

Aronoff, Stan (1989) Geographic Information Systems: A Management Perspective. WDL Publications, Ottawa, Canada

Arora, A.N. and Goyal, R. (2003) Conceptual Groundwater Modeling using GIS. GIS India 2003, National Conference on GIS/GPS/RS/Digital Photogrammetry and CAD, Jaipur.

Ashraf, A & Ahmad, Z (2008) Regional groundwater flow modeling of Upper Chaj Doab of Indus Basin, Pakistan using finite element model (Feflow) and geoinformatics Geophysical Journal International, (GJI), 173, 17–24. doi:10.1111/j.1365-246X. 2007.03708.x.

Ashraf, A.; Ahmad, Z. and Akhtar, G. (2005) A Composite study of Spatial Analysis and Demarcation of Potential Recharge/discharge zones of upper Chaj doab in Indus basin of Pakistan, using Remote Sensing, GIS Techniques and Finite Element Model (Feflow). Int. Conf. Model Care 2005 "Calibration and Reliability in Groundwater Modeling from Uncertainty to Decision Making", The Hague (Scheveningen), the Netherlands (June 6-9, 2005)

Choubey, V.K. (1996) Assessment of waterlogged area in IGNP Stage-I by remotely sensed and field data. Hydrology Journal, Vol. XIX (2), pp. 81-93.

Dafny, E.; Burg, A. and Gvirtzman, H. (2010) Effects of Karst and geological structure on groundwater flow: The case of Yarqon-Taninim Aquifer, Israel. Journal of Hydrology, 389(3–4): 260-275.

Doherty, J. (1995) PEST ver. 2.04. Water mark Computing, Corinda, Australia

Ehlers, M.; Edwards, G. and Bedard, Y. (1989) Integration of Remote sensing with Geographic Information Systems: A Necessary Evolution. Photogrammetric Engineering and Remote sensing, 55, no. 11: 1619-1627.

Indo-Dutch Network Project-IDNP (2002) A Methodology for Identification of Waterlogging and Soil Salinity Conditions Using Remote sesning. CSSRI, Kamal and Alterra-ILRI, Wageningen: pp. 78

Lillesand, T.M. and Kiefer, R.W. (2004) Remote Sensing and Image Interpretation, 5th Edition, John Wiley & Sons.

Middlemis (2000) Groundwater Flow Modeling Guidelines, Murray-Darling basin commission. Aquaterra Consulting Pvt, Ltd, S. Perth, Western Australia

Peleg, N. and Gvirtzman, H. (2010) Groundwater flow modeling of two-levels perched karstic leaking aquifers as a tool for estimating recharge and hydraulic parameters. Journal of Hydrology, 388(1-2): 13–27.

PPSGDP Report (2000) Groundwater Flow and Solute Transport Numerical Models For SCARP II and Shahpur Area of Chaj Doab, Tech. Rep. No. 39, Punjab Private Sector Groundwater Development Project Consultants, Project Management Unit, Irrigation and Power Department, Govt. of Punjab, Lahore

Qureshi, M.K.A.; Sheikh, M.I. and Khan, A. (2003) A Report on the Geology and Environmental concerns of the Rasul-Mandi Bahaudin Quadrangle, Punjab, Pakistan, Ministry of Petroleum and Natural Resources, Geological Survey of Pakistan, Quetta, Pakistan.

Russo, S.L. and Civita, M.V. (2009) Open-loop groundwater heat pumps development for large buildings: A case study. Geothermics, 38(3): 335–345.

Sarwar, A. (1999) Development of a conjunctive use model, an integrated approach of surface and Groundwater modeling using a Geographic Information System (GIS). PhD thesis, University of Bonn, Germany.

Shelton, R.L. and Estes, J.E. (1980) Remote Sensing and Geographic Information Systems: An Unrealized Potential. Geo-processing, 1, no. 4: 395-420.

Siddiqi, J. (1999) Effect of El-Nino on Agriculture of Pakistan, Pakistan Agricultural Research Council, Islamabad.

Soil Survey Report (1967) Reconnaissance soil survey of Gujrat district, Directorate of Soil Survey of Pakistan, Lahore.

Thangarajan, M. (2004) Regional Groundwater Modeling, Capital Publishing Company, New Delhi, India.

UNESCO (1999) Water Resources of Hard rock Aquifers in Arid and Semi-arid Zone. Ed. J W Lloyd.

WAPDA (1993) Pakistan Drainage Sector Environmental Assessment – National Drainage Programme Vol. 4: (Data (water, Soil and Agriculture), NESPAK and Mot MacDonald Int. Ltd., WAPDA, Lahore, Pakistan.

Yang, J.W. and Radulescu, M. (2006) Paleo-fluid flow and heat transport at 1575 Ma over an E–W section in the Northern Lawn Hill Platform, Australia: Theoretical results from finite element modeling. Journal of Geochemical Exploration, 89(1-3): 445–449.

Zaisheng, H.; Wang Hao and Chai Rui (2006) Transboundary Aquifers in Asia with Special Emphasis to China, UNESCO Report

11

Percolation Approach
in Underground Reservoir Modeling

Mohsen Masihi[1] and Peter R. King[2]

[1]*Department of Chemical and Petroleum Engineering, Sharif University of Technology,*
[2]*Earth Science and Engineering Department, Imperial College London,*
[1]*Iran*
[2]*UK*

1. Introduction

The spatial distribution of the underground heterogeneities which may be appeared on various scales can affect the flow and transport of fluids (e.g. underground spread of pollution or recovery from hydrocarbon reservoirs). Reservoir modeling is conventionally used to incorporate the effect of uncertainty in the spatial distribution of geological heterogeneities (such as fractures as appeared on various scales as emphasized by Odling et al, 1999) and to assess the reservoir performance parameters such as the reservoir connectivity and conductivity. It needs detail data to make geological model, then upscale it to coarser grid and run flow simulation. Since the natural variability of geological formations extends over many length scales, this approach requires very fine modeling schemes. In addition, the predictions may not be achievable in the early life of reservoirs because of the sparse data. Moreover, in stochastic modeling in order to make a reliable underground prediction or for sensitivity purposes, it is necessary to construct a number of possible reservoir models. The problem with this approach is that it is computationally very expensive and time consuming. Hence, there is a great incentive to produce much simpler physically-based model to predict the underground reservoir performance very quickly.

In this chapter, we used the idea that the flow characteristics in porous media are mainly controlled by the continuity of conductivity contrasts (e.g. high hydraulic conductive streaks) to model stochastically the underground reservoir flow. There are many cases for both hydrologist and petroleum engineers where the geological formations consists of a mixture of good sandstones with high hydraulic conductivity (i.e. flow units) and poorer rock types (e.g. siltstones, mudstones and shales) with negligible permeability. Good sandstones with a higher permeability and porosity are the main bodies containing fluid within their pores. Obviously the volume of fluid in place and recoverable fluid between a pair of wells depends on the connected fraction of these good fluid bearing zones. Examples include fluvial sediments containing paleo-channels (high conductivity zones embedded in a low conductive background), shale/sandstone sequences (non-conductive inclusions embedded in a conductive matrix), fractured formations (with fractures as the connected high conductive zones), and coastal deposits (deltaic systems representing the conductive media).

As the mathematical model of connectivity and conductivity in complex geometries is given by percolation theory (Stauffer and Aharony 1992) we use "percolation approach in reservoir modeling" for the title of the presented method. Imagine a typical reservoir model to be constructed with an object based technique. Assume a simple permeability model of the reservoir (black and white model for sand/shale or fracture/matrix). Then consider sandbodies/fractures as simple geometrical objects located in space with simple statistics and use percolation theory to estimate the uncertainty in reservoir performance. In particular, we can estimate the probability that wells are in connection via the sands, the fraction of the total sands which is in contact with these wells, the reservoir conductivity, potential oil recovery, sweep efficiency, breakthrough time and post breakthrough behavior within a particular well configuration. Throughout the field life many decisions have to be made based on the technical factors (see Table 1) which can be addressed by using percolation approach.

Production Phase	Business Decision	Technical Factors
Exploration/ appraisal	Number and location of wells Reserves calculation Production rates	Connected oil volume fractions connected by wells Swept volumes Effective permeability
Plateau	End of plateau	Breakthrough time and post breakthrough behavior
Decline	Decline rate or increase of water or gas Infill locations	Flow path tortuosity and connected volume of oil Distribution of remaining oil pools

Table 1. A summary of the key decisions and technical factors during underground hydrocarbon reservoir life

Moreover, the percolation approach is able to estimate the uncertainty in reservoir parameters which is not possible with a single realization reservoir model. The advantage is that the effects of the complex geometry which influence the flow can be easily estimated in a fraction of a second on a spreadsheet by simple algebraic transformations (Masihi et al. 2007). Clearly the disadvantage is that much of the flow physics and subtleties of the heterogeneity distribution are missed.

2. Percolation theory

The mathematical analysis of percolation theory was first developed by Broadbent & Hammersley in 1957 and since then the topic has been intensively studied [Stauffer & Aharony, 1992]. It has been applied in many fields from the spread of forest fires and the spread of diseases to the flow in porous media and fractured rocks [Sahimi 1994]. This theory links the global physical properties such as connectivity and conductivity to the number density of objects placed randomly in space and reduces many results to simple power laws from which all outcomes are predicted by simple transformations. A simple percolation model is an infinite size lattice of squared sites (so called *site* percolation) that are occupied randomly and independently to an occupancy fraction "p". Neighboring sites are in the same cluster if they are both occupied (Fig. 1). This can be checked by an algorithm proposed by Hoshen and Kopelman 1976. Consider that sites are filled with a fluid then the cluster is a region accessible to the same well. Percolation theory describes mathematically how the number and the size of these clusters vary as a function of the "p". As this occupancy probability is increased, the finite size clusters bridged and grow in size. At one particular value (called percolation threshold,p_c) one large cluster (so called the spanning, percolating or infinite cluster) spans the whole region (the blue clusters in figures below). However, there exist also other small clusters which get absorbed as p further increases leaving a few isolated sites. From flow studies point of view across the region (as between two wells) we are interested in the spanning cluster. However, to analyse connectivity to a point (or a well) the contribution of the finite clusters then must be considered.

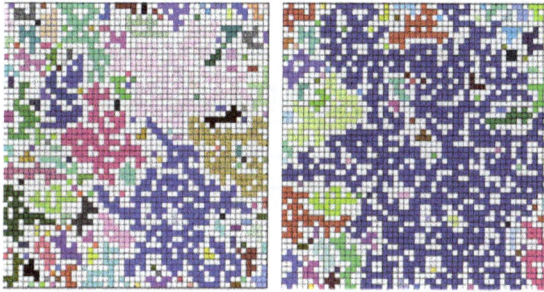

Fig. 1. Illustration of (left) various clusters in different colors before the threshold, and (right) spanning cluster in blue at the threshold

Other extensions of this simple site percolation model include using other lattice shapes (e.g. triangular, hexagonal or cubic lattices), other connection criteria (e.g. next nearest neighbors), directed percolation (e.g. contact is only in left-right direction), bond percolation, continuum percolation (i.e. geometrical objects are distributed randomly in space and correlated percolation (i.e. objects may not be distributed randomly). These do not alter the essential ideas of percolation theory; however, may alter some numerical values of the parameters used to describe the percolation. One of these adjusting parameters is percolation threshold which is dependent on the type/detail of the system. The percolation threshold p_c^∞ value depends on the details of the system. The threshold value for some geometries in two and three dimensions are given in Table 1 [Stauffer and Aharony, 1992; Berkowitz and Balberg, 1993, Baker et al 2002]. As can be seen in Table 2 the threshold values in three dimensions are much lower than the values in two dimensional lattices. One fundamental percolation property is the connectivity $P(p)$ which is defined as a probability that any point in space belongs to the spanning cluster which shows the strength of the percolating cluster and scaled as [Stauffer and Aharony, 1992]:

$$P(p) \propto (p - p_c^\infty)^\beta \tag{1}$$

where p_c^∞ is the percolation threshold of an infinite system and the β is called the connectivity exponent. Above the threshold, the connectivity, $P(p)$, increases rapidly. The other property is the correlation length $\xi(p)$ defined as $\xi^2 = \sum r^2 g(r)/\sum g(r)$ (with $g(r)$ as the two point correlation function or the probability that two sites separated by a distance r are in the same cluster) which represent a typical size of finite clusters and used in analysis of connectivity to a point (or well) [Stauffer&Aharony, 1992],

$$\xi(p) \propto |p - p_c^\infty|^{-v} \tag{2}$$

With v as the correlation length exponent. Moreover, the mean cluster size $S(p)$ follows a simple power law as,

$$S(p) \propto |p - p_c^\infty|^{-\gamma} \tag{3}$$

Where the γ is called the mean cluster size exponent.

Shape	Site	Bond	Continuum
Square	0.593	0.500	
Triangular	0.500	0.347	
Hexagonal	0.696	0.653	
Simple Cubic	0.312	0.249	
Disks			0.676
Aligned Squares			0.666
Spheres			0.289
Aligned Cubes			0.277

Table 2. Threshold values of different shapes for site, bond or continuum percolation

Next property is the effective permeability. From percolation theory the effective permeability of an infinite system $K(p)$ has a power law behavior [Stauffer and Aharony, 1992]:

$$K(p) \propto (p - p_c^\infty)^\mu \tag{4}$$

where the universal exponent μ is called the conductivity exponent. Above the threshold p_c^∞, the effective permeability increases slowly as compared with the connectivity emphasizing that a part of the spanning cluster (called the backbone) contribute to the flow (Fig. 2). The backbone consists of blobs (i.e. the multiple connected part of the backbone) and the red links i.e. the link between two blobs (Fig. 2).

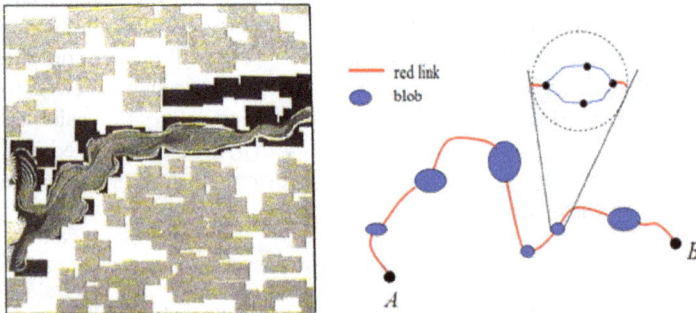

Fig. 2. Illustration of (left) the tortuous conducting path with many dead ends and (right) red links and blobs between two points on the backbone

Backbone between points i and j are the bonds that belong to at least one self-avoiding walk between i and j. The burning algorithm described by Stanley et al 1984 is used to identify the backbone. The backbone fraction of the percolating cluster is the amount of oil that can be swept during water/gas injection. The critical exponents β, v, μ, γ in these power laws are universal. By universality, we mean that their values are independent of the small scale structure (i.e. the kind of lattice, site or bond percolation, lattice or continuum systems). They only depend on the dimensionality of space (i.e. two or three dimensional systems). The generally accepted literature values for these exponents for two and three dimensional systems are given in Table 3 [Stauffer and Aharony, 1992; Berkowitz and Balberg, 1993]. The principle of universality is a very powerful concept as it enables us to study and understand the behaviour of a very wide range of systems without needing to worry so much about the fine scale details.

Exponent	2D	3D
β	5/36	0.4
v	4/3	0.88
γ	43/18	1.8
μ	1.3	2

Table 3. Selected percolation exponents in two and three dimensions.

The basics percolation has been developed on lattices; while, in many natural systems the objects are not restricted to fixed points or the objects can be of any shape with variable length, direction and number of interconnected bonds. In such cases we use continuum percolation, where the geometrical objects are distributed independently and randomly in a region. Examples include fracture systems where there is theoretically no end to the degree of fracturing [Sahimi, 1995] or overlapping sandbodies with various shapes and sizes [Masihi and King 2008]. In continuum percolation models, the principal of universality allows us to use the same scaling laws with the same numerical values of the critical exponents as in the case of lattice percolation. However, we need to the percolation threshold as it depends on the details of the system. Examples of applying percolation theory to uncorrelated (or even correlated) continuum systems that check the universality and determination of the percolation threshold of different models can be found elsewhere [Gawlinski and Stanley, 1981; Lee and Torquato., 1990; King, 1990; Berkowitz, 1995; Lorenz and Ziff, 2001; Baker et al., 2002].

It should be emphasized that the percolation cluster and its other sub-networks have fractal properties. Mass of the percolating cluster (as defined as total number of the sites L) scale as, $M \sim L^{D_c}$ where the fractal dimension D_c of the percolating cluster is given by $D_c = d - \beta/v$ with d as the space dimension. A sub-network of percolating cluster is the Backbone which is a fractal object with its mass M_B scaled with total number of the sites L as $M_B \sim L^{D_B}$ (with $D_B = d - \beta_B/v$ where D_B is the fractal dimension of the Backbone cluster and β_B is the connectivity exponent of the backbone cluster). Another sub-network is called the red bonds which is also a fractal object with its mass M_{red} scaled with total number of the sites L as $M_{red} \sim L^{D_{red}}$ (with $D_{red} = 1/v$ where D_{red} is the fractal dimension of the red bonds). The numerical values of the fractal dimensions of various percolation networks are given in Table 4 [Stauffer and Aharony, 1992; Sahimi, 1995].

Fractal dimension	2D	3D
D_c	91/48	2.52
D_B	1.64	1.8
D_{red}	3/4	1.14

Table 4. The numerical values of the fractal dimensions of various percolation networks

The fractal behavior of percolation-based reservoirs can be observed during well testing results analysis. Conventional build up test for radial flow in a single well homogenous reservoir gives the pressure change Δp over the time period Δt as $\Delta p \sim ln(\Delta t)$. However, for some field data the conventional model matches only a part of the data not being improved by considering other effects including formation damage along the fracture face or nondarcy flow near the wellbore. In such cases, the alternative model which gives a reasonable match is to use solution of pressure transient equations with rock properties have fractal behavior (Chang & Yortsos 1990).

3. Two basic models

Two main applications of percolation approach include low-to-intermediate net to gross reservoirs and fractured reservoirs. Geologists have used the idea that a reservoir consists of the geometrically complex connected and disconnected sandbodies to be able to qualitatively describe it by means of outcrop studies and subsurface and for the reliable estimation of accessible hydrocarbons in place and the expected recoverable oil [Haldorsen et al., 1988]. Hence, sandbodies within a reservoir are assumed to be flow units connected in a complicated way through the sedimentary processes along with and poorer siltstones, mudstones and shales (of lower permeability) [Bridge and Leeder, 1979]. An example is a meandering river which deposits sand layers over the time in its flowing bed results in formation of a system of embedded sandbodies in an impermeable background (Fig. 3). Moreover, outcrop studies often show a network of connected fractures (Fig. 3). In this study, we use overlapping sandbodies and fracture network models within the framework of continuum percolation to determine the reservoir connectivity and conductivity.

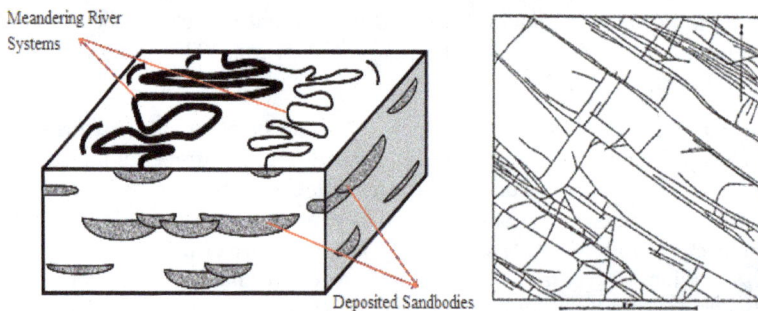

Fig. 3. Illustration of (left) Overlapping sandbodies deposited by a meandering river over the ages (Nurafza et al 2006). (right) 2-D fracture model (Odling et al 1999)

3.1 Overlapping sandbody model

We assume that conductive bodies represented by squares of side size a (or circles) in two dimensional square region of side size X (or by cubes/spheres in 3D) are distributed randomly and independently in an impermeable background. The effective size of the system is defined by the dimensionless length $L = \frac{X}{a}$. The net to gross ratio p, equivalent to the occupancy probability, is defined as the total area of sands divided by the total area of the region. The connectivity or connected sand fraction $P(p, L)$ that is a function of both p and L is defined as the probability of a point on the sandbodies area, belonging to the percolating cluster. As each sandbody is put down through the model build up, it is given a color/cluster number showing a cluster that the body belongs to it. Moreover, we need an efficient computational algorithm to check the sandbody overlapping and to get the statistics of p, P, K for each realization.

3.2 Fracture network model

Many hydrocarbon reservoirs are naturally fractured. By fractures we mean any discontinuity within a rock mass that developed as a response to stress. Conductive fractures form a network of interconnected fractures with a distribution for the fracture orientation, length or aperture. The nature of fluid flow in fractured reservoirs with low matrix permeability depends strongly on the spatial distribution of the fractures. We represent fractures with line segments embedded in a 2D impermeable background (or squares/rectangles in 3D) and assume that they are the only pathways for the flow. Consider line segments representing the fractures in 2D are distributed randomly within a squared region of side size L. Fractures have the same length l with orientations that are distributed uniformly on the interval $-90° \leq \theta \leq 90°$. Equivalent term to the occupancy probability p is defined based on the probability density function that a point is in the effective area of a fracture,

$$p = 1 - e^{\frac{-N\langle a_{ex}\rangle}{4L^2}} \tag{5}$$

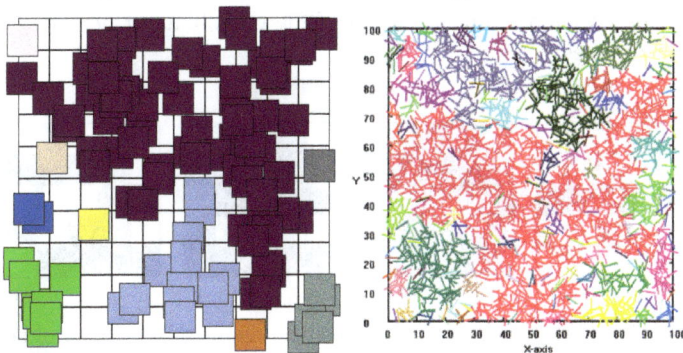

Fig. 4. A realization of (left) sandbody model at the threshold point with the number of fractures N=100 and p=0.6 with connected sand fraction P=0.60 (right) fracture network at the threshold with total number of fractures $n_w = 2066$ and p=0.56. Spanning cluster (in red) consists of $n_s = 897$ fractures

Where N is the number of fractures in the region and the term $\langle a_{ex} \rangle$ is the average excluded area over the distribution of the fracture orientations (e.g. as shown in Masihi et al 2007 for random fracture orientation this is $\langle a_{ex} \rangle = \frac{2l^2}{\pi}$). Moreover, an equivalent to the percolation probability P (or connected fraction) is $P = n_s/n_w$ where n_s and n_w are respectively the number of fractures in the spanning cluster and in the whole fracture network. Moreover, an appropriate computational algorithm is necessary to check the fracture intersections and get the statistics of p, P, K for each realizations.

Two realizations of overlapping sandbody model and fracture network model are shown in Fig. 4.

4. Finite size effects

In reality, each underground reservoir has a finite size. In such as case, instead of a single infinite cluster as occupancy increases the clusters get larger in size until one (or more) cluster spans from one side of the system to the other. For simplicity we describe this on lattice, although the argument is general and the equations describe here are applicable to other continuum models such two basic models described in the previous section. Fig. 5 shows that the connectivity from left to right may occur at very low occupancy (p=0.2), which is much less than the infinite percolation threshold for a 2d square lattice, but not get it at a much higher occupancy probability e.g.$p = 0.8$. As the size increases, the probability of getting such rare configurations decreases until we return to the unique single threshold value for infinite systems. In practice, for a finite system an apparent threshold $\tilde{p}_c(L)$ which depends on the system size can be used. A survey of literature [King, 1990; Stauffer and Aharony, 1992; Berkowitz, 1995; Adler and Thovert 1999; Harter, 2005] suggests that the threshold can be defined as:

- the point at which the connectivity is first non-zero.
- the intercept of the tangent to the connectivity curve taken from its inflexion.
- the occupancy probability at which half of the realizations percolate.
- the point of intersection of all the curves of the probability of percolation obtained at various domain sizes.

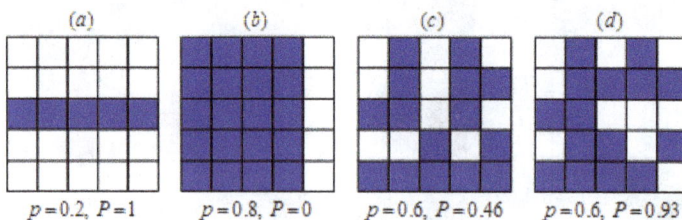

(a)	(b)	(c)	(d)
$p=0.2, P=1$	$p=0.8, P=0$	$p=0.6, P=0.46$	$p=0.6, P=0.93$

Fig. 5. Connection from left to right can occur well (a) below infinite threshold or (b) above infinite threshold. (c, d) Different realizations at the same occupancy give different connectivity, P

We should notice that all these definitions have the same behaviour as the system size goes to infinity. The apparent threshold has the scaling [Stauffer and Aharony, 1992],

$$\tilde{p}_c(L) - p_c^\infty \propto L^{-1/v} \tag{6}$$

From which the infinite percolation threshold p_c^∞ can be estimated. Let describe the finite size scaling for connectivity even though it is applicable to conductivity as well. For finite size systems, we define the connectivity $P(p, L)$ as the fraction of occupied sites belonging to the spanning clusters. Again figure below shows that different realizations at the same occupancy give different values for the connectivity $P(p, L)$.

If we plot the connected fractions $P(p, L)$ as a function of occupancy probability "p" over a large number of realizations for a particular system size we get a scatter of points showing no longer a sharp transition in the connectivity behaviour near the threshold so the main effect of finite boundaries is to *smear out* the percolation transition. Fig. 6 shows a considerable spread in connectivity results for basic sandbody model with percolation threshold of $p_c^\infty = 0.668$ due to the finite size effects.

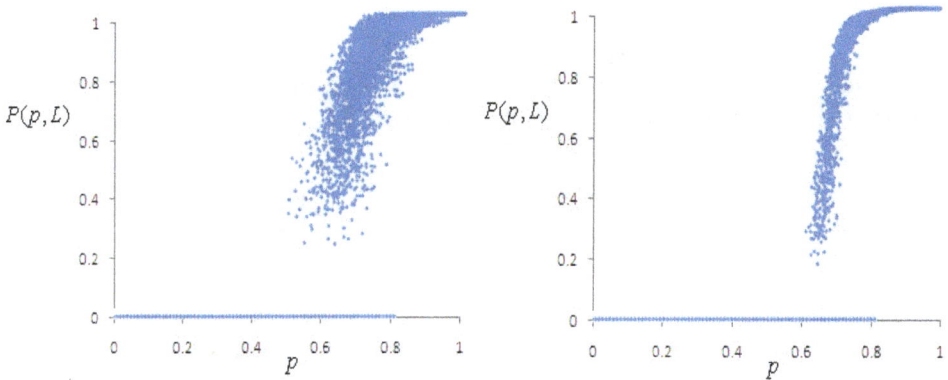

Fig. 6. The typical scatter plots of connectivity, P versus net-to-gross, p at system size L= 16 and 64 showing the smearing out of the percolation transition in the basic two dimensional sandbody model

Instead of working with the whole scatter we can determine the average value of connectivity $P(p, L)$ and standard deviation of connectivity $\Delta(p, L)$ obtained over all realizations at the same occupancy probability, p. As the system size increases the amount of smearing of the percolation transition decreases and the standard deviation of connectivity curves becomes narrower and its peak becomes lower (Fig. 7).

Finite size scaling law, based on the principle that all the properties depend only on the ratio of the system size L to the correlation length ξ, is a way of relating all of the connectivity curves (i.e. P and Δ) obtained for different system sizes to a single curve through the scaling equations (Stauffer and Aharony, 1992):

$$P(p, L) = L^{-\beta/v}\Im\big[(p - p_c^\infty)L^{1/v}\big] \tag{7}$$

$$\Delta(p, L) = L^{-\beta/v}\Re\big[(p - p_c^\infty)L^{1/v}\big] \tag{8}$$

Where functions $\Im$ and $\Re$ are respectively the master function for the mean and standard deviation of connectivity results. Fig. 8 shows the data collapse presented in Figure 7 by using these scaling transformations.

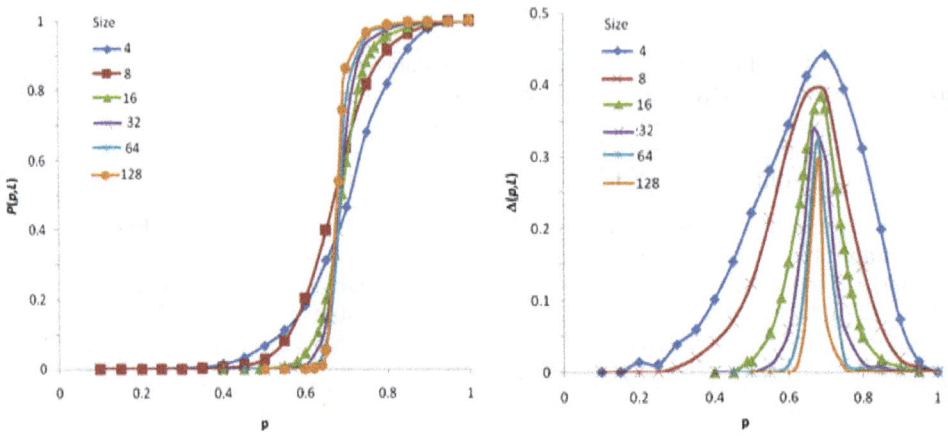

Fig. 7. The average value of connectivity $P(p, L)$ and standard deviation of connectivity $\Delta(p, L)$ results obtained over all realizations at the same occupancy probability, p at six system sizes for squared overlapping sandbody model

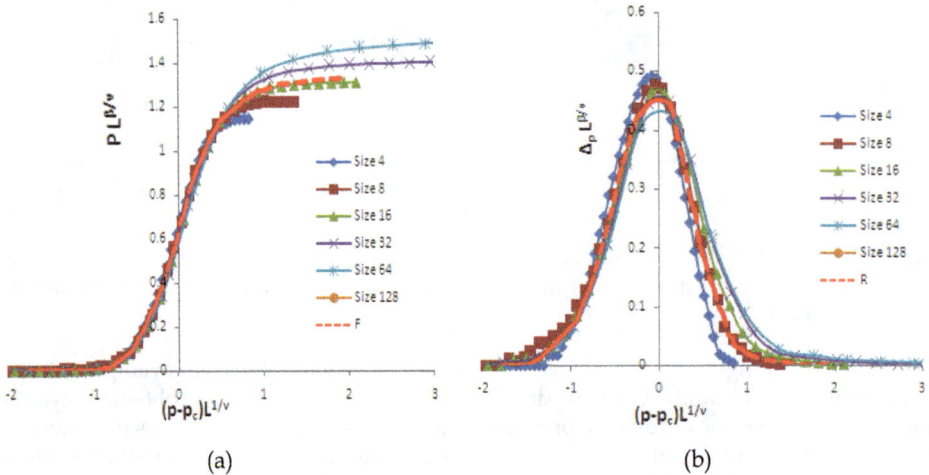

(a)   (b)

Fig. 8. Illustration of the finite size scaling laws for squared sandbodies in two dimensions for (a) the average connectivity and (b) standard deviation in connectivity results (Sadeghnejad et al. 2011)

These provide the master curves for the mean connected sand fraction and the standard deviation of connectivity results ($\mathfrak{I}$ and $\mathfrak{R}$) for the case of overlapping sandbodies model from which one can predict the mean connectivity and its associated uncertainty for any system size very quickly, without performing any explicit realizations.

Example: Consider a typical reservoir of 50 m thick and inter-well spacing of 1 km within which sandbodies of 5 m thick and 100 m long are distributed randomely. Assuming the net-to-gross of 0.75 for the reservoir, estimate the connected sand fraction between the two wells.

Solution: Consider a two dimensional cross section reservoir model. Then, $L_x = \frac{50}{5} = 10$, $L_y = \frac{1000}{100} = 10$. Moreover, we use usual values of $p_c^\infty = 0.668, \frac{1}{v} = 0.75, \frac{\beta}{v} = 0.015$. Then by using master curves in Figure 8 with $(p - p_c^\infty)L^{1/v} = 0.46$ we get $PL^{\beta/v} = 1.05$ and $\Delta L^{\beta/v} = 0.26$ from which $P = 0.82$ and $\Delta = 0.2$. Hence, the predicted connected fraction of sands between two wells is 0.82 ± 0.20

A similar analysis can be done for the effective permeability results by considering single phase flow between two wells under a fixed pressure drop which gives the pressure equation $\nabla \cdot (K \nabla P) = 0$ where K is the permeability and P is the pressure and ∇ is the derivative operator. Having solved this on the system, one can then calculate the total flow rate and equate fluxes with those from an equivalent homogenous system to determine the effective permeability K. As before, once we collect the statistics of the effective permeability $K(p, L)$ and net to gross, p for a given system size, we can plot the effective permeability against the net to gross that gives a scatter of points (see Fig. 9).

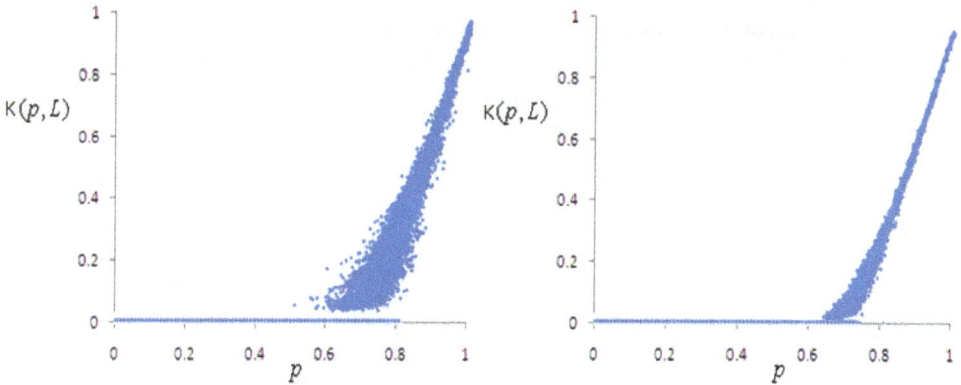

Fig. 9. The typical scatter plots of effective permeability K versus net-to-gross p at system size of $L = 16$ and 64 showing the smearing out of the percolation transition in the basic two dimensional sandbody model

Again, instead of working with the whole scatter one can determine the mean effective permeability, $K(p, L)$, and its associated uncertainty in the effective permeability results, $\Delta_K(p, L)$ which has the scaling: [Stauffer and Aharony, 1992],

$$K(p, L) = L^{-\mu/v} \kappa \left[(p - p_c^\infty) L^{1/v} \right] \tag{9}$$

$$\Delta_K(p, L) = L^{-\mu/v} G \left[(p - p_c^\infty) L^{1/v} \right] \tag{10}$$

Where functions κ and G are two master functions respectively for the mean reservoir effective permeability and its standard deviation. Fig. 10 shows how these scaling laws collapse the effective permeability results of various system size presented in Figure 9 on top of each other.

Using the master curves for the connectivity (i.e. the curve $\mathfrak{F}$ from Figure 8) and effective permeability (i.e. the curve κ from Figure 10) one can replot the connectivity and effective permeability against wellspacing at various net to gross values (Fig. 11). The connected sand fraction shows the fraction of the original oil in place that is in contact with two wells.

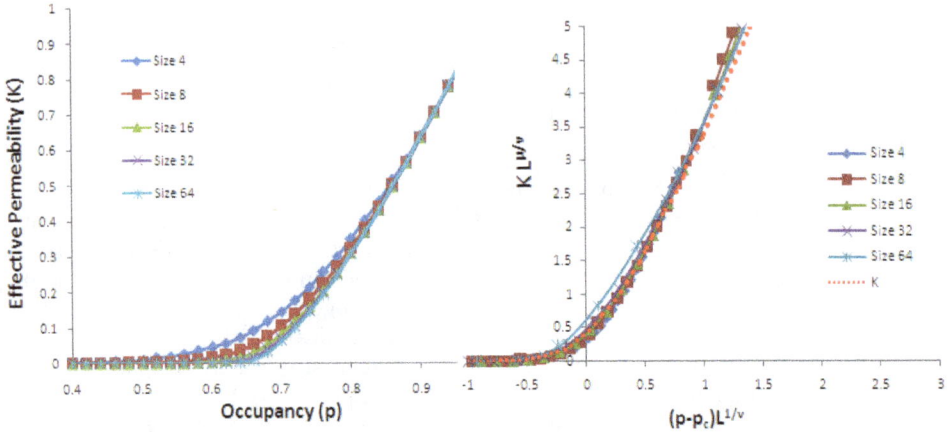

Fig. 10. Illustration of (left) effective permeability results $K(p, L)$ of squared sandbodies in two dimensions at various system sizes (right) the data collapse for conductivity results using finite size scaling

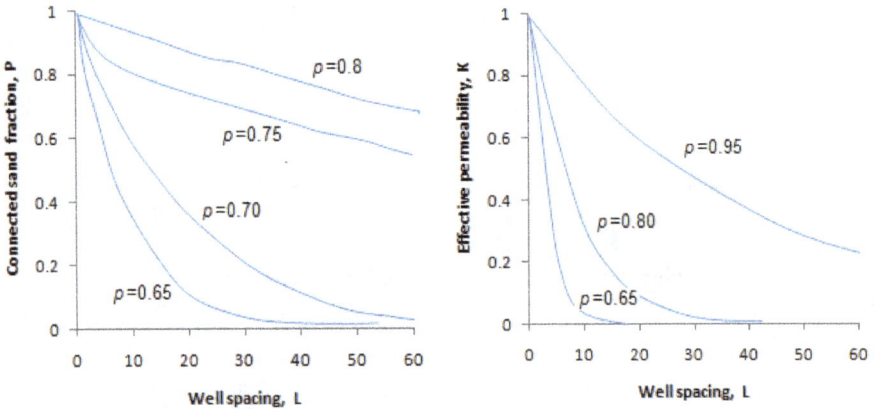

Fig. 11. Illustration of the connectivity results $P(p, L)$ and effective permeability results $K(p, L)$ of squared overlapping sandbodies against well spacing L at various net-to-gross p values

5. Anisotropic reservoirs

In isotropic reservoirs, the horizontal connectivity is the same as the vertical connectivity on average if not for individual realizations. However, for many realistic systems, the objects or their orientation is rarely isotropic. For example in fractured rocks, fracture sets with particular orientations are typically formed as a result of tectonic history [Bear *et al.*, 1993; Sahimi, 1995; Adler and Thovert, 1999; Zhang and Sanderson, 2002] or in overlapping sandbody model the simple assumption of squared sands may not be true and rectangles can be used. This leads to the creation of an *easy* direction for connected paths which is in the short direction and a *difficult* direction which is along the long axis. Monte-Carlo, conformal field and renormalization group theory and duality arguments can be used to incorporate the anisotropic effects in the finite size scaling laws. We define an *apparent* percolation threshold in each direction to be the point where 50% of realizations connect in that direction. Again for simplicity, we describe this on lattice although the results are applicable to other continuum systems such as overlapping sandbody system or fracture system (Masihi et al 2006, 2007; Sadeghnejad et al 2010). Consider an anisotropic system in 2d space which now has different effective size L_x and L_y along the x and y directions and define an aspect ratio $\omega = L_x/L_y$ to represent the anisotropy. Obviously if the effective sizes in two directions are equal (i.e. $L_x = L_y$) then $\omega = 1$ which means that the system is isotropic (see Fig. 12 for a typical realization).

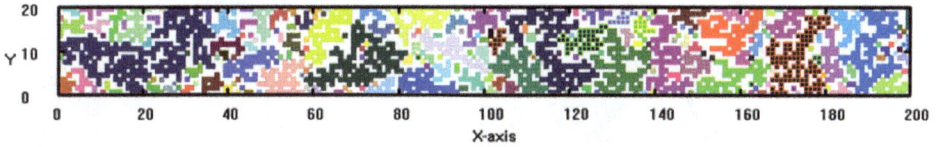

Fig. 12. Illustration of three connected clusters in the Y direction at $\omega = 10$ and $p=0.55$ on a lattice which is globally anisotropic ($L_x = 200$, $L_y = 20$)

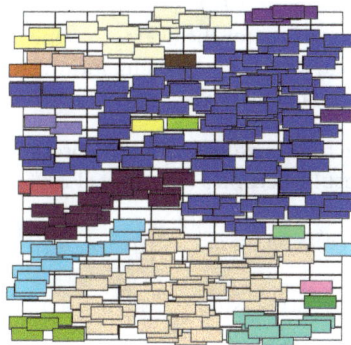

Fig. 13. Illustration of threshold in the horizontal direction for overlapping sandbody model which is globally isotropic but the objects are rectangles with the number of objects N=280, the net to gross $p=0.659$ and the horizontal connected sand fraction of $P_h=0.449$

We investigate finite size scaling at a fixed aspect ratio ω and with the length scale L_x. We expect that both of the apparent x and y percolation thresholds to scale with the length scale as,

$$\tilde{p}_c^i - p_c^\infty = \Lambda_i(\omega)L_x^{-1/v} \tag{11}$$

Where $\tilde{p}_c^i$ is the apparent threshold and Λ_i is simply the constant of proportionality which is dependent on the aspect ratio ω with i labels the coordinate direction. The following simple scaling function has been observed for the proportionality constant Λ_i:

$$\Lambda(\omega) = c(\omega^{1/v} - 1) \tag{12}$$

Where the coefficient c has been estimated numerically about 0.92, 0.45 and 0.58 respectively for the lattice model, overlapping sandbody model and fracture model (King 1990, Masihi et al 2006, 2007, Sadeghnejad et al 2011). It was found that using infinite percolation threshold within the usual finite size scaling law then the anisotropic connectivity in the horizontal and vertical directions (i.e. P_h and P_v) collapse onto the single universal isotropic connectivity curve ($\Im$) as shown in Fig. 14.

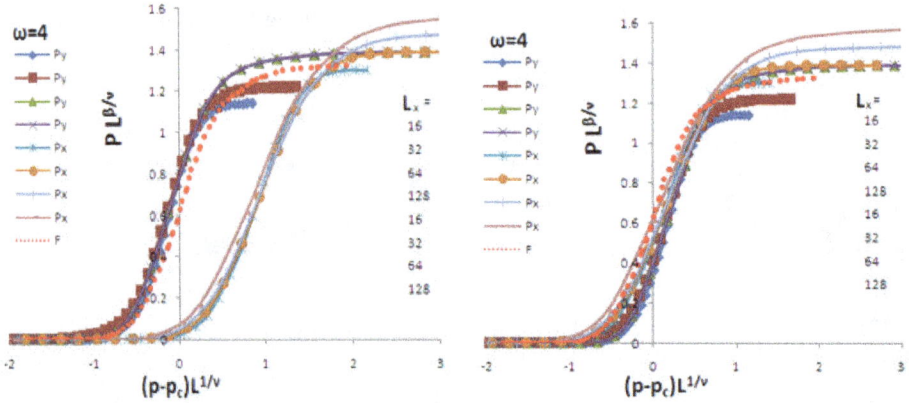

Fig. 14. Illustration of finite size scaling for the average connected fraction (P) of overlapping squares in horizontal and vertical directions by using infinite threshold (left) and apparent threshold (right) in the finite size scaling law.

This led to a way to modify the finite size scaling in the presence of anisotropy. Such scaling, for example for connectivity results becomes (Masihi et al 2006, 2007):

$$P(p, L_x, \omega) = L_x^{-\beta/v} \Im\left[(p - \tilde{p}_c)L_x^{-1/v}\right] \tag{13a}$$

$$\Delta(p, L_x, \omega) = \omega^{1/2} L_x^{-\beta/v} \Re\left[(p - \tilde{p}_c)L_x^{-1/v}\right] \tag{13b}$$

These results enable us to use the same isotropic universal curves given in Figure 8 for predicting the connectivity of anisotropic lattices which is a good enough approximation for engineering purposes.

Example: Plot the variation of the connected sand fraction against well spacing at net-to-gross values of p=0.65, 0.7, 0.75 and 0.80 for a cross section reservoir model of 10m thickness which contains sandbodies of 50 m long and 1 m thick.

Solution: with a well spacing of 500 m then the effective system size is $L_x = 10, L_y = 10$ so the system is isotropic. To calculate the connected sand fraction at a particular net-to-gross (p) we need to invert the finite size scaling with L=10 and the master curve in Fig.8 At a well spacing of 1000 m the effective system size is $L_x = 20, L_y = 10$ so the system is anisotropic with aspect ratio of $\omega = 2$. The horizontal connectivity of this system is the same as the vertical connectivity of a system with $L = 10$ and the aspect ratio of $\omega = 0.5$. With this we can determine the shift in apparent threshold by using Eq. 12 which is $\Lambda(\omega) = -0.18$ and so the apparent threshold from Eq. 11 becomes $\tilde{p}_c = 0.668 + 0.18 \times 10^{-0.75} = 0.7$. Again the connected sand fraction at a particular net-to-gross (p) can be calculated by inverting the finite size scaling with $L=10$ and by using the numerical values from the connectivity master curve. Now by continuing this process we can determine the connectivity as a function of well spacing for various net-to-gross ratios (see Fig. 11 for typical behavior).

6. Size variation and orientation distribution

In reality, the sandbodies in the overlapping sandbody model or fractures in the fracture network model may not have the same size or they do not have a fixed orientation. The distributions for the size and orientation are related to the sedimentological environment which deposited the reservoir. The finite size scaling discussed so far assumes that the sandbodies all have the same lengths and the global behaviour is controlled by two lengths, the system size, L, and the correlation length, ξ. If there is distribution of sizes then this is changed. If the distribution of sizes is a continuous (e.g. uniform or a Gaussian distribution), then it was suggested to replace this variable size system with another system in which the bodies are all with the same effective size. This means that the connectivity of, for example, sandbodies of variable size is identical to the connectivity of sandbodies of the same size. Consider the sandbodies that are represented by squares of variable side size of a. Then it was shown that the effective size $l_{effective}$ which gives the same percolation behavior can be based on the second moment of size distribution as[Balberg et al 1984, Masihi et al 2008]:

$$l_{effective} = \sqrt{\langle a^2 \rangle} \tag{14}$$

Where $\langle \ \rangle$ denotes the average over all fracture lengths. This hypothesis has been verified on the 2d overlapping sandbody model [King 1990], two and three dimensional fracture model [Masihi et al 2008] and 3d overlapping sandbody model [Sadeghnezad et al 2011]. Fig 15 shows the connectivity and conductivity results for a system with size variation as compared with the master curves when effective size is used.

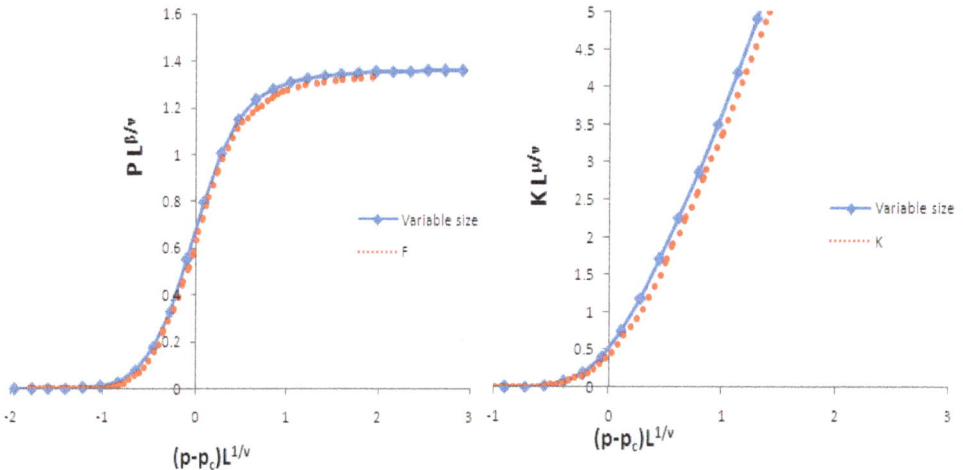

Fig. 15. (left) connected sand fractions and (right) effective permeability for variable sizes squares with uniform length distribution in the range of [0.5-1.5] in a system of size L=24

Moreover, the orientational disorder of the objects (sands or fractures) may enhance the percolation (e.g. connectivity) which means that the infinite threshold has been reduced. In addition to this shift in the threshold there will be a finite size "effective" shift. Consider a system of sandbodies with uniform orientation between $[-\theta_o, \theta_o]$. To account for the shift in the finite threshold a conjecture about the universality of percolation thresholds [Balberg et al 1984, Balberg 1985] saying that there is a universal value of the excluded area (i.e. the area around a geometrical object in which the centre of another object must lie in order for them to overlap) of percolating shapes at the threshold was used [King 1990, Masihi et al 2008] which led to the following infinite threshold:

$$p_c = 1 - e^{-\rho_c} \tag{15}$$

$$\rho_c = \frac{4.41}{2\left(1+\frac{sin^2\theta_o}{\theta_o^2}\right)+\frac{1+\omega^2}{2\omega\theta_o^2}(2\theta_o - sin2\theta_o)} \tag{16}$$

For example, in an overlapping sandbody model with rectangles of an aspect ratio ω and a uniform distribution between $[-\theta, \theta]$ the reduction of the percolation threshold as a function of aspect ratio becomes [Fig. 16],

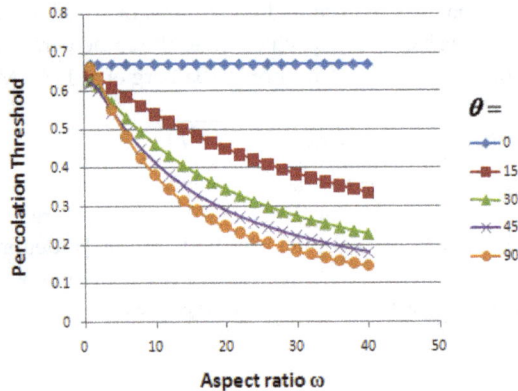

Fig. 16. Illustration of the shift in the threshold due to the orientational disorder of sands

Example: To see the effect of orientational disorder on infinite percolation threshold, one used the Monte Carlo simulation on a simple overlapping sandbody model with rectangles at a fixed aspect ratio $\omega = 2$ and uniform angular dispersion of $\theta_o = \pi/8$ about the vertical axis as shown in Fig. 17. Compare the numerically determined threshold (pc= 0.619) with the value estimated from Eqs. 15 and 16

Solution: The estimated shift in the threshold from Eqs. 15 and 16 for aspect ratio of $\omega = 2$ and distribution range of $\theta_o = \pi/8$ becomes $sp_c = 0.622$ that is in agreement with the numerical results.

To account for the finite size "effective" shift due to the orientational disorder, we use the concept of the effective length which gives the total "reach" of a body. In Fig 18 the effective body size in the x and y directions is the average of all the horizontal and vertical extents of the bodies i.e.,

$$\langle \ell_x \rangle = \frac{1}{n}\Sigma(acos\theta + bsin|\theta|), \quad \langle \ell_y \rangle = \frac{1}{n}\Sigma(bcos\theta + asin|\theta|) \tag{17}$$

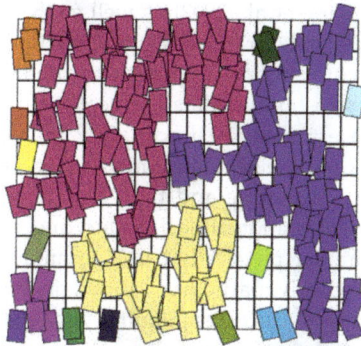

Fig. 17. A realization showing uniform sandbody orientation distribution between $[-\pi/8, \pi/8]$ where $L_x = 20$, $\omega = 2$, N=210, $p_c = 0.619$ and $P_V = 0.316$.

Where sum is over all sandbodies and n is the number of sandbodies within the rectangular system of size X and Y. This shows that the effect of the orientational disorder is to make the sandbodies a bit larger and a bit less elongated. Then, the effective size of the system is defined by two dimensionless lengths $L_x = X/\langle \ell_x \rangle$ and $L_y = Y/\langle \ell_y \rangle$ which implies a new aspect ratio, ω, as $\omega = L_x/L_y$ and so affect the finite size percolation threshold (Eq. 11)

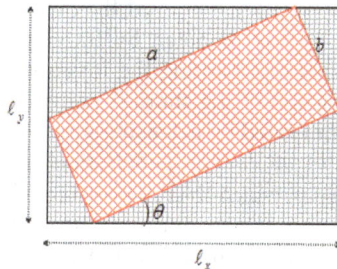

Fig. 18. Illustration of the effective size of a rectangular sandbody of angle θ in two directions

Extensions of the methodology presented in this chapter to three dimensions can be found elsewhere (Masihi et al 2007, Sadeghnejad et al 2010). Moreover, a number of case study including using the data set of a carbonate gas condensate reservoir with tilted fault blocks in the Gulf of Gabes, offshore Tunisia [Nurafza et al 2006], a fractured limestone bed exposed on the southern margin of the Bristol Channel Basin the north Somerset, UK [Masihi et al 2007], and the Burgen reservoir dataset of Nowrooz offshore oil field in the north of Persian Gulf, Iran [Sadeghnejad et al 2011] has been investigated which proved the applicability of percolation approach.

7. Breakthrough and post breakthrough behavior

Percolation concepts can be used to find the probability distribution for the breakthrough time t_{br} between an injector and a producer (Dokholyan et al., 1998; King et al., 1999; Lee et al., 1999; Andrade et al. 2000; King et al., 2002) as well as the post breakthrough behavior. The knowledge of these is important, for example, in water flooding to improve recovery in hydrocarbon reservoirs. The conventional approach to predict the breakthrough time or post breakthrough behavior is to build a detailed geological model, upscale it, and then perform flow simulations. To estimate the uncertainty in such a system, this has to be repeated for various realizations of the geological model. The problem with this approach is that it is computationally very expensive and time consuming. Although there has been some progress to reduce the calculation time [Donato and Blunt, 2004]; there is a great incentive to produce much simpler physically-based techniques to predict these reservoir performance parameters very quickly, particularly for engineering purposes. In this section, we use percolation-based reservoir structures [Stauffer and Aharony 1994] on which we can analyze the breakthrough time and post breakthrough behavior between a injector and a producer [Andrade et al 2000].

Dokholyan et al. [1999] did the pre-elementary study of the breakthrough time. They presented a scaling ansatz for the distribution function of the shortest paths connecting any two points on a percolating cluster. Traveling time and traveling length for tracer dispersion in two-dimensional bond percolation systems have been studied by Lee et al. [1999]. King et al. [1999] used percolation theory to predict the distribution of the shortest path between well pairs and presentd a scaling hypothesis for this distribution which has been confirmed by the numerical simulation. Andrade et al. [2000] concentrated on the flow of fluid between two sites on the percolation cluster. They modelled the medium by using bond percolation on a lattice, while the flow front was modeled by tracer particles driven by a pressure difference between two fixed sites representing the injection and productions wells. They investigated the distribution function of the shortest path connecting these two wells, and proposed a scaling which was confirmed by extensive simulations. Moreover, Andrade et al. [2001] investigated the dynamics of viscous penetration in two-dimensional percolation networks at criticality for the case in which the ratio between the viscosities of displaced and injected fluids is very large. They reported extensive numerical simulations showing that the scaling exponents for the breakthrough time distribution was the same as the previously reported values computed for the case of unit viscosity ratio. Araujo et al. [2002] analysed the distributions of traveling length and minimal traveling time through two-dimensional percolation porous media characterized by relatively long-range spatial correlations. They found that the probability distribution functions follow the same scaling ansatz originally proposed for the uncorrelated case, but with quite different scaling exponents. Using Monte Carlo simulations, comparing the results of the proposed scaling laws from percolation theory with the time for a fluid injected into an oil field to breakthrough into a production well was studied by King et al. [2002]. Lopez et al. [2003] numerically simulated the traveling time of a tracer in convective flow between two wells in a percolation porous media. They analyzed the traveling time probability density function for two values of the fraction of connecting bonds, namely the homogeneous case and the inhomogeneous critical threshold case. Soares et al. [2004] investigated the distribution of shortest paths in percolation systems at the percolation threshold from one injector well to multiple producer wells. They calculated the probability distribution of the optimal path length. Li et al. [2009] applied percolation method to estimate inter-well connectivity of thin intervals for non-communicating stratigraphic intervals in an oil field.

In this part, we characterize the breakthrough time and post breakthrough behaviour between an injector and a producer in a real case by using the Burgan reservoir dataset of Norouz offshore oil field in the south of Iran and applying the propsoed scaling law against the detailed reservoir simulation results. Furthermore, we develop a scaling law for the prediction of post breakthrough behavior based on the probability distribution of breakthrough time. For this purpose, we focus on the time to reach to the water cut of 50% in the production well and introduce a new probability distribution function for this.

It has been shown that the distribution of breakthrough time t_{br} between two wells with well spacing r conditioned to the formation size, L, and net-to-gross ratio, p, has the scaling [Andrade et al 2000]:

$$P(t_{br}|r,L,p) \propto \frac{1}{r^{d_t}}\left(\frac{t_{br}}{r^{d_t}}\right)^{-g_t} f_1\left(\frac{t_{br}}{r^{d_t}}\right) f_2\left(\frac{t_{br}}{L^{d_t}}\right) f_3\left(\frac{t_{br}}{\xi^{d_t}}\right) \tag{18}$$

Where functions $f_1(x) = e^{-ax^{-\phi}}$, $f_2(x) = e^{-bx^{-\psi}}$ and $f_3(x) = e^{-cx^{-\pi}}$. Moreover, t_{br} is the breakthrough time for a given realization, r is the Euclidean distance between the injector and a producer, and ξ is the typical size of reservoir sandbodies. The numerical values of the other exponents were determined through rigorous simulation studies (Table 5).

Exponent	Value	Exponent	Value
d_t	2D: 5.0 3D: 2.3	c	2D: 1.6(−), 2.6(+) 3D: 2.9(−)
g_t	2D: 5.0 3D: 2.3	ϕ	2D: 3.0 3D: 1.6
a	2D: 5.0 3D: 2.3	ψ	2D: 3.0 3D: 2.0
b	2D: 5.0 3D: 2.3	π	2D: 1 3D: 1

Table 5. Numerical values of the exponent in Eq. 18[Andrade et al 2000]

Eq. 18 assumed that the injecting and producing fluids are incompressible, the displacing fluid has the same viscosity and density as displaced fluid (i.e. like passive tracer transport). Pressure field is determined by the solution of the single phase flow equation (i.e. Laplace equation, $\nabla.(K\nabla P) = 0$. Fixed pressure boundary conditions were considered at the two wells. Breakthrough time corresponds to the time when the first tracer reaches the production well for a given reservoir realization. The probability function given by Eq. 1 can be encoded in a spreadsheet from which, using some primary data, the probability distribution of the breakthrough time is determined very quickly. The interested reader is referred to work of Andrade et al 2000 for the idea used in developing Eq. 18

7.1 Application to real field

To validate the approach, we have used the Burgan formation dataset of Norouz offshore oil field in the south of Iran. Core and palynological data have indicated that the Burgan consists of a series of incision-fill sequences occurring in an estuarine/coastal plain/deltaic environment. Consequently, the Burgan formation consists of a thick stack of excellent quality sands incising into each other with a few remaining shalier sediments locally separating these sequences. The excellent formation sands of the Burgan consist mainly of valley-fill deposition sediments, where freshwater fluvial, coarse-grained sediment (high permeable zones) accumulated in the upper and middle reaches of the incised valley system, while tidally influenced sediments accumulated in the marine-influenced middle and lower reaches. As a result of this aggradations, less fluvial-derived sediment reached the lower reaches of the valley system, which resulted in the clean, well-sorted shoreline material being re-worked and deposited in the tidal channels in this area [Huerlimann 2004]. Because of this process, the valley-fill deposition sediments can be considered as high permeable flow units in a background that is essentially impermeable which makes the Burgan reservoir ready to be modeled with the percolation approach.

To have different realizations, various points for the well locations in the entire Burgan reservoir were randomly selected. Nearly 300 flow simulations were run on these wells at the full field scale of the Burgan reservoir formation. To include well spacing effect, two distinct well spacing has been considered as 500m, and 1000m. Then we plot the histogram of the breakthrough times by using the results of 300 simulation runs. Moreover, we encoded the Eq. 18 in an excel spreadsheet for the Burgan reservoir data to produce the breakthrough time probability distribution. The detail of this is as follows: We first make the reservoir sizes dimensionless with the dimension of the sandbody in the appropriate direction (L_x, L_y, L_z) and apply the scaling for the t_{br} with the minimum dimensionless length involved $L_{min} = \min(L_x, L_y, L_z)$. We note that the scaling was for the passive tracer transport; however, the actual displacement is not the same as passive tracer transport. Hence, the Buckley Leverett displacement idea along fastest streamline may be assumed. Moreover, it should be noted that in encoding Eq. 18 with the scaling $t \propto r^{d_t}$ we actually mean $\frac{t}{t_o} \propto \left(\frac{r}{r_o}\right)^{d_t}$ where r_o is a typical length in the system (e.g. sandbody size or r_s) and t_o is the typical time (e.g. the time needed to transit one sandbody). Consider the transit time t (in seconds) for a fluid of viscosity μ (cp) between two wells of radius r_w(in cm) with pressure drop ΔP (in atm) separated by a distance r (in cm) in a homogeneous reservoir with a permeability k (Darcy) is given by $t = \frac{4\mu r^2}{3k\Delta P} \ln \frac{r}{r_w}$. Then, with linear assumption for the pressure drop across the well spacing, the pressure drop across the sandbody becomes $\Delta P \left(\frac{r}{r_s}\right)$. Taking into account the effect of the sandbodies overlapping through the simulation analysis then the typical transit time for a sandbody becomes, $t_o = \frac{12\mu r_s^2}{15k\Delta P\left(\frac{r}{r_s}\right)} \ln \frac{r_s}{r_w}$. These procedures were repeated for two cases with well spacings 500 and 1000 m. The comparison of the result of the scaling law of the breakthrough time (curve) with the results obtained from the conventional numerical simulations (bar chart) for two well spacing is shown in Fig. 19.

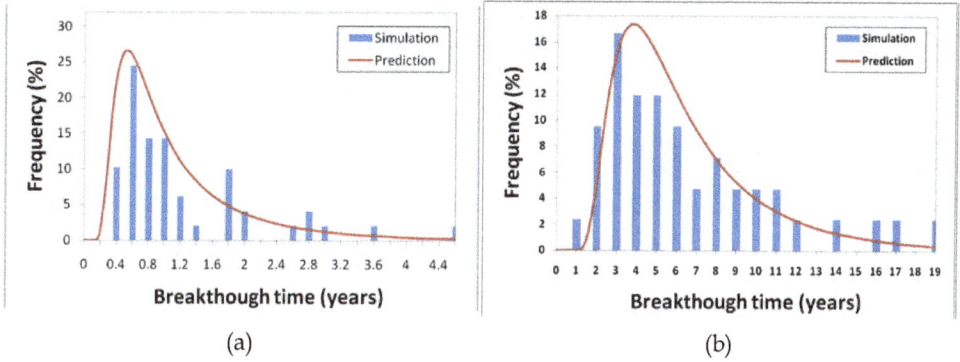

Fig. 19. Comparison of prediction of the breakthrough time using the scaling laws of Eq. 18 (curve) with results from the numerical simulations (bar chart) for well spacing of a) 500m, b) 1000 m.

As it can be seen from Fig. 19, there is a reasonable agreement between the prediction from the scaling law, given by Eq. 18, and the direct numerical simulation results.

7.2 Post breakthrough behavior

The main question in evaluating the post breakthrough behavior is that if the probability distribution of the breakthrough time is available how does this reduce the uncertainty in the post breakthrough behavior. Specifically, we want to check if there is a correlation between the breakthrough time results and the time taken for the oil production to fall by, for example 50%, so called $t_{1/2}$, (or water cut to increase to 50%). To get these statistics, we continue the conventional flow simulations to reach to this water cut value. Then we plot $t_{1/2}$ versus t_{br} on a log-log scale. Fig. 20 shows the results these cross plots at two well spacing values.

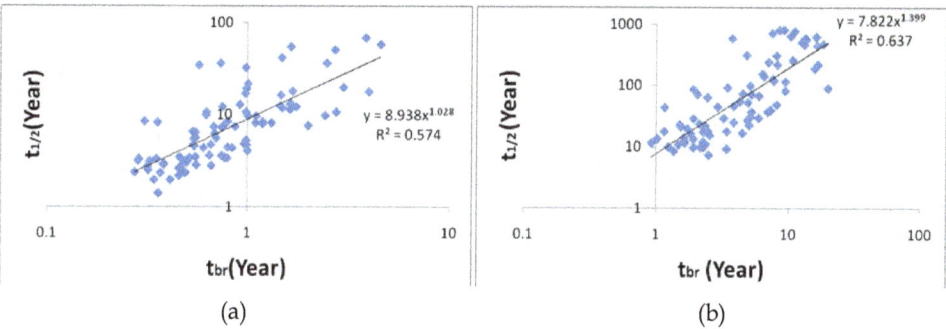

Fig. 20. Illustration of time to reach to the water cut of 50% in production wells versus breakthrough time for different well spacing as a)500m, b) 1000 m.

As it can be seen, there is a positive correlation between the breakthrough time and the time to reach to the water cut of 50% in the production well. The degree of correlation observed in Fig. 20 depends mainly on the heterogeneity of the reservoir. It should be noted that the understudy reservoir is a heterogeneous channel reservoir. By plotting the $t_{1/2}$ versus t_{br} of all simulation results from different well spacing on a log-log scale, one can find a power law scaling between the two times $t_{1/2}$ and t_{br} as:

$$t_{1/2} = 9.343 t_{br}^{1.27} \tag{19}$$

Therefore, for estimation of the probability distribution of the post breakthrough time (defined by the water cut of 50%), one can use the probability distribution of the breakthrough time from Eq.18 and scaled it by using Eq. 19. Fig. 21 compares the simulation results of the post breakthrough behavior with the proposed scaling law.

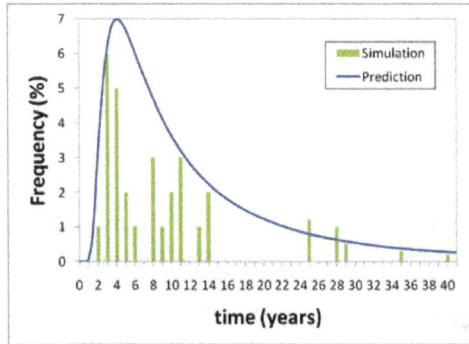

Fig. 21. Comparison of prediction of the time to reach to the water cut of 50% using the scaling laws of post breakthrough behavior (curve) with results from the numerical simulations (bar chart)

8. Summary and conclusion

We have described an approach based on percolation theory to model underground reservoirs and to estimate the reservoir performance parameters very rapidly. This is a field scale approach and it uses simple data acquired from reservoir geology such as reservoir size, net-to-gross ratio within the framework of continuum percolation theory to rapidly evaluate the important parameters of geometrically complex reservoirs, especially in the early reservoir development stage when data are very limited. This approach has been used in estimating the connectivity and conductivity between two or more wells in the reservoir. These can be used for reserve estimation and infill drilling projects. Moreover much details of the methodology have been presented. In particular, the effects of anisotropy in the system, size variations and angular dispersion of sandbodies on the percolation results have been addressed. We observed that the effective length based on the second moment of length distribution can be used to incorporate the effect of variable size. Moreover, the approach to incorporate the effects of angular dispersion of sandbodies within the finite size scaling has been presented.

The methodology has not been limited to estimate the reservoir static data. We have described how to estimate the breakthrough time and post breakthrough behavior between two injection and production wells in terms of percolation concepts. In particular, the proposed scaling law of the breakthrough time probability has been compared with the results obtained from the conventional numerical simulations on the Burgan formation dataset of Norouz offshore oil field in the south of Iran which showed a good agreement between the prediction from the percolation based expression and the numerical simulation results. We have also extend the approach to estimate the post breakthrough behavior (e.g. the time to reach to the water cut of 50% in the production well) given the breakthrough time is available. There was a good agreement between the prediction from the derived correlation and the numerical simulation results. Moreover, the prediction from the scaling law took a fraction of a second of CPU times (as it only needs some algebraic calculations) compared with many hours required for the conventional numerical simulations.

9. References

Adler, P. M. and Thovert, J. F., Fractures and fracture networks, pp 163-218, Kluwer Academic Publishers, London, 1999.

Andrade Jr. J.S., A.D. Araujo, S.V. Buldyrev, S. Havlin, H.E. Stanley, Dynamics of viscous penetration in percolation porous media, Phys. Rev. E 63 (2001), 051403.

Baker, R., Paul, G., Sreenivasan, S. and Stanley, H. E., Continuum percolation threshold for interpenetrating squares and cubes, Physical Review, E, 66: 046136.1-046136.5, 2002.

Balberg, I., Universal percolation threshold limits in the continuum, Physical Review B, 31(6): 4053-4055, 1985.

Balberg, I., Anderson, C. H., Alexander, S. and Wagner, N., Excluded volume and its relation to the onset of percolation, Physical Review B, 30(7): 3933-3943, 1984.

Bear, J., Tsang, C. F. and de Marsily, G., Flow and contaminant transport in fractured rock, pp 169-231, Academic Press, Inc., San Diego, CA, 1993.

Berkowitz, B., Analysis of fracture network connectivity using percolation theory, Mathematical Geology, 27(4): 467-483, 1995.

Berkowitz, B. and Balberg, I., Percolation theory and its application to ground water hydrology, Water Resources Research, 29(4): 775-794, 1993.

Bridge, J.S., Leeder, M.R. (1979) A simulation model of alluvial stratigraphy. J. Sediment. 26(5) 617-644

Broadbent, S.R., Hammersley, J.M. (1957). Percolation processes: I. crystals and mazes. Math Proc Cambridge Philos Soc. Doi:10.1017/S0305004100032680

Chang and Yortsos (SPE Formation Evaluation 5, 1990, 31-38

Dokholyan NV, Lee Y, Buldyrev SV, Havlin S, King PR & Stanley HE, Scaling of the distribution of shortest paths in percolation, J Stat Phys 93: 603-613 1998.

D. Donato, G., M.J. Blunt, Streamline based dual porosity simulation of reactive transport and flow in fractured reservoirs, Water Resour. J.S. Andrade Jr., S.V. Buldyrev, N.V. Dokholyan, S. Havlin, P.R. King, Y. Lee, G. Paul, H.E. Stanley, Flow between two sites on a percolation cluster, Phys. Rev. E 62 (2000) 8270-8281.

Gawlinski, E. T. and Stanley, H. E., Continuum percolation in two dimensions: Monte Carlo tests of scaling and universality for non-interacting discs, Journal of Physics A: Mathematical and General, 14: L291-L299, 1981.

Haldorsen, H.H., Brand, P.J., Macdonald, C.J. (1988) Review of the Stochastic Nature of Reservoirs, In: Edwards, S., King, P.R. (Eds.), Proceed Math Oil Prod, Clarendon Press, Oxford

Harter, T., Finite size scaling analysis of percolation in three dimensional correlated binary Markov chain random fields, Physical Review E, 72: 026120.1-026120.7, 2005.

Hoshen, J. and Kopelman, R., Percolation and cluster distribution, I, cluster multiple labelling technique and critical concentration algorithm, Physical Review, B 14(8): 3438-3445, 1976.

A. Huerlimann, MDP technical report of Shell Company, Appendix seven of Soroosh & Nowrooz Burgan rock properties, NISOC, Iran, 2004.

King, P. R., The connectivity and conductivity of overlapping sandbodies, In: Buller, A. T., North Sea Oil and Gas Reservoirs-II, Graham and Trotman, London, 353-361, 1990.

King, P.R., J.S. Andrade Jr., S.V. Buldyrev, N. Dokholyan, Y. Lee, S. Havlin, H.E. Stanley, Predicting oil recovery using percolation, Phys. Rev. A 266 (1999) 107-114.

King, P.R., S.V. Buldyrev, N.V. Dokholyan, S. Havlin, E. Lopez, G. Paul, H.E. Stanley, Uncertainty in oil production predicted by percolation theory, Phys. Rev. A 306 (2002) 376-380.

Lee, S. B. and Torquato, S., Monte Carlo study of correlated continuum percolation: Universality and percolation thresholds, Physical Review A 41(10): 5338-5344, 1990.

Lee, Y., J.S. Andrade Jr., S.V. Buldyrev, N.V. Dokholyan, S. Havlin, P.R. King, G. Paul, H E. Stanley, Traveling time and traveling length in critical percolation clusters, Phys. Rev. E 60 (1999) 3245-3248.

Li, W., J.L. Jensen, W.B. Ayers, S.M. Hubbard, M.R. Heidari, Comparison of interwell connectivity predictions using percolation, geometrical, and Monte Carlo models, J. Pet. Sci. Eng. 25 (2009) 125-129.

Lopez, E., S.V. Buldyrev, N.V. Dokholyan, L. Goldmakher, S. Havlin, P.R. King, H.E. Stanley, Postbreakthrough behavior in flow through porous media, Phys. Rev. E 67, (2003) 056314.

Lorenz, C. D. and Ziff, R. M., Precise determination of the critical percolation threshold for the three dimensional Swiss cheese model using a growth algorithm, Journal of Chemical Physics, 114(8): 3659-3661, 2001.

Masihi, M. and P R King "Connectivity Prediction in Fractured Reservoirs with Variable Fracture Size; Analysis and Validation", SPE Journal, 13 (1), 88-98, March 2008.

Masihi, M., King, P.R., Nurafza, P.: Effect of anisotropy on finite-size scaling in percolation theory. Phys. Rev. E (2006), doi: 10.1103/PhysRevE.74.042102

Masihi, M., King, P.R., Nurafza, P.: Fast estimation of connectivity in fractured reservoirs using Percolation theory. SPE J. (2007), doi: 10.2118/94186-PA

Nurafza, P., King, P.R., Masihi, M.: Facies connectivity modeling: analysis and field study. Annual Conf. and Exhibition SPE Europec/EAGE, Vienna, Austria (2006)

Odling, N. E., Gillespie, P., Bourgine, B., Castaing, C., Chiles, J-P., Christensen, N. P., Fillion, E., Genter, A., Olsen, C., Thrane, L., Trice, R., Aarseth, E., Walsh, J. J. and Watterson, J., Variations in fracture system geometry and their implications for fluid flow in fractured hydrocarbon reservoirs, Petroleum Geosciences, 5: 373-384, 1999.

Sadeghnejad, S., Masihi, M., King, P.R., Shojaei, A., Pishvaei, M.: Effect of anisotropy on the scaling of connectivity and conductivity in continuum Percolation theory. Phys. Rev. E (2010), doi: 10.1103/PhysRevE.81.061119

Sadeghnejad, S., Masihi, M., King, P.R., Shojaei, A., Pishvaei, M.: Reservoir conductivity evaluation using Percolation theory. Pet. Sci. Technol. (2011), doi: 10.1080/10916460903502506

Sahimi, M., Flow and transport in porous media and fractured Rock, pp 103-157, VCH publication, 1995.

Sahimi, M., Applications of Percolation Theory, 258 pp, Taylor and Francis, 1994.

Soares, R.F., G. Corso, L.S. Lucen, J.E. Freitas, L.R. da Silva, G. Paul, H.E. Stanley, Distributionof shortest paths at percolation threshold: application to oil recovery with multiple wells, Phys. A 343 (2004) 739-747.

Stauffer, D. and Aharony, A., Introduction to percolation theory, 181 pp, Taylor and Francis, London, 1992.

Zhang, X. and Sanderson, D. J., Numerical modelling and analysis of fluid flow and deformation of fractured rock masses, 288 pp, Pergaman, Elsevier Science Ltd., UK, 1st edition, 2002.

12

Simplified Conceptual Structures and Analytical Solutions for Groundwater Discharge Using Reservoir Equations

Alon Rimmer[1] and Andreas Hartmann[2]
[1]Israel Oceanographic and Limnological Research Ltd.,
The Yigal Alon Kinneret Limnological Laboratory,
[2]Institute of Hydrology, Freiburg University,
[1]Israel
[2]Germany

1. Introduction

The approaches to the study of hydrological issues are generally divided into two very different groups: (1) the physical approach; and (2) the system approach (Singh 1988). The physical approach is motivated primarily by scientific study and understanding of the physical phenomena, whereas the practical application of this knowledge to engineering and water resources management is recognized but not always fully required. Unlike detailed physical studies of each hydrological problem, the system approach is driven by the need to establish working relationships between measured parameters for solving practical hydrological problems. This approach simplifies the issue because it is unfeasible to consider the entire physical system. Therefore, a logical approach consists of measuring those variables in the hydrologic cycle, which appear significant to the problem, and establish explicit mathematical relationships between them.

An initial step and a well-recognized part of groundwater flow analysis is the definition of a conceptual model. It is usually a simplified perception of the dominant physical components of the studied groundwater system. The main purpose for constructing a conceptual model is concentrating on the parts relevant for solving the hydrological problem.

Common ways to convert a conceptual model of a groundwater system into mathematical formulations are reservoir (or 'tank') type models (Dooge, 1973; Sugawara, 1995). These model types are often used as a theoretical tool in surface and subsurface hydrology, for water management, control of inflows and outflows in lakes, rivers, reservoirs, and aquifers. The linear reservoir concept is an important component of many widely used hydrological models like the TOPMODEL (Beven & Kirby, 1979), HBV (Lindström et al., 1997) or WaSiM-ETH (Schulla & Jasper, 2007). Reservoir type models are especially useful in karst environments, because the essential information for physical approaches is usually not available (Jukic & Denic-Jukic, 2009). The lack of information and the necessity to use simplified reservoir models become evident in the high number of recently published studies on karst hydrology (Fleury et al., 2007; Geyer et al., 2008; Hartmann et al., 2011; Jukic & Denic-Jukic, 2009; Kessler & Kafri, 2007; Le Moine et al., 2008; Rimmer & Salingar, 2006; Tritz et al., 2011).

In this chapter, a set of typical groundwater modeling problems is described, exemplifying the use of simple reservoir structures to model spring discharge and/or groundwater level during time. In each example, we will explicate the use of the proposed reservoir type system. Moreover, in each case, we will examine an analytical solution associated with the proposed system using simple domain geometries. The advantage of analytical solutions is that their equations offer quick answers to the proposed mechanism based on a few basic parameters. These solutions therefore allow an immediate system understanding and provide a meaning value for each parameter or group of parameters. Given the differential equation that describes the groundwater system, most of the presented analytical solutions can be found using the 'symbolic mathematical toolbox' of MATLAB (http://www.mathworks.com).

2. Examples

Our set of example models include: (1) the classic formation of the linear reservoir problem for an aquifer drained by a single spring; (2) spring discharge potentially fed by two parallel aquifers; (3) spring discharge potentially fed by two serial groundwater aquifers; (4) two parallel aquifers with linear exchange and linear discharges; (5) the discharge from an aquifer with two outlets; (6) the discharge from an aquifer into a lake (submerged springs); and (7) the cases of long-term change of groundwater level and annual spring discharge. Although in most cases the models will be applied with a given set of measured data, it is important to clarify that these types of models are not location-specific, and can be used to model various groundwater flow systems.

2.1 The formation of linear reservoir problem for a single spring discharge

In a traditional hydrology, a spring discharge is often conceptually described and modelled using simple linear reservoirs. We can start the simplification of a system by examining the spring discharge Q (L^3 T^{-1}) according to Darcy's Law:

$$Q(t) = -k_i G \frac{h(t) - H_0}{\Delta x} \tag{1}$$

where h (with units of length, L) represents an equivalent unknown hydraulic head in the aquifer, H_0 (L) is the head at the spring outlet (if an exact head can be evaluated) so that h-H_0 represents the equivalent hydraulic head difference between two points, located at Δx (L) distance one from the other. The k_i (units of length over time, L T^{-1}) is the saturated hydraulic conductivity, and G (L^2) represents an "equivalent" cross section of the flow. For practical purposes, it is assumed that k_i, G and Δx are constant for a given natural aquifer, and therefore Eq. 1 can be simplified to:

$$Q(t) = \alpha \cdot [h(t) - H_0] \quad ; \quad \alpha = -\frac{k_i G}{\Delta x} \tag{2}$$

considering H_0=0 in Eq. 2, further simplification can be conducted by conceptualising the drained aquifer as a reservoir (0) with storage V (L^3) varying in time; constant recharge area A (L^2) and a given effective porosity n (-):

$$V(t) = A \cdot n \cdot h(t) \tag{3}$$

according to Eqs. 2 and 3, in such a reservoir model, spring discharge through the outlet, Q_{out}, is proportional to storage.

$$V(t) = KQ_{out}(t) \quad ; \quad K = \frac{(A \cdot n)}{\alpha} \tag{4}$$

where K (given in units of time, T) is known as the reservoir constant or storage, representing the recharge area, the porosity, the saturated hydraulic conductivity, and the equivalent path and cross section of the flow within the aquifer. Usually, changes of K in time or from one season to another are not physically justified, and it should be independent of both the selected period of modeling, and the boundary conditions (amount of precipitation).

The equation for the continuous water balance in this kind of reservoir is:

$$\frac{dV(t)}{dt} = Q_{in}(t) - Q_{out}(t) \text{ with } Q_{out}(t=0) = Q_0. \tag{5}$$

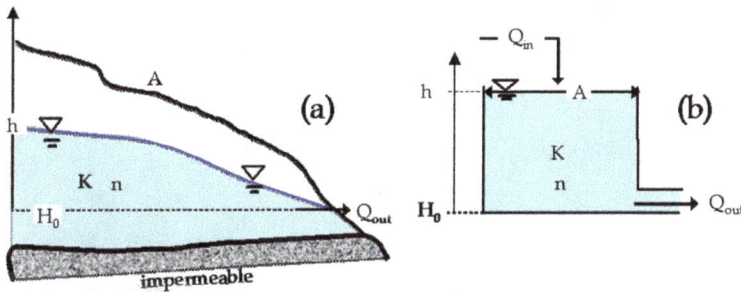

Fig. 1. (a) Schematic description of groundwater system; (b) linear reservoir model

Incorporating Eq. 4 into Eq. 5 results in the linear reservoir differential equation:

$$K \frac{dQ_{out}(t)}{dt} = Q_{in}(t) - Q_{out}(t) \quad ; \quad t_0 \geq t \geq 0 \tag{6}$$

A well-known application in hydrology is the determination of K. This task becomes significantly easier in the dry period that follows the rainy season, since the flow is then a smooth, physical and unidirectional process, with no random processes (such as rainstorms) to be taken into account. At this time $Q_{in}(t)=0$ and the mathematical description of linear reservoir model (Eq. 6) reduces to the homogeneous equation:

$$\frac{dQ_{out}(t)}{dt} + \frac{Q_{out}(t)}{K} = 0 \quad ; \quad Q_{out}(t=0) = Q_0 \tag{7}$$

Eq. 7 is solved analytically by:

$$a. \quad Q_{out}(t) = Q_0 \exp\left(-\frac{t}{K}\right)$$

$$b. \quad V(t) = K Q_{out}(t) = K Q_0 \exp\left(-\frac{t}{K}\right)$$

$$(8)$$

In Eq. 8, V is the volume (assume 10^3 m³), t is the time (day), Q_{out} is the outflow (10^3 m³ day⁻¹), Q_0 is the outflow (10^3 m³ day⁻¹) at the day when Q_{in} vanished, and K is the reservoir constant with units identical to the units of t (day).

Analyzing spring recession in this way is known as Maillet's approach (Maillet, 1905). An application of this fundamental method is presented in Fig. 2, with measured discharge flow from the Carcara Springs in the Western Galilee, Israel, during the dry period starting in March 1981. The springs emerge from the aquifer of the Upper Judea Group formation, which appears to be connected to the aquifer of the Lower Judea Group formations. In this time of the year, the regional groundwater level is usually high. Data from 1950-1985 indicated that the spring had never dried, a situation that changed significantly since the beginning of pumping in 1985 (These changes are discussed in section 2.5).

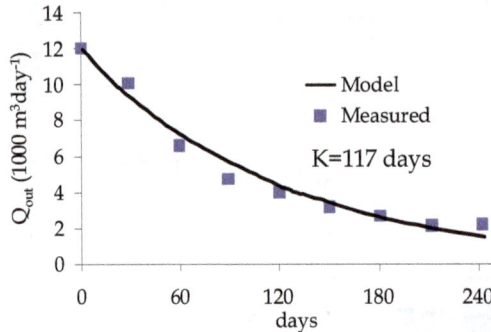

Fig. 2. The discharge of Carcara Spring during the dry period starting in March1981. K was calibrated to 117 days.

2.2 Parallel linear reservoirs

During a dry season that follows a rainy season, the discharge of a spring reduces in time. The shape of the graph discharge vs. time corresponds to the sum of several exponential functions (Bonacci, 1993; Grasso & Jeannin, 1994). Often, such spring discharge is represented as a combination of two parallel linear reservoirs (Fig. 3), mathematically represented by:

$$a. \quad Q_{out}(t) = Q_{out1}(t) + Q_{out2}(t) =$$

$$Q_{01} \exp\left(-\frac{t}{K_1}\right) + Q_{02} \exp\left(-\frac{t}{K_2}\right)$$

$$(9)$$

$$b. \quad V(t) = K_1 Q_{out1}(t) + K_2 Q_{out2}(t)$$

A simple optimization algorithm can be applied to identify the K_1 and K_2 constants, as well as the initial flows Q_{01} and Q_{02} during a recession period. When K is small, the recession of the reservoir will be fast, and its discharge and volume will reach zero within a short time. When K is large, the recession will be slow, and the reservoir outflow will last for a long time. If $K_1 >> K_2$ the discharge from reservoir 2 (second component in the right hand side of Eq. 9) decreases much faster than the discharge from reservoir 1 (first component in the right hand side of Eq.9), which will still be active much longer.

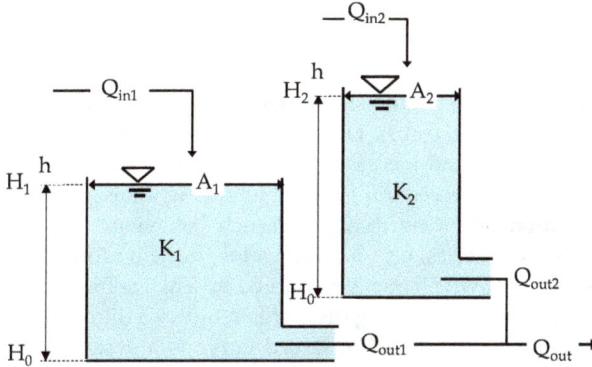

Fig. 3. Two parallel reservoirs. The reservoirs are fed by groundwater recharge originating from the surface and drain simultaneously to the stream.

The parallel groundwater reservoir structure is incorporated in many hydrological models, e.g., the Vensim model (Fleury et al., 2007). Here it is exemplified by applying it to the recession discharge of the Hermon Stream (Israel) during the year of 1996 (Fig. 4). The stream is one of the three main tributaries of the Upper Jordan River (Rimmer and Salingar, 2006). It is fed mainly by the Banias Spring, located at the edge of the karst exposures on the lower parts of the Hermon Mountain, at an altitude of 359 m a.s.l. The Banias annual average discharge is ~67 M m^3 (~2.15 m^3/s). The spring exhibits behaviour of pluvio-nival regime, where discharge is mainly due to precipitation, but also slightly influenced by snowmelt (Gilad & Bonne, 1990; Samuels et al., 2010).

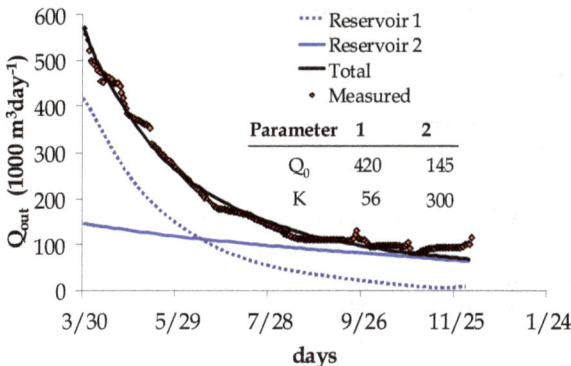

Fig. 4. Stream discharge of the Hermon stream during 1996. The stream is fed by two parallel reservoirs during the recession period: Since $K_2 >> K_1$ the reservoir 1 discharge represents most of the sharp changes following the rainy season, while reservoir 2 represents the more stable component.

The optimization algorithm revealed that K_1 = 56 days and Q_{01} = 420,000 m³day⁻¹, representing the immediate aquifer that contributes to the spring, while K_2 = 300 days and Q_{01} = 145,000 m³day⁻¹, representing the discharge from a large stable aquifer, which also drains into the Dan Spring located nearby. During the recession period, the discharge of reservoir 1 ceases after ~170 days, while the memory of reservoir 2 remains for ~2.5 years (Rimmer & Salingar, 2006).

2.3 Two serial linear reservoirs

Usually, the discharge recession of a karst spring is fast at the beginning of a dry season and slows at its end (see Eqs. 8 and/or 9). However, there are cases in which the recession is rather slow at the beginning, and increases towards the end of the dry season (Rimmer & Salingar, 2006). Moreover, the recession is faster following a low precipitation season, than after a high precipitation one. One reason for such behaviour can be explained by the interplay of two systems in series, e.g., a large vadose zone on top of the phreatic zone or two groundwater systems of which one is recharged by leakage from the other (Fig. 5a). The pattern analysis of such measured spring discharge requires a different setup. The proposed mechanism for examining this type of observed curve is a system with two serial linear reservoirs (Fig. 5b).

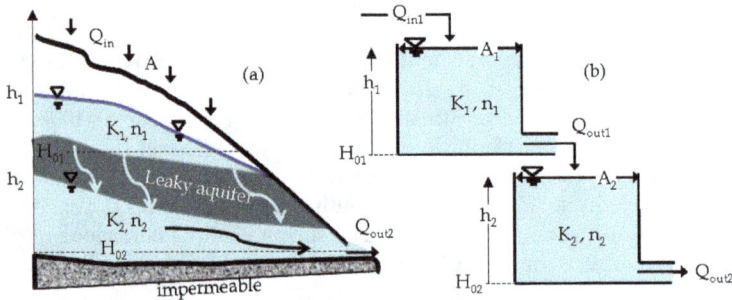

Fig. 5. (a) schematic description of the proposed groundwater system; (b) the system represented by two serial reservoirs. Excess saturation flow from the earth surface feed the upper reservoir (1), which recharges the lower reservoir (2).

In this example, the simplified system is described by an upper linear reservoir, contributing to a lower reservoir, draining through a spring outlet. Similarly to the previous case, we are particularly interested in determining the system storage coefficients K_1 and K_2, and the initial conditions (flow at the beginning of the dry season) Q_{01} and Q_{02}. By defining the input to the upper reservoir (1) during the dry season as zero, the input to the lower reservoir (2) is an exponential recession with time, typical to a linear reservoir system (Section 2.1; see also Nash 1957; Huggins & Burney, 1982). We therefore write the differential equation for the lower reservoir (2) for a single dry season as follows:

$$\frac{dQ_{out2}(t)}{dt} + \frac{Q_{out2}(t)}{K_2} = \frac{Q_{out1}(t)}{K_2} \quad ; \quad t \geq 0$$

$$\text{with } 1.\, Q_{out1}(t=0) = Q_{01} \qquad 2.\, Q_{out2}(t=0) = Q_{02} ;$$

(10)

where Q_{01} and Q_{02} are the initial conditions, yet to be determined from the measured data of each season. In Eq. 10 the contribution from the upper to the lower groundwater reservoir and the upper reservoir volume are determined by:

$$Q_{out1}(t) = Q_{01} \exp\left(-\frac{t}{K_1}\right)$$

$$V_1(t) = K_1 Q_{out1}$$

(11)

and Eq. 10 can be solved analytically so that the discharge from the lower groundwater reservoir and its volume are determined by:

$$Q_{out2}(t) = Q_{out2/1} + Q_{out2/2}$$

$$Q_{out2/1} = \frac{Q_{01} K_1}{(K_1 - K_2)}\left[\exp\left(\frac{-t}{K_1}\right) - \exp\left(\frac{-t}{K_2}\right)\right]$$

$$Q_{out2/2} = Q_{02} \exp\left(\frac{-t}{K_2}\right)$$

$$V_2(t) = K_2 Q_{out2}$$

(12)

Here, the outflow from the lower reservoir is combined of the contribution from the upper reservoir $Q_{out2/1}$ and the self-discharge of the lower reservoir $Q_{out2/2}$. With an optimization algorithm, Eq. 12 may be used to evaluate K_1, K_2, Q_{01} and Q_{02} for each season, so that it

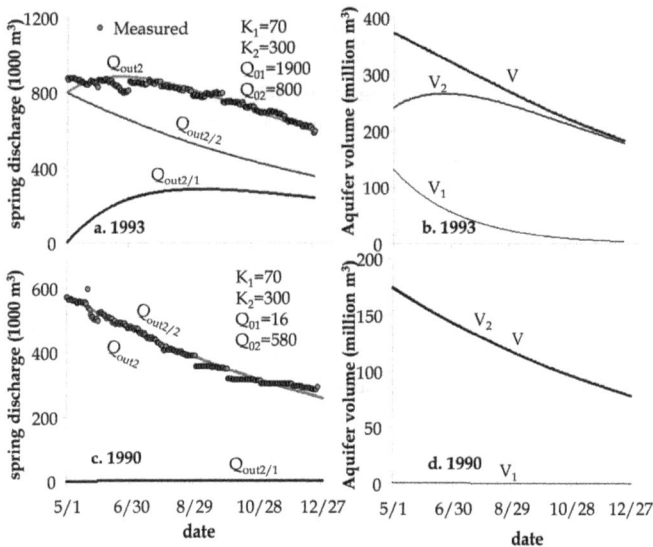

Fig. 6. Illustration of the terms in Eq. 12- the upper reservoir contribution $Q_{out2/1}$ and the self-discharge of the lower reservoir $Q_{out2/2}$ combine the total outflow Q_{out2}. The $K_1 = 70$ days and $K_2 = 300$ days are identical for both rainy (1993) and dry (1990) years, while the two initial conditions Q_{01} and Q_{02} are different. (a, b): 1993; (c, d): 1990. (a, c): spring discharge; (b, d): Aquifer volume.

would match the measured spring discharge. Two restrictions should be imposed on the calibration procedure in order to take into account the physical conditions of the entire system. First, the same K_1 and K_2 must be used for all seasons, and second, there should be a good correlation between Q_{01} and Q_{02} and the annual precipitation during the years, since the entire system is driven by the same precipitation.

Fig. 6 illustrates the curve fitting of Eq. 12 to the discharge of the Dan Spring, Israel, during the dry season that followed two different rainy seasons. The $K_1 = 70$ days and $K_2 = 300$ days were evaluated as the best fit. The initial conditions of $Q_{01}=1900$ m^3 day^{-1} and $Q_{02} =800$ m^3 day^{-1} were valid following a very rainy season (1992-1993), while $Q_{01}=16$ m^3 day^{-1} and Q_{02} =580 m^3 day^{-1} were valid for extremely dry season (1989-1990). Following the rainy season, both reservoirs were partly filled according to the amount of precipitation. However, while the recession of the lower reservoir follows the same rate $(\exp(-t/K_2))$ under any initial condition, the additional recharge from the upper reservoir changes significantly the $Q_{out2}(t)$ curve during the dry season. Consequently, the flow rate of the spring may increase first, following a very rainy year (1992-1993) or reduce immediately following a very dry year (1989-1990). Similar applications can be found in Kiraly, (2003) or Rimmer & Salingar, (2006).

2.4 Two reservoirs with linear exchange

The karst environment is often described as a system with dual porosity (Goldscheider & Drew, 2007), including the fast flow component within the preferential flow paths (karstic conduits), and the slower Darcian groundwater flow within the fissure matrix (Fig. 7a). This process can be conceptualized by dividing the groundwater system in two reservoirs, one representing the conduits and the other representing the fissure matrix (Fig. 7b). Similar to section 2.2, the water exchange between the reservoirs is controlled by the difference in their levels and with similar considerations as in Eqs. 1-4 the spring discharge Q_1 (L^{3}T^{-1}) (or the conduit outflow) may be derived by:

$$Q_1(t) = \frac{V_1(t)}{K_1} \tag{13}$$

Applying the same procedure to the exchange flow between the fissure matrix and the conduit reservoir (reservoirs 2 and 1 in in Fig. 7), and aggregating all constant parameters in a single exchange constant $K_E(T)$, the exchange flow Q_E (L^{3}T^{-1}) can be described as a simple linear relation between water level differences (Fig. 7),

$$Q_E(t) = k_{12}G_{12}\frac{h_2(t)-h_1(t)}{\Delta x_{12}} = \frac{V_2(t)-f_P V_1(t)}{K_E} \tag{14}$$

Where:

$$h_2(t) = \frac{V_2(t)}{An_2} \qquad \frac{1}{K_E} = \frac{k_{12}G_{12}}{An_2\Delta x_{12}} \qquad f_P = \frac{n_2}{n_1} \tag{15}$$

Similar to Eqs. 1-4, h_2 (L) is the water table elevation of the fissure matrix, k_{12} (L T^{-1}) is a representative saturated hydraulic conductivity, G_{12} (L^2) is an equivalent cross-section, and

Δx_{12} (L) an average flow distance; all are parameters representing the interface between conduits and fissure matrix.

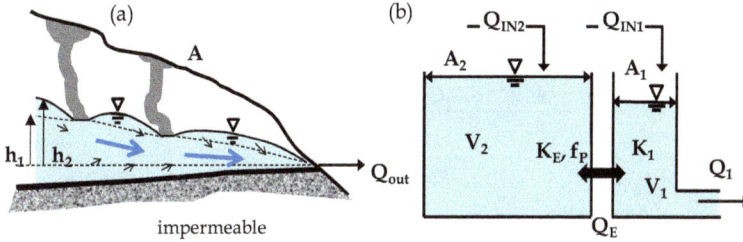

Fig. 7. (a) Schematic description of the groundwater system with karstic conduits and fissure matrix; (b) the system represented by a reservoir combination with a fissure matrix reservoir (left, 2) and conduit reservoir (right, 1).

With the effective porosity of the fissure matrix n_2, and the area A_2, the relation between water level and stored water volume V_2 (L³) can be established. Note that in Eqs. 13-15 the same area A is used because the simplification approach assumes that the conduits section is embedded within the fissure matrix (double porosity approach) and that the porosity differences between the conduits and fissure matrix were taken into account by the porosity factor f_P. With it A, A_1 and A_2 in Fig. 7 are related to each other as follows:

$$A = A_1 + A_2 = A_1 \left(1 + f_P \right) \tag{16}$$

Having defined the flow processes of the conduit and the fissure matrix, water balance for both reservoirs can be established:

$$\frac{dV_1(t)}{dt} = Q_{IN1} + \frac{V_2(t) - f_P V_1(t)}{K_E} - \frac{V_1(t)}{K_1}$$
$$\frac{dV_2(t)}{dt} = Q_{IN2} - \frac{V_2(t) - f_P V_1(t)}{K_E} \tag{17}$$

Rearranging Eq. (17) results in a linear system of inhomogeneous differential equations:

$$\begin{pmatrix} D + \dfrac{1}{K_1} + \dfrac{f_P}{K_E} & -\dfrac{1}{K_E} \\ -\dfrac{f_P}{K_E} & D + \dfrac{1}{K_E} \end{pmatrix} \cdot \begin{pmatrix} V_1(t) \\ V_2(t) \end{pmatrix} = \begin{pmatrix} Q_{IN1} \\ Q_{IN2} \end{pmatrix} \tag{18}$$

Hereby, D is the differential operator d/dt. Assuming constant inflows Q_{IN1} and Q_{IN2}, Eq. (18) can be solved analytically with standard methods (Kramer's rule, variation of constants; e.g. Boyce and DiPrima, 2000) and yield:

$$V_1(t) = B_1 \exp(A_1 t) + B_2 \exp(A_2 t) + C_1$$
$$V_2(t) = B_3 \exp(A_1 t) + B_4 \exp(A_2 t) + C_2 \tag{19}$$

With the constants

$$A_{1,2} = -\frac{1}{2}\left(\frac{1}{K_1} + \frac{1+f_P}{K_E}\right) \pm \sqrt{\frac{1}{4}\left(\frac{1}{K_1} + \frac{1+f_P}{K_E}\right)^2 - \frac{1}{K_1 K_E}} \tag{20}$$

$$B_1 = \frac{V'_{10} - A_2 V_{10} + A_2 C_1}{A_1 - A_2} \qquad\qquad B_2 = V_{10} - B_1 - C_1$$

$$B_3 = \frac{V'_{20} - A_2 V_{20} + A_2 C_2}{A_1 - A_2} \qquad\qquad B_4 = V_{20} - B_3 - C_2 \tag{21}$$

$$C_1 = K_1\left(Q_{IN1} + Q_{IN2}\right)$$
$$C_2 = K_E Q_{IN2} + K_1 f_P\left(Q_{IN1} + Q_{IN2}\right) \tag{22}$$

Where V_{10} and V_{20} are the reservoir volumes and V'_{10} and V'_{20} the storage change at t= 0. V'_{10} and V'_{20} can be obtained by Eq. 18):

$$V'_{10} = \frac{dV_1(t=0)}{dt} = Q_{IN1} + \frac{V_{20} - f_P V_{10}}{K_E} - \frac{V_{10}}{K_1}$$

$$V'_{20} = \frac{dV_2(t=0)}{dt} = Q_{IN2} - \frac{V_{20} - f_P V_{10}}{K_E} \tag{23}$$

Except for $A_{1,2}$, the constants, as well as the initial conditions refer to a single time step, and have to be calculated each time step again. For instance at time step t V_{10} would be equal to $V_1(t-1)$ and V_{20} equal to $V_2(t-2)$, respectively.

Methods that consider the exchange between fissure matrix and conduits can be found in Cornaton & Perrochet (2002) and Sauter (1992). In Fig. 8, the exchange reservoirs solution was applied to the last recharge event and the dry season recession in 1998 at the Banias Spring (see section 2.2). The exchange between the conduit and the fissure matrix reservoir resulted in a buffering of the recharge signal and a slow increase in fissure matrix storage. Exchange flow was negative, indicating flow towards the fissure matrix. Around the end of

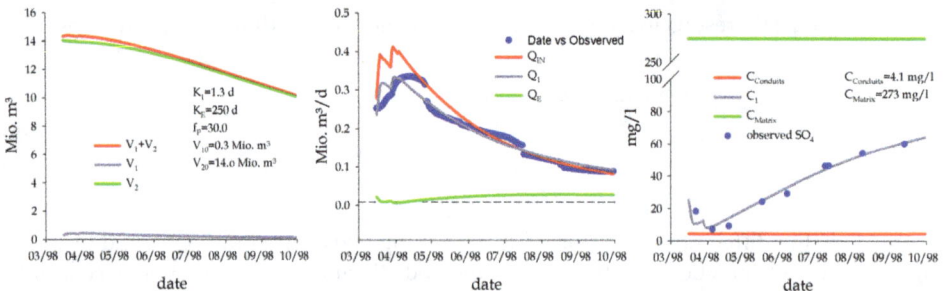

Fig. 8. Left: stored water in the conduit reservoir V_1, the fissure matrix reservoir V_2 and total storage V_1+V_2; Middle: total recharge Q_{IN}, spring discharge Q_1 and exchange flow Q_E vs. observed spring flow; Right: conduit and matrix water SO_4 concentrations, $C_{Conduits}$ and C_{Matrix}, discharge concentrations C_1 vs. observed SO_4 concentrations

April, it changed its direction, which means that parts of the stored water in the fissure matrix were released again to the springs. This switch of direction of the exchange flow was nearly insignificant in terms of flow rates but had immense impact on the water quality. This is exemplified by the SO_4 variations observed at the same spring during the same time: by simply attributing constant SO_4 concentrations to the conduit and matrix flows their mixing at the spring outlet resulted in an acceptable agreement with the observations.

2.5 The linear reservoir with two outlets

In this section, the case of the effect of additional outlet is discussed. Consider the case of a spring discharge, which differs from the basic case (section 2.1) in two important elements: (1) the spring may dry out completely, so that the exponential recession (Eq. 8) is not valid for low flow rates; and (2) From the water mass balance calculations, it is assumed that the groundwater recharge is larger than the spring discharge, and therefore part of the water continues to flow downstream the aquifer to deeper layers. When these two conditions are valid, the linear reservoir with upper and lower outlets (Fig. 9) may represent the system rather well.

With similar considerations, we can handle the problem in Eq. 5 with two outlets, and no recharge, described as follows:

$$\frac{dV(t)}{dt} = \begin{cases} Q_{out0} + Q_{out1} & ; \quad 0 \leq t \leq t_1 \\ Q_{out0} & ; \quad t \geq t_1 \end{cases} \tag{24}$$

where t_1 is the time in which the upper outlet (Q_{out1}) is drying, leaving only the flow in the lower outlet (Q_{out0}).

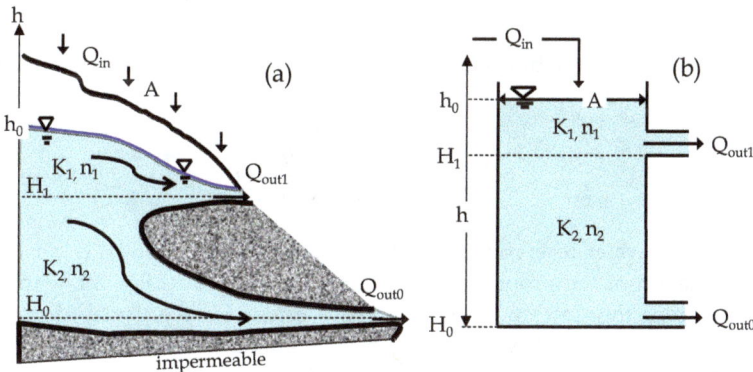

Fig. 9. Linear reservoir with two outlets at different levels

If it is assumed that outlet (1) changes the pressure field only locally, we can consider each outflow separately as a linear function of the head above it, so that:

$$Q_{out1}(t) = -\alpha_1 \cdot (h(t) - H_1)$$
$$Q_{out0}(t) = -\alpha_2 \cdot (h(t) - H_0) \tag{25}$$

With Eq. 25 incorporated into Eq. 24, assuming no inflow and $H_0=0$ the problem is defined as follows:

$$\frac{dh(t)}{dt} = \begin{cases} -\left(\dfrac{1}{K_1}+\dfrac{1}{K_2}\right)h(t)+\left(\dfrac{1}{K_1}H_1\right) & ; \quad 0 \le t \le t_1 \\[4mm] -\dfrac{1}{K_2}h(t) & ; \quad t_1 \le t \end{cases} \qquad ; \quad with \quad h(t=0)=h_0 \qquad (26)$$

The analytical solution to the problem in Eq. 26 is:

$$h(t)=\frac{C}{q}+\left(h_0-\frac{C}{q}\right)\exp(-qt) \quad ; \quad 0 \le t \le t_1$$

where

$$C=\frac{H_1}{K_1} \quad ; \quad q=\frac{1}{K_1}+\frac{1}{K_2} \qquad (27)$$

and

$$h(t)=C\exp\left(-\frac{1}{K_2}(t-t_1)\right) \quad ; \qquad t \ge t_1$$

In order to keep continuous recession curve, the head at time t_1 should be equal to H_1, and identical for the two problems, therefore:

$$h(t_1)=\frac{C}{q}+\left(h_0-\frac{C}{q}\right)\exp(-qt_1)=H_1 \qquad (28)$$

From Eq. 28 we can define the time t_1 in which the flow from the upper outlet vanishes. It is a function of h_0, H_1, K_1 and K_2.

$$t_1=t|_{h=H_1}=-\ln\left(\frac{H_1-C/q}{h_0-C/q}\right)\Big/q \qquad (29)$$

That type of groundwater reservoir is also included in the HBV model (Lindström et al., 1997). An application of the proposed mechanism is presented in Fig. 10, with measured discharge flow from the Carcara Springs in the Western Galilee, Israel, during the dry period starting in March 2002. Note that this spring is similar to the one presented in section 2.1 and therefore K_1 was calibrated to 117 days. However, there is a major difference between section 2.1 and 2.5; since the beginning of pumping in 1985, water levels have been dropping significantly, so that the spring has been drying completely almost every dry season since 1995. The drying requires analysing the spring discharge with Eq. 26 rather than with Eq. 7, and additional calibration of $K_2=100$ days, which, as expected, turned to be nearly similar to K_1. On the regular scale (Fig. 10a) the difference between a drying and not drying spring is not easily perceived, but it becomes clear, and the value of $t_1 = 161$ days is obvious when plotted on a logarithmic scale (Fig. 10b).

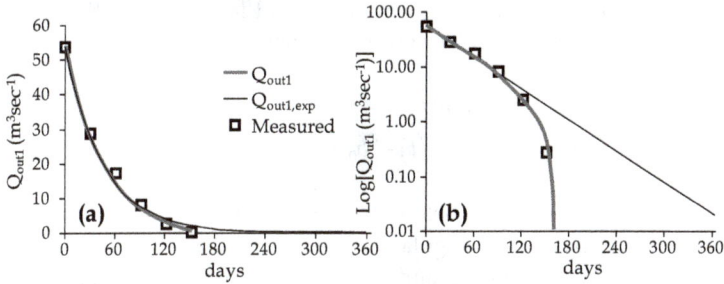

Fig. 10. The outflows through the upper outlet in a linear reservoir with two outlets in different level Q_{out1} and, for comparison, the outflow from a regular linear reservoir without lower outlet $Q_{out1,exp}$. (a) regular scale; (b) logarithmic scale.

2.6 Aquifer drainage to submerged springs

The same physical factors were considered in modelling the process of groundwater discharge into springs onshore and offshore a lake, or a river (Fig. 11). Unlike in previous cases, here, the analytical solution was applied to the entire annual cycle, in order to exemplify the case where the spring outflows are dictated by the downstream head at the lake or river.

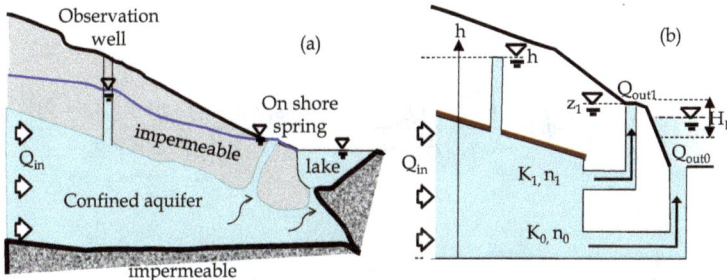

Fig. 11. (a) Schematic description of groundwater system that discharges simultaneously from confined aquifer into springs onshore and offshore a lake. (b) A model where the reservoir drains through a constant level (z_1) onshore spring $Q_{out1}(t)$, and a time varying boundary $H_L(t)$ representing offshore spring $Q_{out0}(t)$. The hydraulic head within a short distance (up to several hundred meters) from the lake is $h(t)$.

The proposed simplified model aims to link the time-dependent spring discharge to the hydraulic heads in the contributing aquifer under the fluctuations of the lake level. These fluctuations are independent of the aquifer system, and affect the spring flow as a close boundary condition. Here we assume constant recharge Q_{in}, and time dependent discharge from the aquifer to the onshore $Q_{out1}(t)$ and offshore $Q_{out0}(t)$ springs:

$$
\begin{aligned}
Q_{out1}(t) &= \alpha_1 \cdot \left[h(t) - z_1 \right] \\
Q_{out0}(t) &= \alpha_0 \cdot \left[h(t) - H_L(t) \right]
\end{aligned}
\tag{30}
$$

We regard the elevation of the onshore spring $z_1=0$, and lake H_L level fluctuating as a sin function around an average level. Under these conditions:

$$Q_{out1}(t) = \alpha_1 \cdot h(t)$$
$$Q_{out0}(t) = \alpha_0 \cdot \left(h(t) - \left[b_0 + b_1 \sin(\beta_1 t + \gamma) \right] \right) \tag{31}$$

Here b_0 (m) is the average lake level below the spring outlet (b_0 is negative); b_1 (m) is the lake fluctuations amplitude; β_1 is the angular frequency in radians which for yearly rotations has a set value of $\beta_1 = 2 \times \pi / 365.25$; and ω is the phase shift (radians). With similar considerations as in earlier problems, we can handle the reservoir mass balance with two outlets, and constant recharge:

$$\frac{dV(t)}{dt} = Q_{in} - Q_{out0} - Q_{out1} \tag{32}$$

Incorporating Eq. 31 into Eq. 32 and rearranging using Eqs. 2, 3 and 4 results in:

$$\frac{dh(t)}{dt} + \left(\frac{1}{K_1} + \frac{1}{K_0} \right) h(t) = \frac{Q_{in}}{A \cdot n} + \frac{b_0}{K_0} + \frac{b_1}{K_0} \left[\sin(\beta_1 t + \gamma) \right] \tag{33}$$
$$\text{with}: \quad h(t=0) = H_0$$

This equation is solved analytically by:

$$h(t) = H_0 \exp(-qt) + \frac{b_0}{K_0 q} + \frac{Q_{in}}{A \cdot n \cdot q} - b_1 \frac{\beta \cos(\beta_1 t + \gamma) + q \sin(\beta_1 t + \gamma)}{K_0 \left(\beta_1^2 + q^2 \right)} \tag{34}$$

$$q = \frac{1}{K_0} + \frac{1}{K_1}$$

An initial test of this solution reveals that if $t \to \infty$, the lake level assumed to be steady with no fluctuations ($b_1=0$), and the inflow $Q_{in}=0$, then:

$$h(t) = \frac{b_0}{K_0 q} = \frac{b_0}{\left(1 + \dfrac{K_0}{K_1} \right)} \tag{35}$$

From Eq. 35 it can be concluded that if the connection between the aquifer and the lake is significantly stronger than the connection to the onshore spring ($K_0 << K_1$), then the aquifer hydraulic head assumes the level of the lake $h(t) \to b_0$, but if $K_0 >> K_1$ the aquifer hydraulic head adapts to the level of the spring outlet $h(t) \to 0$. If $Q_{in}>0$ then h(t) increases by Q_{in} resulting higher discharge through the spring outlets. Discharge of an onshore spring $Q_{out1}(t)$ is straight forward to measure. Therefore, we can evaluate it according to Eq. 31 and calibrate α_1. However the offshore spring discharge $Q_{out2}(t)$ is usually difficult to measure resulting in infinite possibilities to evaluate it since α_0 is also unknown.

As an example, the analytical solution is applied to the Fuliya Springs (Fig. 12) onshore and offshore lake Kinneret Israel. The case of the Fuliya saline springs was classified as confined carbonate aquifer, interacting with the lake through fractures and faults (Goldshmidt et al., 1967; Gvirtzman et al., 1997; Bergelson et al., 1998). The carbonate aquifer system of these springs overlays deep-seated brine, from which saline flux mixes with the fresh groundwater. Diluted saline water drains through fracture springs to both onshore and offshore springs (Rimmer et al., 1999, Abbo et al., 2003). Hydrogeological studies of this natural group of springs, as well as their intensive monitoring (Rimmer et al., 1999) allow us to analyze the simultaneous discharge processes of both onshore and offshore springs in more detail. The observations show that the measured hydraulic head of the aquifer and the discharge to the onshore springs follows the fluctuations (increase or decrease) of the measured lake level (Fig. 12). Discharge to offshore springs could not be measured directly. There is however clear evidence (Simon & Eizik, 1991) that it behaves as a "mirror" picture of the lake level. These results were previously verified by a partial analytical solution proposed by Rimmer et al., (1999) and later by a detailed numerical model (Abbo et al., 2003). With the current analytical solution in Eq. 34 we can test the offshore and the total discharge in time by changing α_0 (Fig. 13). The 'real' value of $Q_{out2}(t)$ remain however unknown.

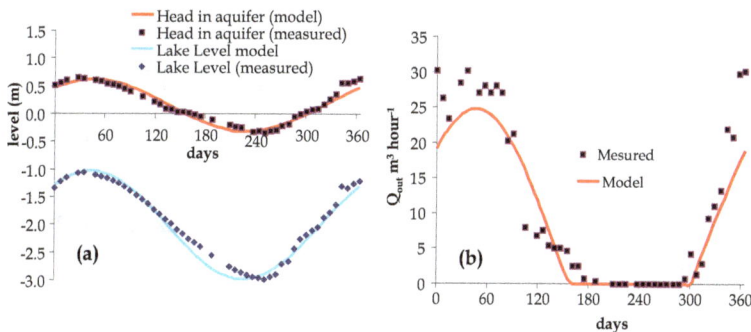

Fig. 12. Application of the analytical solution (Eq. 34) to a. the measured Lake Kinneret level and the measured hydraulic head in the aquifer ~100 m from the lake, and to b. Fuliya Spring discharge through onshore spring. Discharge to the onshore spring vanishes when the hydraulic head drops below the level of the spring outlet.

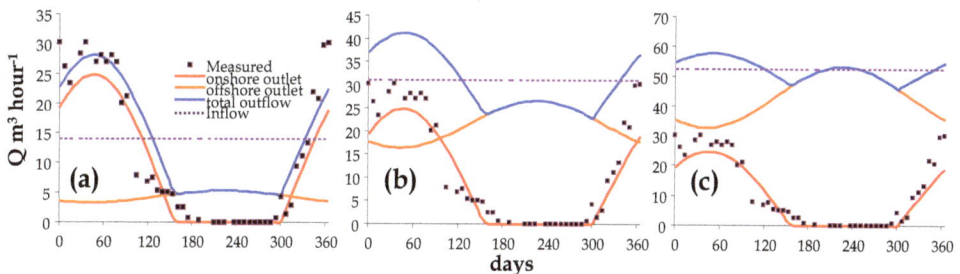

Fig. 13. Application of the analytical solution (Eq. 34 and 31) to Fuliya Spring discharge through both onshore and offshore springs, with three different values of α_0 (a: α_0=-2, b: α_0=-10, c: α_0=-20).

2.7 Long term reduction of groundwater level and spring discharge

The same physical considerations may be used to examine the process of long-term changes of groundwater level and annual spring discharge (Fig. 14). Unlike previous cases, the time scale here is much larger than a daily scale. The analytical solution is applied here for multi annual changes of hydrological variables such as groundwater level and spring discharge, to test whether the aquifer storage is affected by the initiation of large changes upstream. Such changes are for example the initiation of pumping wells, or construction of large water storage reservoirs, which started at a certain point in time.

We consider an average constant annual inflow Q_{in0} to the reservoir that represents the aquifer storage. The outflow is similar to Eq. 2, where elevation of the spring outlet is set to $H_0=0$. A constant flow Q_p represents an outflow from the reservoir in addition to the spring outlet, such as pumping wells or reduction of inflows due to significant land use changes. Under these definitions:

$$Q_{in}(t) = Q_{in0}$$
$$Q_{out}(t) = \alpha_0 \cdot \left[h(t) - H_0 \right] \tag{36}$$
$$Q_p(t) = Q_p$$

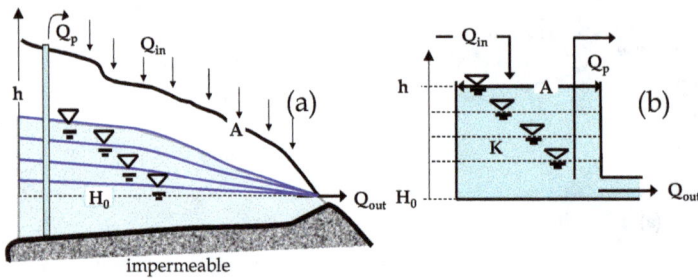

Fig. 14. (a) Schematic description of groundwater system; (b) linear reservoir model -the water flux through the outlet is proportional with storage.

With similar considerations as in the problems described above, the reservoir mass balance is controlled by two outflows (Fig. 14) – one is constant in time Q_p, whereas the other is time-dependent spring discharge $Q_{out}(t)$. On the annual time scale, the natural recharge Q_{in0} is considered as constant. The time when the change occurred (pump, land use change) is considered as t=0. The reservoir equation is therefore:

$$\frac{dV(t)}{dt} = Q_{in0} - \alpha \left[h(t) - H_0 \right] - Q_p \tag{37}$$
$$t < 0 \rightarrow Q_p = 0 \quad ; \quad t \geq 0 \rightarrow Q_p > 0$$

Rearranging the problem results in

$$\frac{dh(t)}{dt} = -\frac{1}{K} h(t) + \frac{Q_{in0} - Q_p}{A \cdot n} \quad ; \quad with \quad h(t=0) = h_0 \tag{38}$$
$$t < 0 \rightarrow Q_p = 0 \quad ; \quad t \geq 0 \rightarrow Q_p > 0$$

It should be emphasized that K in this case represents a timescale by far larger than the seasonal timescale. Eq. (38) is solved analytically with

$$h(t) = \left[h_0 - \frac{K}{A \cdot n}(Q_{in0} - Q_p)\right] \exp\left(\frac{-t}{K}\right) + \frac{K}{A \cdot n}(Q_{in0} - Q_p)$$

$$Q_{out}(t) = \frac{A \cdot n}{K} h(t)$$

(39)

It is assumed that prior to t=0, a steady state had been reached with Q_p=0 and therefore h(t)=h_0 =(K/A·n)Q_{in0}. At t→∞ the system is set on a new steady state h(t)= (K/A·n)·(Q_{in0}-Q_p). The expression [(h_0- K/A·n)·(Q_{in0}-Q_p)] is the aquifer system long-term full response to the change Q_p in water inflows and outflows. If this expression is zero, aquifer level and spring discharge will remain unchanged in time. If the expression is positive, hydraulic head decrease from one steady state to another, and vice versa.

As an example, this analytical solution is applied to the groundwater level in the Lower Judea Group Aquifer near the Uja spring, located in the Eastern Basin of the Judea-Samaria Mountains, ~12 km north-west of the town of Jericho. According to water level measurements and the stratigraphic analysis in this region (Guttman, 2007; Laronne Ben-Itzhak & Gvirtzman, 2005), the Judea Group aquifer, with a thickness of about 800 to 850 m, is comprised of two sub-aquifers: the upper and the lower aquifers. The upper and lower sub-aquifers are separated by relatively low permeability formations, causing groundwater levels in the upper aquifer to be significantly higher than those in lower aquifer do.

Near the Uja Spring there are four wells. (Mekorot Uja-Na'aran wells 1,2,3,4) drilled into the lower aquifer (Guttman, 2007). The first well (Uja 1) was drilled in 1964 by the Jordanian authority to a depth of 288 m and later was deepened by the Israeli authorities to 536 m. This well pumped from the upper part of the lower aquifer. In 1974, a new well (Uja 2) was drilled to a depth of 615 m in order to replace the Uja 1 well. At the beginning of the 1980s, two more wells were drilled (Uja 3, to a depth of 738 m and Uja 4 to 650.5 m) a few kilometres south of the other two wells. The three wells (Uja 2,3,4) currently pump ~3×10^6 m³ annually from the lower aquifer of the Judea Group.

It is assumed that the steady state of groundwater levels in the wells stood at 100 m below sea level (bsl.) prior to the significant pumping in 1974, whereas currently, the new steady state is ~280 m bsl. The long-term measured reduction of groundwater level is nearly exponential from 1974 to 1991 (Fig. 15). Following the extremely rainy season of 1991-1992 the levels increased to ~220 bsl, but since the year 2000 it returned to the steady state of ~280 m bsl. The proposed solution for this case was reached assuming K=1980 days; Q_{in}=8200 m³ day⁻¹ (3×10^6 m³ annually); Q_p=8200 m³ day⁻¹; A×n=90,000 m²; and reduction of level (t=0) initiated in 1974.

The physical interpretation of these results is that prior to the year 1974 a flux of ~3×10^6 m³ passed through the local Lower Judea Group aquifer annually (both Q_{in} and Q_{out} were ~ 8200 m³ day⁻¹). The continuous pumping caused a significant reduction of groundwater level, and brought the system to a new steady state in which the natural flow of groundwater is reduced. The artificial deployment replaced the natural groundwater outflow, which originally travelled downstream following the hydraulic gradient.

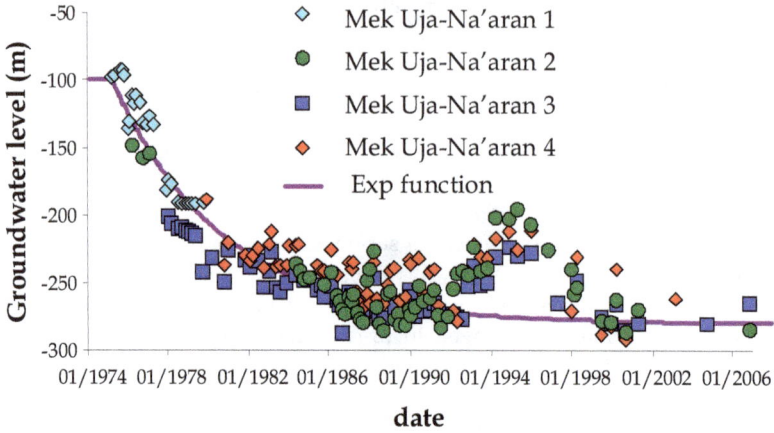

Fig. 15. Measured and modeled multiannual ground water level in the Lower Judea Group aquifer near Uja Spring from 1974 to 2007.

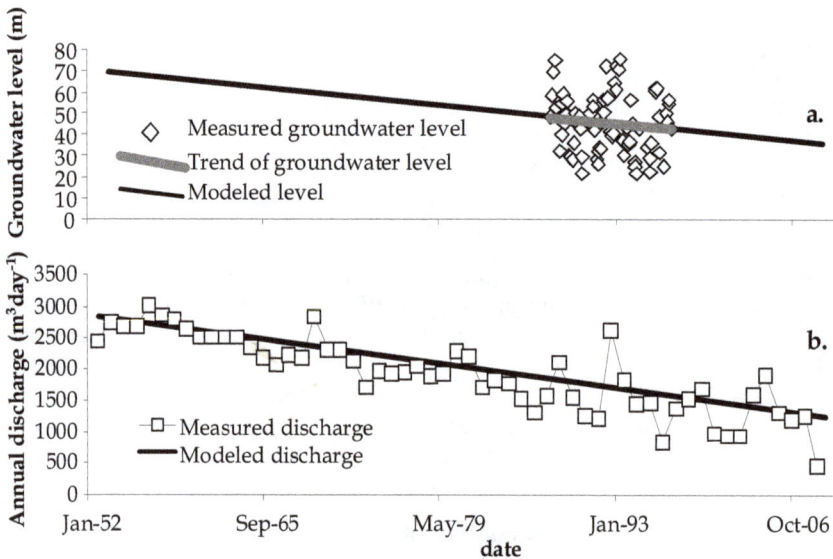

Fig. 16. Measured and modeled multiannual trends of (a) groundwater level (monthly values from 1988 to1999), and (b) average annual discharge of the Masrefot Spring from 1952 to 2008.

Another simplified solution can be obtained for the special case in which long-term regional groundwater level and spring discharge is being constantly reduced. One possible explanation for such reduction is the increasing deployment of the aquifer. From the modelling point of view this is a case where Q_p in Eqs. 36-39 follows a linear change in time (Q_p = a·t+ b). We consider an average constant annual inflow Q_{in} and outflow Q_{out} similar to Eq. (2), but the additional outflow Q_p, representing local pumping, is evolving and increasing linearly in time. When Q_p = a·t+ b is implemented in equations (36-39) the analytical solution is:

$$h(t) = \left(h_0 + \frac{(Kb - KQ_{in})}{A \cdot n} \right) \cdot \exp\left(-\frac{t}{K} \right) + \frac{K}{A \cdot n}\left(Q_{in} - (at + b) \right)$$

$$Q_{out}(t) = \frac{A \cdot n}{K} h(t)$$

(40)

In this case, the exponential term (first term in the right hand side of Eq. 40) may approach zero very quickly, while most of the reduction of groundwater level and spring discharge depends on the decrease of recharge expressed in the second term in Eqs. 40 by $Q_{in}-(a·t+b)$. At large t the system continuously reduces in time as expected.

As an example, this analytical solution is applied to the Masrefot Spring (Fig. 16), which is affected by the hydraulic heads of the Lower Judea Group aquifer in the Western Galilee, Israel. This spring was selected for this case since on the one hand, according to Kafri (1970), its seasonal changes in discharge are hardly noticed due to the large storage of water that feeds them. On the other hand, the long-term history of measured spring discharge (~60 years, Fig. 16) may reflect the reduction of regional groundwater level. The regional water supply system includes dozens of pumping wells. Analysis of the actual annual pumping in these wells revealed a nearly linear increase of pumped water (r^2=0.87) at least between 1960 and 1990, with an average increase of ~350,000 m³ year⁻¹. This is probably the reason for the systematic linear decrease of groundwater levels and Masrefot Spring discharge.

3. Summary

The steps towards modeling groundwater usually include (1) definition of the modeling domain; (2) definition of the hydrogeological structure; and (3) evaluation of initial and boundary conditions. The objective of this paper is to suggest an additional step (4), estimating the dominant parts that define the timely response of the hydrological system. This step is particularly important in developing conceptual models that simplify the hydrological problem to its relevant processes. We showed that using an analytical solution with this methodology could result in some important understanding of the system in question. Although the analytical solution can sometimes be the entire required modeling, usually the usage of analytical solution is only the first idea that we have on the time-dependent system. The cases described in section 2.1 and 2.2 are indeed well known, and often found in the literature. However, the cases described in sections 2.3-2.7 are less familiar, but can be used for creating other new types of models. In section 2.3, we showed that a system of two serial reservoirs might be used to characterize the flow instead of parallel reservoirs. In section 2.4, the possibility of exchanging parallel reservoirs was discussed. Section 2.5 proposed that the recession curve might not fall into the well-known exponential shape due to downward flow to lower outlet, while section 2.6 showed how spring flow could change in time due to nearby dominant boundary condition (such as lake). Finally, section 2.7 suggested simple modeling for multiannual groundwater levels and spring discharge under reduction in water availability, a specific environmental problem that is occurring now, and expected in the future. Altogether, our set of examples can help in developing new process-based models for better system understanding and forward prediction.

4. References

Abbo, H., U. Shavit, D. Markel, and A. Rimmer, (2003). A Numerical Study on the Influence of Fractured Regions on Lake/Groundwater Interaction; the Lake Kinneret Case. Journal of Hydrology, 283/1-4 pp. 225-243.

Bakalowicz, M., 2005. Karst groundwater: a challenge for new resources. Hydrogeology Journal, 13: 148-160.

Bergelson, G., Nativ, R. & Bein, A. 1998. Assessment of hydraulic parameters in the aquifers sorrounding and underlying Sea of Galilee. GroundWater 36:409-417.

Beven, K.J., Kirby, M.J., 1979. A physically based, variable contributing area model of basin hydrology. Hydrological Sciences Bulletin, 24: 43-69.

Bonacci, O., 1993. Karst springs hydrographs as indicators of karst aquifers / Les hydrogrammes des sources karstiques en tant qu'indicateurs des aquifÃ¨res karstiques. Hydrological Sciences Journal, 38(1): 51 - 62.

Boyce, W.E., DiPrima, R.C., 2000. Elementary Differential Equations. Wiley, New York, 700 pp, ISBN 9780470039403

Cornaton, F., Perrochet, P., 2002. Analytical 1D dual-porosity equivalent solutions to 3D discrete-continuum models. Application to karstic spring hydrograph modelling. Journal of Hydrology, 262: 165-176.

Dooge, J.C.I. 1973. Linear theory of hydrologic systems. US Dept. Agric. Tech. Bull. No. 1468, pp. 267–293.

Fleury, P., Plagnes, V., Bakalowicz, M., 2007. Modelling of the functioning of karst aquifers with a reservoir model: Application to Fontaine de Vaucluse (South of France). Journal of Hydrology, 345: 38-49.

Geyer, T., Birk., S., Liedl, R., Sauter, M., 2008. Quantification of temporal distribution of recharge in karst systems from spring hydrographs. Journal of Hydrology, 348: 452-463.

Goldscheider, N., Drew, D., 2007. Methods in Karst Hydrogeology. Taylor & Francis Group, 264 p., ISBN 9780415428736

Goldshmidt, M.J., Arad A., Neev, D., 1967. The mechanism of the saline springs in the Lake Tiberias depression. Min. Dev. Geol. Surv., Jerusalem, Hydrol. Pap. #11, Bull. 45. 19 pp.

Guttman, J. 2007. The Karstic Flow System in Uja Area – West Bank: An Example of two Separated Flow Systems in the Same Area. Chapter6 in: Shuval H. and Dweik, H. [Eds.] Water Resources in the Middle East. Springer. ISBN 9783540695097.

Grasso, D.A., Jeannin, P.-Y., 1994. Etude critique des methods d'analyse de la réponse globale des systèmes karstiques. Application au site du Bure (JU, Suisse). Bulletin d'Hydrogéologie, 13: 87-113.

Gvirtzman, H., Garven, G., Gvirtzman, G., 1997. Hydrogeological modeling of the saline hot springs at the Sea of Galilee, Israel. Water Resources Research, 33(5): 913-926.

Hartmann, A., Kralik, M., Humer, F., Lange, J., Weiler, M., 2011. Identification of a karst system's intrinsic hydrodynamic parameters: upscaling from single springs to the whole aquifer. Environmental Earth Sciences, DOI: 10.1007/s12665-011-1033-9.

Jukic, D., Denic-Jukic, V., 2009. Groundwater balance estimation in karst by using a conceptual rainfall-runoff model. Journal of Hydrology, 373(3-4): 302-315.

Kafri, U., 1970. The hydrogeology of the Judea Group Aquifer in the western and central Galilee, Israel. Geological Service of Israel (GSI). Report Hydro/1/70 (in Hebrew).

Kessler, A., Kafri, U., 2007. Application of a cell model for operational management of the Na'aman groundwater basin, Israel. Israel Journal of Earth Science, 56(1): 29-46.

Kiraly, L., 2003. Karstification and Groundwater Flow. Speleogenesis and Evolution of Karst Aquifers, 1(3): 1-24.

Laronne Ben-Itzhak L., and H. Gvirtzman, 2005. Groundwater flow along and across structural folding: an example from the Judean Desert, Israel. Journal of Hydrology 312 (2005) 51–69.

Le Moine, N., Andréassian, V., Mathevet, T., 2008. Confronting surface- and groundwater balances on the La Rochefoucauld-Touvre karstic system (Charente, France). Water Resources Research, 44(W03403).

Lindström, G., Johannson, B., Perrson, M., Gardelin, M., Bergström, S., 1997. Development and test of the distributed HBV-96 hydrological model. Journal of Hydrology, 201: 272-288.

Maillet, E., 1905. Essais d'hydraulique souterraine et fluviale. In: Hermann, A. (Ed.), Mécanique et Physique du Globe, Paris.

Rimmer, A., Hurwitz, S., Gvirtzman, H., 1999. Spatial and temporal characteristics of saline springs: Sea of Galilee, Israel. Ground Water, 37(5): 663-673.

Rimmer, A., Salingar, Y., 2006. Modelling precipitation-streamflow processes in karst basin: The case of the Jordan River sources, Israel. Journal of Hydrology, 331: 524-542.

Sauter, M., 1992. Quantification and Forecasting of Regional Groundwater Flow and Transport in a Karst Aquifer (Gallusquelle, Alm, SW. Germany. Tuebinger Geowissenhschaftliche Arbeiten, C13. Institut und Museum für Geologie und Pläontologie der Universität Tübingen, Tuebingen.

Schulla, J., Jasper, K., 2007. Model Description WaSiM-ETH (Water balance Simulation Model ETH), ETH Zurich, Zurich, CH.

Simon, e. and A. Eizik, 1991. Hydrological observations in Lake Kinneret for the year 1989-1990. -Tahal Report, 01/91/19, Tahal, Tel-Aviv, 23 pp (In Hebrew).

Singh, V.P., 1988. Hydrologic systems, rainfall-runoff modeling. Prentice Hall, NJ.

Sugawara, M., 1995. Tank Model, in "Computer Models of Watershed Hydrology". Singh V.P. [Ed.]. Water Resources Publications, Colorado, pp. 165–214.

Tritz, S., Guinot, V., Jourde, H., 2011. Modelling the behaviour of a karst system catchment using non linear hysteretic conceptual model. Journal of Hydrology, Journal of Hydrology 397(3-4): 250-262.

13

Quantity and Quality Modeling of Groundwater by Conjugation of ANN and Co-Kriging Approaches

Vahid Nourani and Reza Goli Ejlali
University of Tabriz,
Iran

1. Introduction

Today, groundwater is a major source of supply for domestic and agricultural purposes; especially in arid and semi arid regions. More water is being consumed to meet of a society whose population increases steadily. Worldwide, irrigated land has increased from 50 million ha in 1900 to 267 million ha in 2000 (Cay and Uyan, 2009). The climatic changes stemming from global warming also have negative effects on water resources. Both over exploitation from aquifers, and drought events have caused severe water table level drop in many areas. However, the level of groundwater has reduced remarkably in many areas, as a result of unconscious and excessive irrigation. Depletion of groundwater supplies, conflicts between groundwater and surface water users and potential for groundwater contamination are the main concerns that will become increasingly important as further aquifer development takes place in any basin.

The natural chemical composition of groundwater is influenced predominantly by type and depth of soils and subsurface geological formations through which groundwater passes. Groundwater quality is also influenced by contribution from the atmosphere and surface water bodies. Quality of groundwater is also influenced by anthropogenic factors. For example, over exploitation of groundwater in coastal regions may result in sea water ingress and consequent increase in salinity of groundwater and excessive use of fertilizers and pesticides in agriculture and improper disposal of urban/industrial waste can cause contamination of groundwater resources.

Groundwater systems possess features such as complexity, nonlinearity, being multi-scale and random, all governed by natural and/or anthropogenic factors, which complicate the dynamic predictions. Therefore many hydrological models have been developed to simulate this complex process. Models based on their involvement of physical characteristics generally fall into three main categories: black box models, conceptual models and physical based models (Nourani and Mano, 2007). The conceptual and physically based models are the main tools for predicting hydrological variables and understanding the physical processes that are taking place in a system. In these models, the internal physical processes are modeled in a simplified way. Even if not applying the exact differential laws of conservation, conceptual models attempt to describe large scale behavior of hydrological processes in a basin. However, these models require a large quantity of good quality data, sophisticated programs for calibration using rigorous optimization techniques and a detailed understanding of the underlying physical process. Because of the recognized limitations of these models and the growing need to properly manage overdeveloped groundwater systems, significant researches have been devoted to improve their predictive capabilities. Despite large investments in time and resources, prediction accuracy attainable with numerical flow models has not improved satisfactorily for many types of groundwater management problems. Studies on groundwater levels reveal spatial and temporal information on aquifers and aquiferous systems and help us to take appropriate measures. For management of groundwater resources, traditional numerical methods, with specific boundary conditions, are able to depict the complex structures of aquifers including complicated prediction of groundwater levels. However, the vast and accurate data required to run a numerical model are difficult to obtain owing to spatial variations and the unavailability of previous hydrogeology surveys. As a result, numerical methods have been restricted in their use in remote, sparsely monitored areas. If sufficient data are not available, and accurate predictions are more important than understanding the actual physics of the situation, black box models remain a good alternative method and can provide useful predictions without the costly calibration time (Daliakopoulos et al., 2005).

In recent years, Artificial Neural Network (ANN) as a black box model has been widely used for forecasting in many areas of science and engineering. ANNs are proven to be effective in modeling virtually any nonlinear function to an arbitrary degree of accuracy. The main advantage of this approach over traditional methods is that the method does not require the complex nature of the underlying process under consideration to be explicitly described in mathematical form. This makes ANNs attractive tools for modeling water table fluctuations.

The development of ANNs began approximately 70 years ago (McCulloch and Pitts, 1943), inspired by a desire to understand the human brain and emulate its behavior. Although the idea of ANNs was proposed by McCulloch and Pitts, the development of these techniques has experienced a renaissance only in the last decades due to Hopfield's effort (Hopfield, 1982) in iterative auto-associable neural networks. A tremendous growth in the interest of this computational mechanism has occurred since Rumelhart et al. (1986) rediscovered a mathematically rigorous theoretical framework for neural networks, i.e., back propagation algorithm. Consequently, ANNs have found applications in many engineering problems.

Since the early nineties, ANNs have been successfully used in environmental and hydrology-related areas such as rainfall-runoff modeling, stream flow forecasting, groundwater modeling, water quality, water management policy, precipitation forecasting, and reservoir operations (ASCE, 2000a,b). Also, ANN models have been used for rainfall-runoff modeling (Tayfur and Singh, 2006), precipitation forecasting and water quality modeling (Govindaraju and Ramachandra Rao, 2000). In the water level modeling context, Tayfur et al. (2005) presented an ANN model to predict water levels in piezometers placed in the body of an earthfill dam in Poland considering upstream and downstream water levels of the dam as input data. Neural networks have also been applied with success to temporal prediction of groundwater level (Coulibaly et al., 2001). Two researches have been carried out for forecasting floods in a karestic media (Beaudeau et al., 2001) and determining aquifer outflow influential parameters, and simulating aquifer outflow in a fissured chalky media (Lallahem and Mania, 2003). ANNs have been successfully used for identifying the temporal data necessary to calculate groundwater level in only one piezometer (Lallahem et al., 2005). ANNs were also employed to solve complex groundwater problems and for predicting transient water level in a multilayer groundwater system under variable pumping states and climate conditions (Coppola et al., 2003). Coppola et al. (2005) developed an ANN model for accurately predicting potentiometric surface elevations in alluvial aquifers. Relationships among lake levels, rainfall, evapotranspiration and groundwater levels were determined by Dogan et al. (2008) using ANN-based models. Nourani et al. (2008) employed ANN approach for time-space modeling of groundwater level in an urbanized basin.

In spite of reliable ability of the ANNs in temporal and time series predictions, they could not find notable application for the spatial modeling of the environmental processes. Instead, geostatistics powerful interpolating tools are extremely used for unbiased estimation of the spatial variables at a given point. Geostatistics has made rapid advances in recent years since it first developed by Matheron (1963). Recently, the term geostatistics has been used more generally to describe all applications of statistics in hydrogeology in which the attributes is a random field in space. The heterogeneity of the subsurface often is difficult to characterize adequately for use in deterministic models; therefore, geostatistical techniques often are used to generate estimates of parameters in deterministic mathematical models where parameters are random variables in space. For groundwater flow problems, attributes such as water levels are sampled at a limited number of sites whereas values at un-sampled sites usually are needed for analysis. Geostatistical techniques such as Kriging and Cokriging can be applied to estimate the values of attributes at un-sampled sites (Ma et al., 1999). For examples, various forms of geostatistical tools have been used to map potentiometric surfaces from water level data alone (Delhomme, 1978; Aboufirassi and Marino, 1983; Neuman and Jacobsen, 1984). A comprehensive review of the applications of geostatistics to hydrogeology can be found in the ASCE Task Committee report (ASCE, 1990). Also, a few applications of the geostatistics tools in groundwater level predictions can be found in the literature (e.g. Ma et al., 1999; Finke et al., 2004; Gundogdu and Guney, 2007; Barca and Passarella, 2008; Cay and Uyan, 2009; and Taany et al., 2009).

Nourani et al., (2010) proposed a hybrid model (ANNG) for spatiotemporal forecasting of groundwater level in coastal aquifers. The basic idea of the models combination in the forecasting is the use each model's unique feature to capture different pattern in the data. Both theoretical and empirical findings suggest that the combining different methods can be efficient way to improve forecasting (Zhang and Dong, 2001). Therefore, the developed hybrid model employs the ability of ANN in time series modeling and capability of Kriging in spatial estimation in a unique framework and may be considered as a more general groundwater level modeling tool. According to the inherent capability of ANNs in temporal forecasting and geostatistics tools in spatial estimating, a new modified hybrid ANN-Geostatistic (MANNG) black box model is proposed in this text and its potential is evaluated for spatio-temporal prediction of groundwater level and salinity in a coastal aquifer located in Iran.

2. Study area and data

The data used in this study are from the Shabestar plain (Figure 1) which is located in northwest Iran at Azerbaijan province (between 45° 26' and 46° 2' north latitude and 38° 3' and 38° 23' east longitude). The plain area is 1300 km² and its main channel is Daryanchai which discharges to Urmieh Lake. The headwaters of the river are situated in the Misho Mountain. Plain elevation is varying between 1278 m to 3135 m above sea level and its longest waterway has 15 km length.

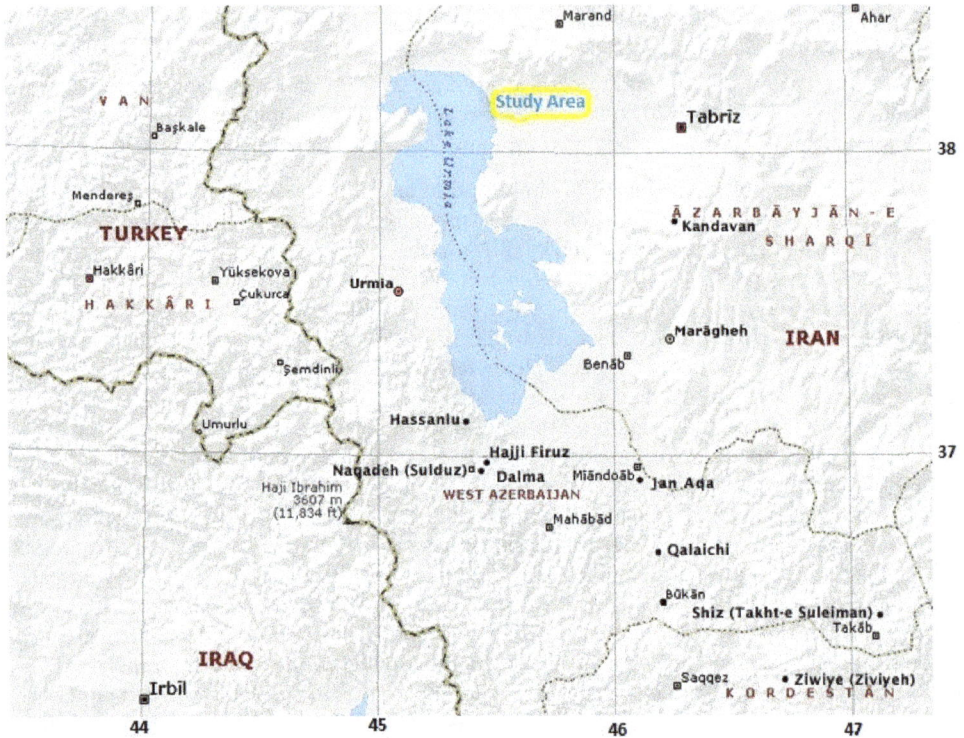

Fig. 1. Study area

The mean daily temperature varies from -19°C in January up to 42°C in July with a yearly average of 11°C and the average annual rainfall is about 250 mm.

Urmieh Lake, located in northwestern Iran, is an oligotrophic lake of thalassohaline origin and the 20th largest, and the second hyper saline lake in the world with a total surface area between 4750 and 6100km² and a maximum depth of 16 m at an altitude of 1250 m. The lake is divided into north and south parts separated by a causeway in which a 1500 m gap provides little exchange of water between the two parts. Due to drought and increased demands for agricultural water in the lake's basin, the salinity of the lake has risen to more than 300g/l during recent years, and large areas of the lake bed have been desiccated. The possible causes of rising salinity are likely to be surface flow diversions, groundwater extractions and unsuitable climate condition.

Fluctuation of Urmieh Lake water levels has tremendous environmental impacts, especially on the adjoining groundwater resources. About 4.4 million people live in the Urmieh Lake basin, whose irrigation economy is strongly dependent on existing surface and groundwater resources in the area. Accordingly, human population growth in the lake's basin has seriously increased the need for agricultural and potable water in recent years, all of which are supplied from surface and groundwater sources in the area. These issues, together with poor weather conditions, have reduced significantly the volume of water entering the lake so that, at present, Urmieh Lake has shrunk significantly and large areas of the former lake bed have been exposed. According to the interaction between the water depth of the lake and groundwater level of the plain, decreasing of the water depth of the lake leads to decrease of groundwater level of the plain and also increase the groundwater salinity. In this research, it is tried to utilize the ANN and geostatistic concepts in order to investigate the effects of the lake's water depth and other hydro-meteorological parameters on the groundwater level and salinity via a spatiotemporal modeling.

The data utilized in this study were collected over 13years (from April 1994 to March 2006) with one month time interval. Table 1 shows the statistical analysis of the observed groundwater levels of piezometers.

Piez. No.	X(UTM) (m)	Y(UTM) (m)	Piezometer Elevations (m)	Mean (m)	Min. (m)	Max. (m)	Variance	Standard deviation (m)	Skewness coefficient
P1	586050	4238025	1401.48	1390.0	1389.6	1391.1	0.069140	0.262944	1.224652
P2	562800	4230450	1583.24	1547.8	1540.9	1553.8	8.785094	2.963966	-0.12895
P3	561450	4217350	1277.70	1333.7	1331.4	1336.8	1.211814	1.100824	0.132759
P4	562250	4221350	1322.79	1272.0	1268.1	1276.2	3.930377	1.982518	0.026029
P5	576925	4223350	1309.97	1297.7	1295.4	1303.4	3.523323	1.877052	1.473395
P6	577600	4222950	1303.96	1302.5	1301.7	1303.6	0.235159	0.484932	0.611962
P7	584800	4229250	1325.98	1299.1	1298.1	1301.3	0.399340	0.631933	1.321099
P8	546600	4223900	1301.86	1321.8	1319.5	1323.7	1.302649	1.141337	-0.374780
P9	551700	4220350	1292.05	1282.2	1279.0	1284.4	3.338116	1.827051	-0.346380
P10	554550	4220050	1289.02	1284.2	1282.4	1285.8	0.970875	0.985330	0.031911
P11	555050	4220250	1288.98	1285.9	1283.6	1287.3	0.805980	0.897764	-0.651490

Table 1. Statistical analysis of observed data in piezometers

The monthly data collected consist of the following categories:

1. Observed water levels and salinities of piezometers located within the Shabestar plain (P1, P2, P3,..., P11 for training and TP1, TP2, and TP3 for cross-validation purposes). Figure 2 shows positions of the piezometers in the study area.
2. Rainfall in Sharafkhaneh station,
3. Average discharge of Daryanchai in Daryan station,
4. Urmieh Lake level,
5. Temperature in Sharafkhaneh station.

3. Artificial Neural Network

ANNs offer an effective approach for handling large amounts of dynamic, non-linear and noisy data, especially when the underlying physical relationships are not fully understood. This makes them well suited to time series modeling problems of a data-driven nature. In general the advantages of an ANNs over other statistical and conceptual models can be classified as (Nourani et al., 2008):

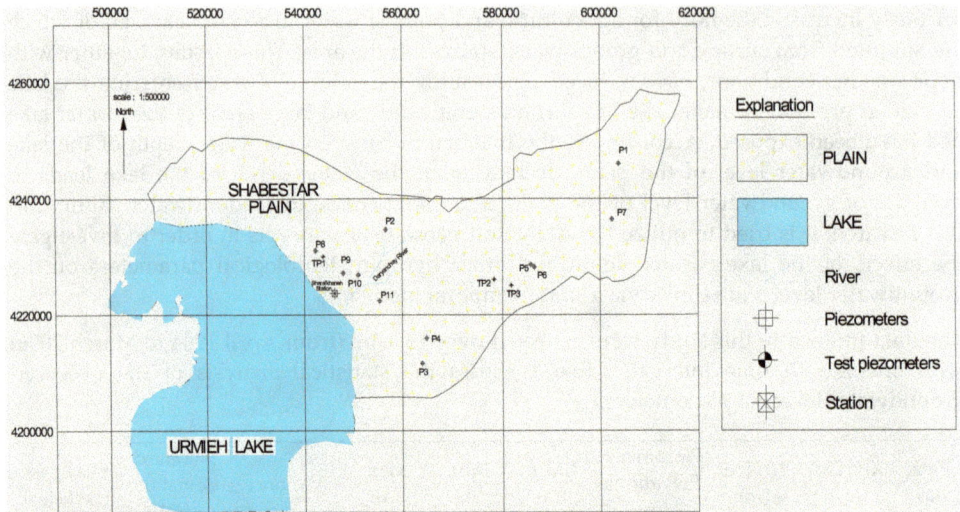

Fig. 2. Piezometers positions

1. The application of ANN does not require a prior knowledge of the process because ANNs have black-box properties,
2. ANNs have the inherent property of nonlinearity since neurons activate a nonlinear filter called an activation function,
3. ANNs can have multiple input having different characteristics, which can represent the time-space variability,
4. ANN has been proven to be effective in modeling virtually any nonlinear function to an arbitrary degree of accuracy. The main advantage of this approach over traditional methods is that it does not require the complex nature of the underlying process under consideration to be explicitly.

ANN is composed of a number of interconnected simple processing elements called neurons or nodes with the attractive attribute of information processing characteristics such as nonlinearity, parallelism, noise tolerance, and learning and generalization capability. Among the applied neural networks, the feed forward neural networks (FFNN) with back-propagation (BP) algorithm are the most common used methods in solving various engineering problems (Nourani et al., 2009).

FFNN technique consists of layers of parallel processing elements called neurons, with each layer being fully connected to the preceding layer by interconnection strengths, or weights. Initial estimated weight values are progressively corrected during a training process that compares predicted outputs with known outputs. Learning of these ANNs is generally accomplished by Back Propagation (BP) algorithm (Hornik et al., 1989). The objective of the BP algorithm is to find the optimal weights, which would generate an output vector, as close as possible to the target values of the output vector, with the selected accuracy.

The network is determined by architecture of the network, the magnitude of the weights and the processing element's mode of operation. The neuron is a processing element that takes a number of inputs, weights them, sums them up, adds a bias and uses the results as the argument for a singular valued function called the transfer function. The transfer function results in the neuron's output. At the start of training, the output of each node tends to be small. Consequently, the derivatives of the transfer function and changes in the connection weights are large with respect to the input. As learning progresses and the network reaches a local minimum in error surface, the node outputs approach stable values. Consequently the derivatives of the transfer function with respect to input, as well as changes in the connection weights, are small.

The Back Propagation (BP) neural network is the most widely used ANN in hydrologic modeling and is also used in this study. A typical BP neural network model is a full-connected neural network including input layer, hidden layer and output layer.

Back-propagation (BP) algorithms use input vectors and corresponding target vectors to train ANN. The standard BP algorithm is a gradient descent algorithm, in which the network weights are changed along the negative of the gradient of the performance function. There are a number of variations in the basic BP algorithm that is based on other optimization techniques such as conjugate gradient and Newton methods (Hornik et al., 1989).

For properly trained BP networks, a new input leads to an output similar to the correct output. This ANN property enables training of a network on a representatives set of input/target pairs and achieves sound forecasting results. A clear systematic document about the BP algorithm and the methods for designing the BP model are given by Basheer and Hajmeer (2000) and Jiang et al. (2008). Some researchers claim that networks with a single hidden layer can approximate any continuous function to a desired accuracy and is enough for most forecasting problems (Hornik et al., 1989).

In this study, at first step by using a three-layer neural network via a sensitivity analysis the effective data sets are chosen. All input values are standardized to a specific range separately after data division. Input and output variables are normalized by scaling between zero and one to eliminate their dimensions and to ensure that all variables receive equal attention during training of the models. Finally, the training and testing data sets are selected, and the network is trained.

The Levenberg-Marquardt (LM) method is a modification of the classic Newton algorithm for finding an optimum solution to a minimization problem. Levenberg-Marquardt has large computational and memory requirement and thus it can only be used in small networks (Maier and Dandy, 1998). It is faster and less easily trapped in local minima than other optimization algorithms (Coulibaly et al., 2001a, b, c; Toth et al., 2000).

In this study, among the many training methods, the Levenberg- Marquuardt training algorithm was selected, considering its fast convergence ability (Sahoo et al., 2005). Also a Tangent Sigmoid transfer function was used for hidden layer and a linear transfer function for the output layer according to Qu et al. (2004). The numbers of hidden layer nodes and training epochs are determined using trial and error in the test scenarios.

4. Geostatistics

Since detailed information about geostatistics and geostatistical techniques such as Kriging and Cokriging can be found in the scientific literature (e.g., Isaaks and Srivastava, 1989), only a brief description of this methods which is employed in this research is provided.

Kriging technique is a spatial interpolation estimator $Z(x_0)$ used to find the best linear unbiased estimator of a second-order stationary random field with an unknown constant mean:

$$\underline{Z}(x_0) = \sum_{i=1}^{n} \lambda_i Z(x_i) \tag{1}$$

Where $\underline{Z}(x_0)$ is Kriging estimate at location x_0; $Z(x_i)$ is sampled value at x_i; λ_i is weighting factor for $Z(x_i)$; and $i = 1, \ldots, n$ in which n denotes to the numbers of samples. The estimation error can be written as:

$$R(x_0) = \underline{Z}(x_0) - Z(x_0) = \sum_{i=1}^{n} \lambda_i . Z(x_i) - Z(x_0) \tag{2}$$

Where $Z(x_0)$ is unknown true value at x_0; and $R(x_0)$ is estimation error. For an unbiased estimator, the mean of the estimates must be equal to the true mean, therefore (Ma et al., 1999):

$$E(R(x_0)) = 0 \tag{3}$$

Where E is expected value and then:

$$\sum_{i=1}^{n} \lambda_i = 1 \tag{4}$$

The best linear unbiased estimator must have minimum variance of estimation error. The minimization of the estimation error variance under the constraint of unbiasedness leads to a set of simultaneous linear algebraic equations for the weighting factors as follows (Ma et al., 1999):

$$E\left[\left(\sum_{i=1}^{n} \lambda_i Z(x_i) - Z(x_0)\right)^2\right] = Var\left[\sum_{i=1}^{n} \lambda_i Z(x_i) - Z(x_0)\right] \tag{5}$$

Where Var, is the abbreviation of variance function. The weighting factors λ_i can be determined by solving a nonlinear optimization problem involving the minimization of the foregoing function subject to the constraint in (4) by using the Lagrange multiplier μ as:

$$L(\lambda_i, \mu) = Var\left[\sum_{i=1}^{n} \lambda_i Z(x_i) - Z(x_0)\right] - 2\mu\left(\sum_{i=1}^{n} \lambda_i - 1\right) \tag{6}$$

The necessary conditions for optimal λ_i and μ values involve setting the first derivative of Equation (6) to zero; therefore, the system of simultaneous linear algebraic equations for λ and μ can be expressed in matrix form as (Ma et al., 1999):

$$
\begin{bmatrix}
\gamma_{11} & \gamma_{12} & \cdots & \gamma_{1n} & 1 \\
\gamma_{21} & \gamma_{22} & & \gamma_{2n} & 1 \\
\vdots & & \ddots & \vdots & \\
\gamma_{n1} & \gamma_{n2} & \cdots & \gamma_{nn} & 1 \\
1 & 1 & \cdots & 1 & 0
\end{bmatrix}
\begin{bmatrix}
\lambda_1 \\ \lambda_2 \\ \vdots \\ \lambda_n \\ \mu
\end{bmatrix}
=
\begin{bmatrix}
\gamma_{01} \\ \gamma_{02} \\ \vdots \\ \gamma_{0n} \\ 1
\end{bmatrix}
\tag{7}
$$

The Variogram γ can be derived from sampled data as follows:

The presence of a spatial structure where observations close to each other are more alike than those that are far apart (spatial autocorrelation) is a prerequisite to the application of geostatistics. The experimental Variogram measures the average degree of dissimilarity between un-sampled values and a nearby data value and thus can depict autocorrelation at various distances. The value of the experimental Variogram for a separation distance of h (referred to as the lag) is half the average squared difference between the value at $z(x_i)$ and the value at $z(x_{i+h})$ as (Ma et al., 1999):

$$
\gamma(h) = \frac{\left\{ \sum_{i=1}^{n} \left[Z(x_i) - Z(x_i + h) \right]^2 \right\}}{2n}
\tag{8}
$$

Where n is the number of data pairs within a given class of distance and direction. If the values of $z(x_i)$ and $z(x_{i+h})$ are auto correlated the results of Equation (8) will be small, relative to an uncorrelated pair of points. From analysis of the experimented Variogram, a suitable model (e.g., spherical, exponential) is then fitted, usually by weighted least squares and the parameters (e.g., range, nugget and sill) are then used in the Kriging procedure (Isaaks and Srivastava, 1989).

The "co-regionalization" (expressed as correlation) between two variables, i.e. the variable of interest, groundwater salinity in this case and another easily obtained and inexpensive variable, can be exploited to advantage for estimation purposes by the Cokriging technique. In this sense, the advantages of Cokriging are realized through reductions in costs or sampling effort. The cross semivariogram is used to quantify cross-spatial auto-covariance between the original variable and the covariate. The cross-semivariance is computed through the equation:

$$
\gamma_{uv}(h) = E[\{Z_u(x) - Z_u(x+h)\}\{Z_v(x) - Z_v(x+h)\}]
\tag{9}
$$

Where $\gamma_{uv}(h)$ is cross-semivariance between u and v variable, $Z_u(x)$ is primary variable and $Z_v(x)$ is secondary variable.

5. Proposed conjugated model and results

By combining the artificial neural network capability in modeling complicated and non-linear systems and geostatistical ability in linear estimation with low estimation error, a new hybrid model (MANNG) of spatiotemporal groundwater level and salinity forecasting in coastal aquifers has been proposed in this paper which uses both of mentioned models in unique framework. Figure 3 shows the proposed model scheme.

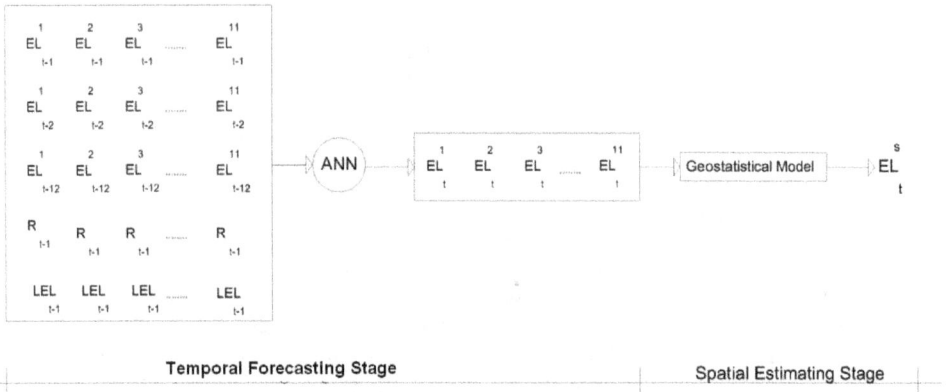

Fig. 3. Diagram of new modified proposed hybrid model (MANNG)

The proposed model contains two separated stages. At the first stage, an ANN is trained for all of the piezometers (P1, P2, ..., and P11) for time series modeling of the water level. The model predicts the preceding month ground water level of the piezometers based on quantity of present month rainfall in study area (R_{t-1}), Urmieh Lake water surface level at that month (LEL_{t-1}) and groundwater levels in present, first and twelfth previous months (EL_{t-1}, EL_{t-2}, EL_{t-12}) in order to handle the seasonality of the process as well as the auto regressive characteristics. A sensitivity analysis was employed in order to select the mentioned input parameters from the all available data, as it will be discussed in the next section.

At the second stage, the predicted values of water levels at different piezometers are imposed to a calibrated geostatistics model in order to estimate groundwater level and salinity at any desired point in the plain. Finally, as a cross-validation process the proposed spatio-temporal model is evaluated by the data of piezometers TP1,TP2, and TP3 which are not contributed in the calibration step of the model. The details and results of the stages are presented in the following sections.

5.1 Temporal forecasting stage

In order to ensure good generalization ability by an ANN model, some empirical relationships between the number of training samples and the number of connection weights have been suggested in the literature. However, network geometry is generally highly problem dependent and these guidelines do not ensure optimal network geometry, where optimality is defined as the smallest network that adequately captures the relationships in the training data (principle of parsimony). In addition, there is quite a high variability in the number of input and hidden nodes suggested by the various rules. While research is being conducted in this direction by the scientists working in ANNs, it may be noted that traditionally, optimal network geometries have been found by trial and error (Maier and Dandy, 2000). Consequently, in the current application the number of hidden neurons in the network, which is responsible for capturing the dynamic and complex relationship between various input and output variables, was identified by several trials. Also, this trial and error procedure with domain knowledge was explored for general guidance in the number of inputs selected.

The trial and error procedure started initially with two hidden neurons, and the number of hidden neurons was increased up to fifty with a step size of one in each trial. For each set of input and hidden neurons, the network was trained in batch mode to minimize the mean square error at the output layer. In order to check any over-fitting during training, a validation was performed by keeping track of the efficiency of the fitted model. The training was stopped when there was no significant improvement in the efficiency. The parsimonious structure that resulted in minimum root mean squared error (Equation 10), and maximum efficiency coefficient (Equation 11) during training as well as testing was selected as the final form of the ANN model for all piezometers.

The variables are scaled to a limit between zero and one as the activation function warrants. The total available data were divided into two sets, calibration and validation sets. In the training step the models were trained using data of ten years (1994-2003) and then validated on the rest of the data (2004-2006).

The Root Mean Squared Error (RMSE) and coefficient of efficiency (CE) were used in order to assess the effectiveness of each model and its ability to make precise predictions. The RMSE calculated by

$$RMSE = \sqrt{\frac{\sum_{i=1}^{N}(y_i - \hat{y}_i)^2}{N}} \tag{10}$$

Where y_i and $\hat{y}_i$ are the observed and predicted data respectively and N is the number of observations. RMSE indicates the discrepancy between the observed and calculated values. The lowest the RMSE, the more accurate the prediction is. Nash and Sutcliffe (1970) proposed the non-dimensional coefficient of efficiency (CE) criterion on the basis of standardization of the residual variance with initial variance, which provides a measure for the proportion of the variance explained by the model. It can be used to compare the relative performances of the models which are developed by different methods. It is estimated as (Nash and Sutcliffe, 1970).

$$CE = 1 - \frac{\sum_{i=1}^{N}(y_i - \hat{y}_i)^2}{\sum_{i=1}^{N}(y_i - \overline{y}_i)^2} \tag{11}$$

Where $\overline{y}_i$ is the average of observed values and the CE represents the initial uncertainty explained by the model. The CE is varying between -∞, 1 and the best fit between observed and calculated values would have CE=1. The quality of the fit statistics is measured by RMSE and CE between the computed and observed data. The sensitivity analysis showed that present month rainfall, lake water surface level at that month and groundwater levels in first, second and twelfth previous months are the most dominant parameters in forecasting the groundwater level in the most of piezometers and these parameters were considered as the input neurons for ANNs (Nourani et al.,2010).

The results of temporal modeling of groundwater levels in piezometers P1,P2,...,P11 ,as the first stage of the hybrid modeling have been briefly shown in table 2.

Piezometer	UTM		Networks Parameters		Calibration		Validation	
	x	y	Structure	Epoch	CE	RMSE(m)	CE	RMSE(m)
P1	586050	4238025	(5,6 ,1)	40	0.85	0.08	0.78	0.11
P2	562800	4230450	(5, 6,1)	40	0.95	0.07	0.89	0.10
P3	561450	4217350	(5,6 ,1)	40	0.95	0.06	0.88	0.09
P4	562250	4221350	(5,6,1)	40	0.96	0.07	0.86	0.10
P5	576925	4223350	(5,6,1)	40	0.89	0.05	0.83	0.11
P6	577600	4222950	(5, 6,1)	40	0.90	0.05	0.83	0.10
P7	584800	4229250	(5,6 ,1)	40	0.88	0.06	0.81	0.09
P8	546600	4223900	(5,6,1)	40	0.96	0.02	0.88	0.04
P9	551700	4220350	(5,6 ,1)	40	0.96	0.03	0.89	0.05
P10	554550	4220050	(5,6 ,1)	40	0.95	0.04	0.92	0.06
P11	555050	4220250	(5,6 ,1)	40	0.97	0.03	0.89	0.06

Table 2. ANN results for temporal forecasting stage

5.2 Spatial estimation stage

Groundwater has become one of the important sources of water for meeting the requirements of various sectors in the world in the last few decades. It plays a vital role in countries economic, development and in ensuring them food security. The rapid pace of agriculture development, industrialization and urbanization has resulted in the over exploitation and contamination of groundwater resources in the world, resulting in various adverse environmental impacts and threatening its long-term sustainability.

Salinity is the saltiness or dissolved salt contents of a water body. Salt content is an important factor in water use. Salinity can be technically defined as the total mass in grams of all the dissolved substances per kilogram of water (TDS). Different substances dissolve in water giving it taste and odor.

Salinity always exists in groundwater but in variable amounts ($100<TDS<50000$ mg/lit). It is mostly influenced by aquifer material, solubility of minerals, duration of contact and factors such as the permeability of soil, drainage facilities, quantity of rainfall and above all, the climate of the area.

The salinity of groundwater in coastal areas may be due to air borne salts originating from air water interface over the sea and also due to over pumping of fresh water which overlays saline water in coastal aquifer system.

Unlike Ordinary Kriging dealing with the primary variable alone, Cokriging utilized not only the primary variable (e.g., salinity) but also cross-correlated secondary variables (e.g., groundwater level). Cokriging is thus a linear interpolator of both primary and secondary data values. If only a limited number of observations are available for the primary variable in concern, knowledge of secondary variables that are correlated with the primary variable can be used to reduce the estimation error and to improve the estimation. The estimation error is thereby reduced since more information is being used for the estimation of the primary parameter; a twofold reduction in error of estimation would be typical. Improvement of Cokriging over Ordinary Kriging with the primary variable alone is greatest when the primary variable is under sampled, as we often encounter in salinity sampling.

In this study we apply the common geostatistical method of Cokriging to estimate groundwater salinity in the study area.

The Variogram measures dissimilarity, or increasing variance between points (decreasing correlation) as a function of distance. In addition to helping us assess how values at different location vary over distance, the Variogram provide a way to study the influence of other factors which may affect whether the spatial correlation varies only with distance (the isotropic case) or with direction and distance (the anisotropic case).Variogram map provides a visual picture of semivariance in every compass direction. If there is anisotropy, this allows one to easily find the appropriate principal axis for defining the anisotropic Variogram model. In this map, the surface (z-axis) is semivariance, and the x and y axes are separation distances in E-W and N-S directions, respectively. The center of the map corresponds to the origin of the Variogram $\gamma(h)=0$ for every direction.

At stage two of the current modeling which deals with spatial prediction of groundwater level, estimated groundwater level of following month at the location of each piezometer was firstly corrected via bedrock elevation at the same location because of termination of existing trend (see Figure 4). Afterward, the Variogram map of the study area was plotted using the temporally averaged values of the groundwater levels at different piezometers.

Figure 5 shows that, the isotropic spatial modeling of the groundwater levels could be taken in use.

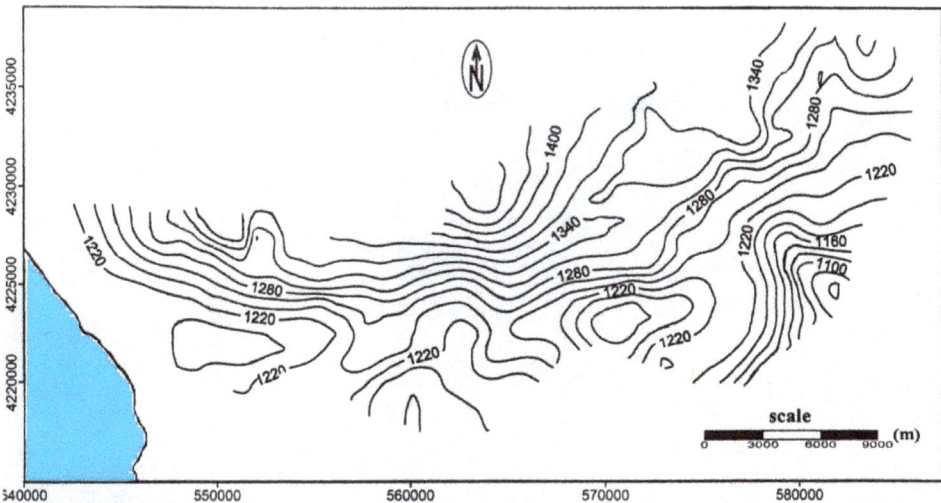

Fig. 4. Bedrock elevations in study area (units in meters)

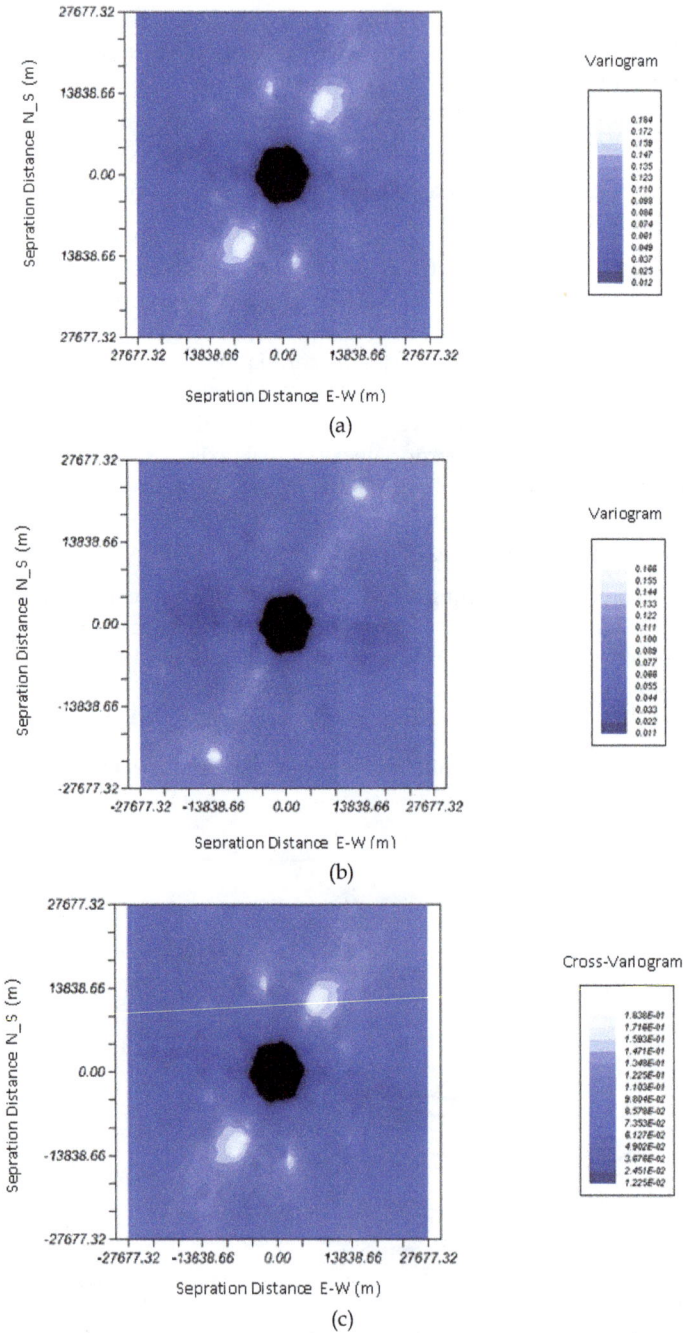

Fig. 5. Variogram maps : (a) Primary variogram (TDS); (b) Covariate variogram (EL) and (c) Cross variogram (TDS and EL)

Thereafter, a suitable Variogram model was determined by fitting some well-known Variogram models (i.e., spherical, exponential, Gaussian) to the experimental Variogram using weighted least squares method (Myers, 1982).

The geostatistical model having the least error was selected by comparing the observed water-table and salinity values with the values estimated by Variogram models. (Gundogdu and Guney,2007).

According to table 3 the best fitted models were Spherical, Gaussian and Spherical models for first, second and co-variables respectively, and their parameters (i.e., range, nugget and sill) were then used in the CoKriging procedure.

The results of the modeling have been presented in Figure 6.

RSS	Variogram for TDS	Variogram for Elevation	Cross Variogram
Gaussian model	9.22E-03	**5.09E-03**	5.98E-03
Exponential model	8.32E-03	9.20E-03	7.80E-03
Spherical model	**7.41E-03**	6.41E-03	**5.93E-03**

Table 3. Results of the different Variogram models

Based on the mentioned Variogram models spatial ground water level and salinity estimation of the area has been carried out using CoKriging method. The calibrated CoKriging method was then verified via a cross validation technique. Cross validation is a process for checking the compatibility between a set of data, the spatial model and neighborhood design. In cross validation, each point in the spatial model is individually removed from the model, and then its value is estimated by a covariance model. In this way, it is possible to complete estimated versus actual values. Figure 7 shows the results of cross validation procedure as a scatter plot, which denotes to the reliability of the proposed geostatistical modeling.

At this moment both stages of the hybrid model have been completed and the model can be used for spatio-temporal modeling of groundwater level within the Shabestar plain.

Finally, the proposed new modified hybrid model was validated using the verification data set (2004-2006, 3 years) of piezometers TP1, TP2, and TP3 which have not been utilized neither for training the ANNs nor for the calibration of the geostatistices model. For this purpose, the forecasted values of the water level time series at different piezometers (P1, P2,..., and P11) via the trained ANNs models for the verification data set (2003 to 2006) were imposed to the calibrated geostatistical model in order to estimate the water level and salinity of piezometers TP1,TP2, and TP3, time step by time step.

The results of the modeling have been presented and compared with the previous model results (ANNG) in Figure 8 which demonstrates the capability of the new proposed time-space hybrid model (MANNG).

Gussian Model
0.0003+0.09[1-exp(-h/9500)^2)]
r^2=0.86

Spherical model
8.989E-8+0.076[1.5(h/9763)-0.5((h/9763)^3)]
r^2=0.90

Exponential Model
1.229E-7+0.115[1-exp(-h/9432)]
r^2=0.95

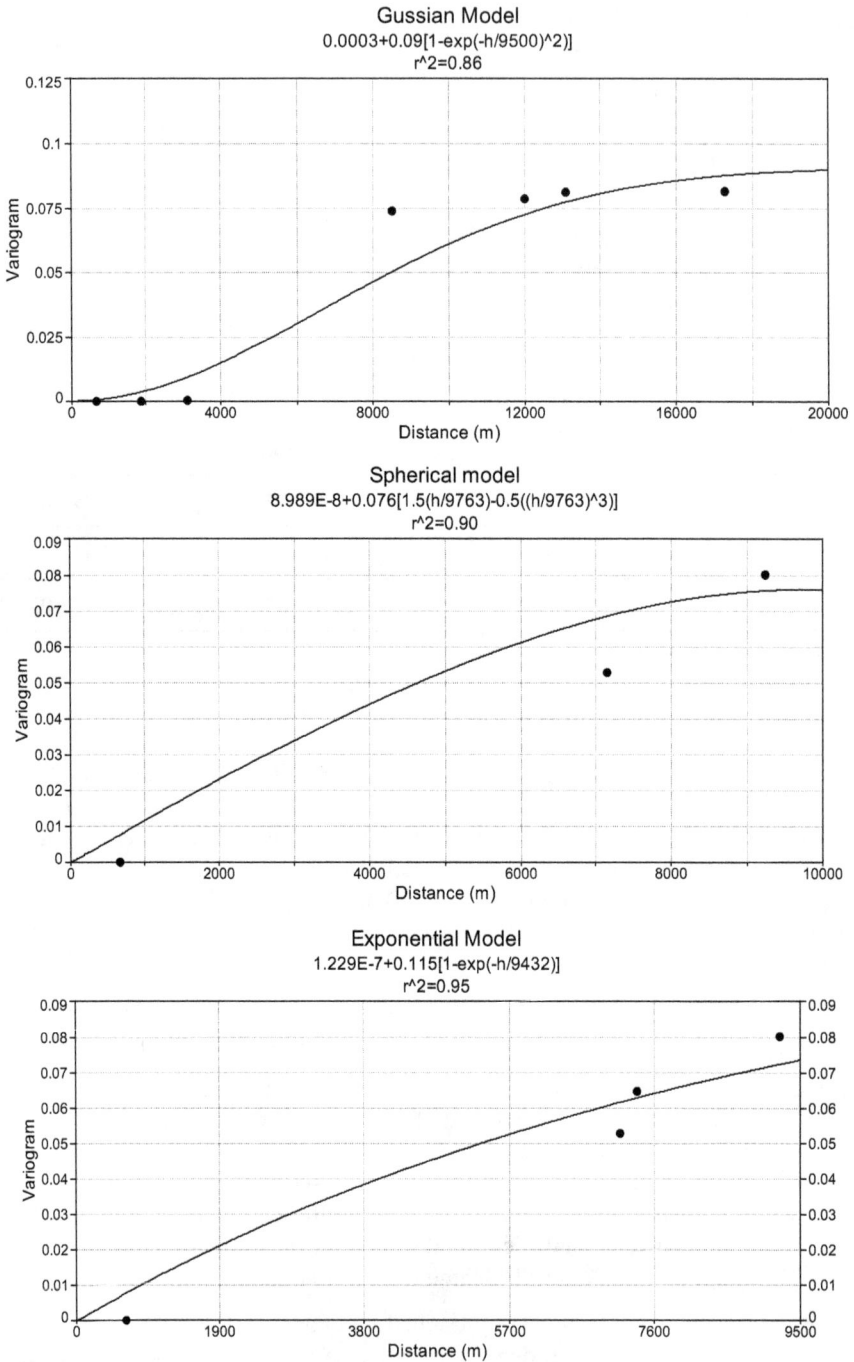

Fig. 6a. Variogram models for TDS.

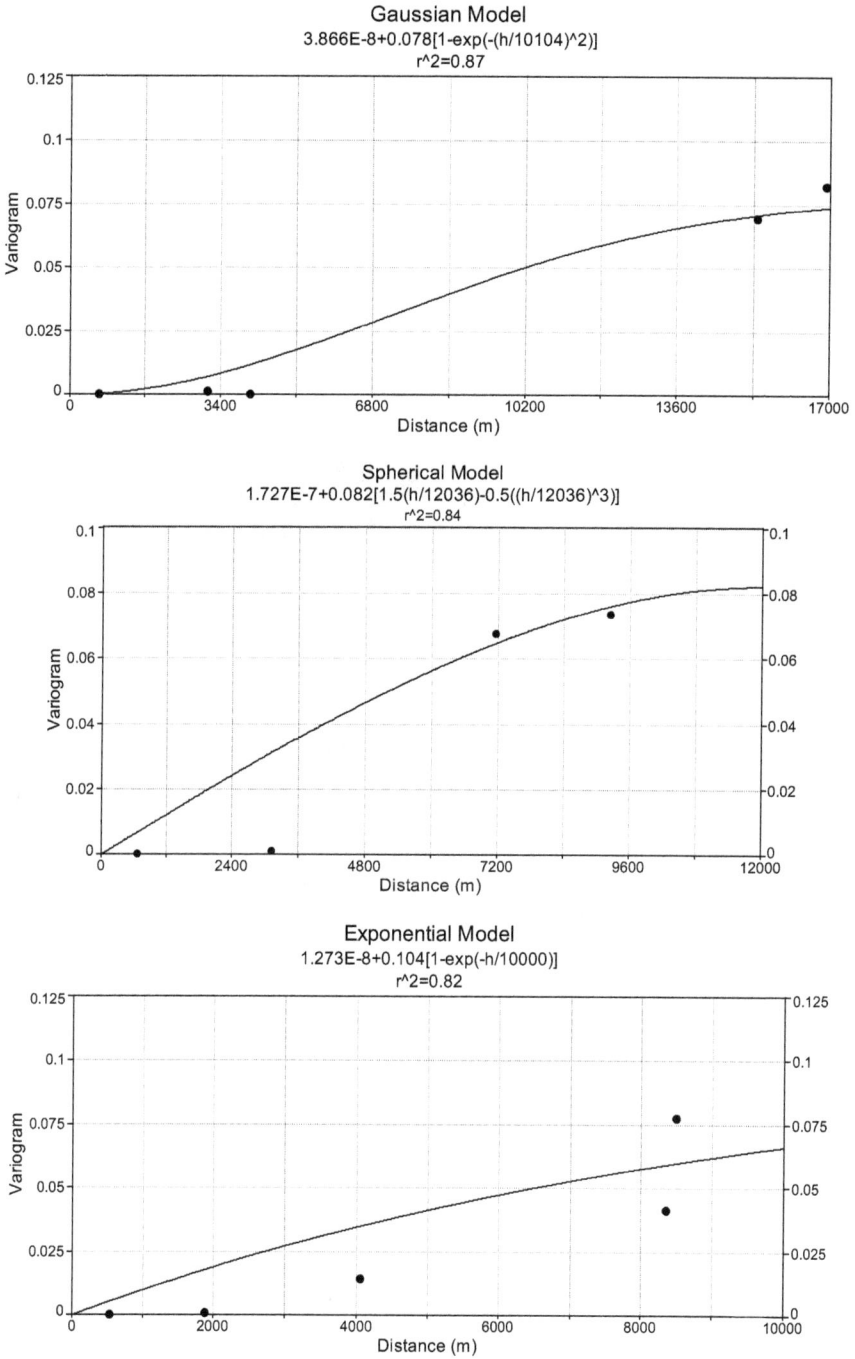

Gaussian Model
$3.866E-8+0.078[1-exp(-(h/10104)^2)]$
$r^2=0.87$

Spherical Model
$1.727E-7+0.082[1.5(h/12036)-0.5((h/12036)^3)]$
$r^2=0.84$

Exponential Model
$1.273E-8+0.104[1-exp(-h/10000)]$
$r^2=0.82$

Fig. 6b. Variogram models for groundwater level.

Gaussian Model
-0.008-0.07[1-exp(-(h/9173)^2)]
r^2=0.83

Spherical Model
-0.001-0.075[1.5(h/10774)-0.5((h/10774)^3)]
r^2=0.94

Exponential Model
-0.029-0.075[1-exp(-h/10000)]
r^2=0.8

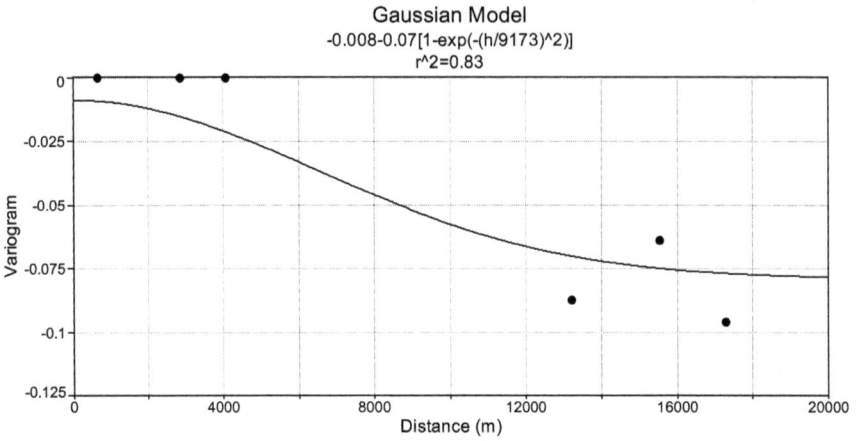

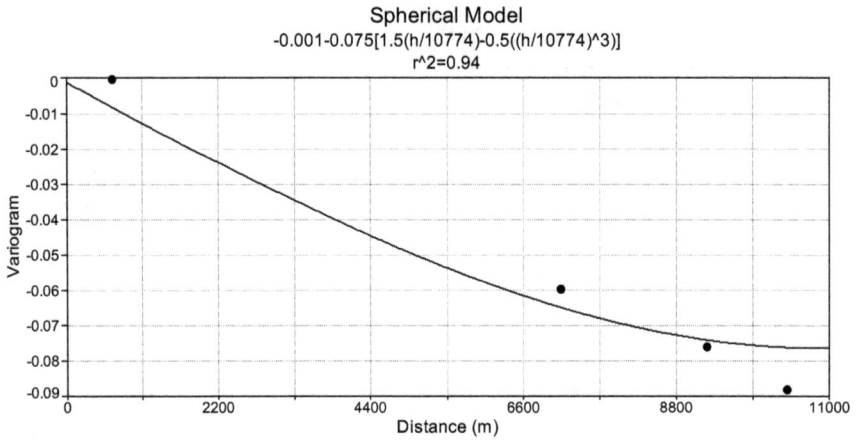

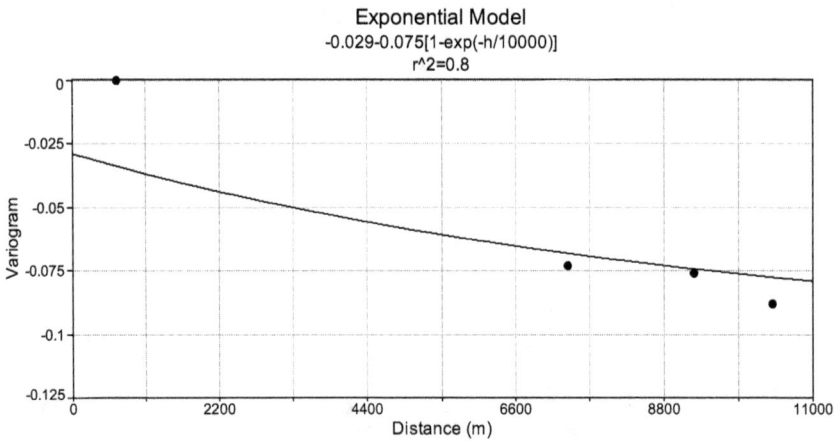

Fig. 6c. Variogram models for Cross.

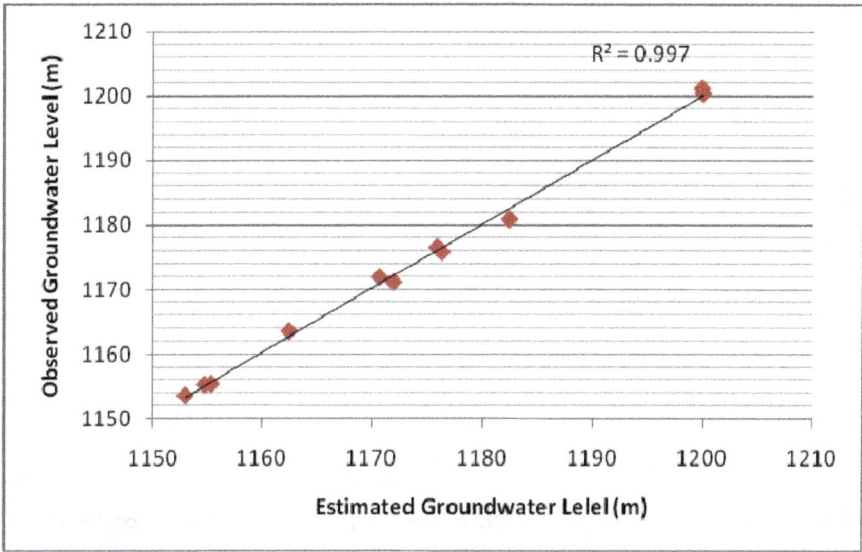

(a)

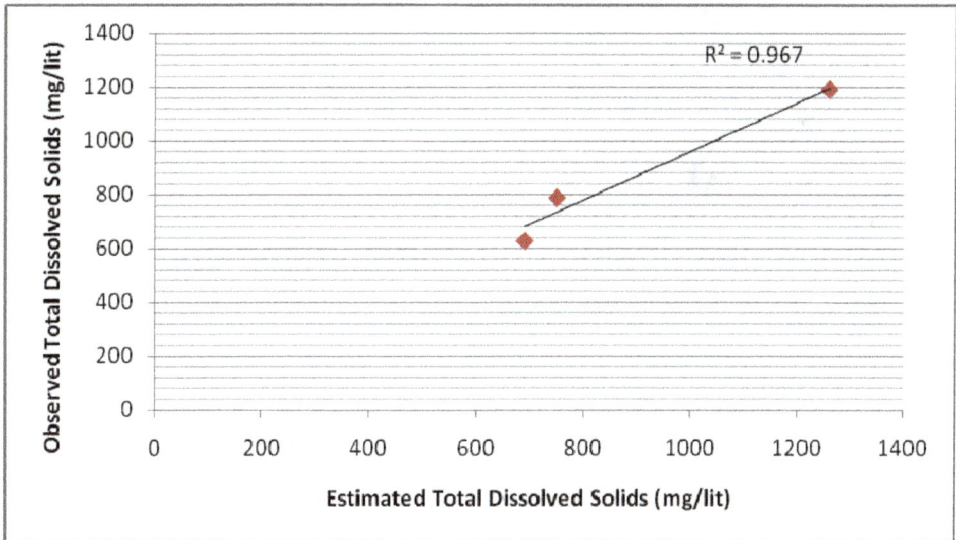

(b)

Fig. 7. Cross validation results: (a) for Groundwater level and (b) for TDS.

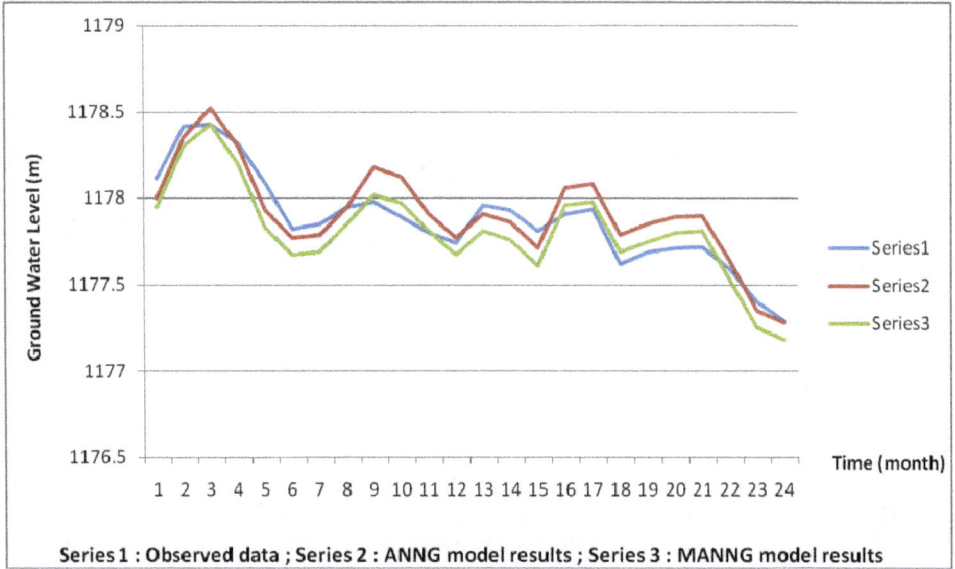

(a)
[R²=0.75 for ANNG & R²=0.83 for MANNG]

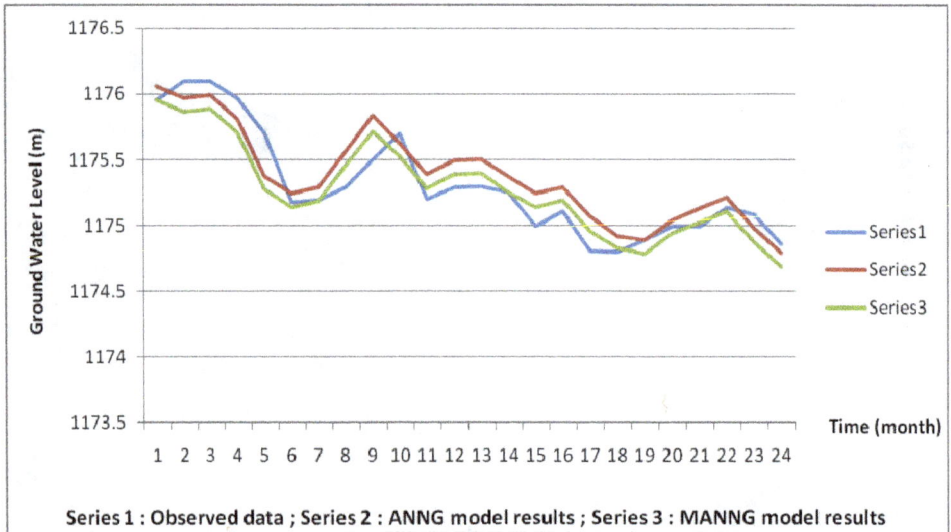

(b)
[R²=0.78 for ANNG & R²=0.84 for MANNG]

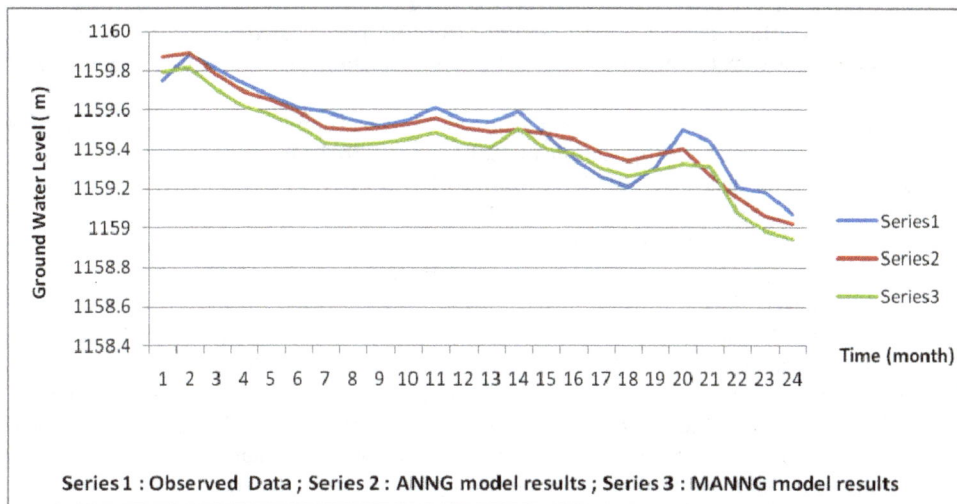

Series 1 : Observed Data ; Series 2 : ANNG model results ; Series 3 : MANNG model results

(c)

[$R^2=0.87$ for ANNG & $R^2=0.91$ for MANNG]

Fig. 8. Results of spatiotemporal modeling for piezometers; a) TP1, b) TP2 and c) TP3.

According to the obtained results it can be clearly seen that the model is more capable to estimate the groundwater levels and salinity where are close to the lake. Since the water depth of lake is considered as a input variable to the ANNs, the proposed model could simulate the groundwater level of the near region to the lake were accurate than the far points.

7. Concluding remarks

There are many hydrological variables that can be viewed as spatiotemporal phenomena. For example, monthly rainfalls or piezometric readings exhibit random aspects both with respect to time and space. The estimation of such variables at un-sampled spatial locations or un-sampled times requires the adequate techniques into space-time domain. In this study, according to inherent capability of artificial neural networks in temporal forecasting and geostatistics in spatial estimating, the potential of the proposed hybrid empirical model (MANNG) was evaluated for the purpose of spatio-temporal prediction of groundwater levels and salinity in a coastal aquifer in Iran.

Monthly groundwater levels data from eleven piezometer (P1, P2,... P11), rainfall and lake water surface elevations in the 13 years are the inputs of multilayer feedforward neural network. CoKriging was applied to the outputs from ANN model to estimate groundwater levels and salinity in un-sampled locations such as coordinates of three selected piezometers (TP1, TP2, and TP3).

This modeling framework is applied for the Shabestar plain which is located in northwest Iran at Azerbaijan province. The major results of the study are summarized as follows:

- The results of the research reported in the paper shows high efficiency of three-layer back propagation artificial neural network (BPANN) with Levenberg-Marquardt (LM) training algorithm for groundwater elevation prediction in the case study for coastal aquifer.

- Because of spatial structure between groundwater levels and salinity in adjacent points of this coastal aquifer, application of CoKriging with isotropic adequate Variogram geostatistical models have been led to appropriate results.

- In general, the results of the case study are satisfactory and demonstrate that the proposed hybrid model (MANNG) is a promising spatio-temporal prediction tool for groundwater modeling and may be also employed to fill the temporally and/or spatially missed data.

- According to Fig 8, application of the new modified hybrid model (MANNG) respect to previous model (ANNG) presented by Nourani et al. (2010), was led to exact results. In the other word, the results of cross validation procedure of the new model were 3 percent better than the old model. Generally, the proposed modified hybrid empirical model (MANNG) was used for the purpose of spatio-temporal prediction of groundwater levels and salinity in a coastal aquifer in Iran, efficiently.

8. References

Aboufirassi, M. & Marino, MA. (1983). Kriging of water level in the Souss aquifer, Morocco. *Math. Geol.* 15:537.

ASCE Task Committee (1990). Review of geostatstics in geohydrology, Part II: Applications. *J. Hydraulic Eng.* 116: 633-658.

ASCE Task Committee (2000a). Artificial neural networks in hydrology, Part I: Preliminary concepts. *J. Hydrol. Eng.* 5: 115-123.

ASCE Task Committee (2000b). Artificial neural networks in hydrology, Part II: Hydrologic application. *J. Hydrol. Eng.* 5: 124-136.

Barca, E. & Passarella, G. (2008). Spatial evaluation of the risk of groundwater quality degradation. A comparison between disjunctive kriging and geostatistical simulation. *Environ. Monit. Assess.* 137:261.

Basheer, IA. & Hajmeer, M. (2000). Artificial neural networks : fundamentals, computing, design, and application. *J. Micro. Methods* 43:3.

Beaudeau, P.; Leboulanger, T.; Lacroix, M.; Hanneton, S. & Wang, HQ. (2001). Forecasting of turbid floods in a karstic drain using an artificial neural network. *Ground Water* 39:139.

Cay, T. & Uyan, M.,(2009). "Spatial and temporal groundwater level variation geostatistical modeling in the city of Konya, Turkey". *Water Environ. Res.* 12:2460.

Coppola, E.; Poulton, M.; Charles, E.; Dustman, J. & Szidarovszky, F. (2003). Application of artificial neural networks to complex groundwater management problems. *Natural Resour. Res.* 12:303.

Coppola, E.; Rana, AJ.; Poulton, M.; Szidarovszky, F. & Uhi, VW. (2005). A neural networks model for prediction aquifer water level elevation. *Ground Water* 43:231.

Coulibaly, P.; Anctil, F.; Aravena, R.; & Bobee, B. (2001a). Artificial neural network modeling of water table depth fluctuation. *Water Resour. Res.* 37:885.

Coulibaly, P.; Anctil, F.; & Bobee, B. (2001b). Multivariate reservoir inflow forecasting using temporal neural networks.*J. Hydrol. Eng.* 9-10:367.

Coulibaly, P.; Bobee, B.; & Anctil, F. (2001c). Improving extreme hydrologic events forecasting using a new criterion for artificial neural network selection. *Hydrol. Process.* 15:1533.

Daliakopoulos, I.; Coulibaly, P.; & Tsanis, I.K. (2005). Groundwater level forecasting using artificial neural networks. *J. Hydrol.* 309:229.

Delhomme, JP. (1978). Kriging in hydrosciences. *Adv. Water Resour.* 1:251.

Dogan, A.; Demirpence, H.; & Cobaner, M. (2008). Prediction of groundwater levels from lake levels and climate data using ANN approach. *Water SA* 34:199.

Finke, PA.; Brus, DJ.; Bierkens, MFP.; Hoogland T.; Knotters, M.; & Vries, F. (2004). Mapping groundwater dynamics using multiple sources of exhaustive high resolution data. *Geoderma* 123:23.

Goovaerts, P. (1997). Geostatistics for natural resources evaluation. Oxford University Press.

Goovaerts P. (1999). "Geostatistics in soil science: state-of-the-art and perspectives". *Geoderma* . 89:1.

Govindaraju, RS. & Ramachandra, RA. (2000). Artificial neural networks in hydrology. Kluwer Academic Publishing, Netherlands.

Gundogdu, K. & Guney, I., (2007). "Spatial analysis of groundwater levels using universal kriging". *J. Eearth Syst. Sci.* 116:49.

Hopfield, JJ. (1982). Neural networks and physical systems with emergent collective computational abilities. *Proc. Natl. Acad. Sci.* 79:2554.

Hornik, K.; Stinchcombe, M. & White, H. (1989). Multilayer feed forward networks are universal approximators. *Neural Netw.* 2:359.

Isaaks, E.H. & Srivastava, R.M. (1989). Applied Geostatistics. Oxford University Press, New York.

Jiang, SY.; Ren, ZY.; Xue, KM. & Li, CF. (2008). Application of BPANN for prediction of backward ball spinning of thin-walled tubular part with longitudinal inner ribs. *J. Materials Process. Tech.* 196:190.

Lallahem, S. & Mania, J. (2003). Evaluation and forecasting of daily groundwater inflow in a small chalky watershed. *Hydrol. Process.* 17:1561.

Lallahem, S.; Mania, J.; Hani, A. & Najjar, Y. (2005). On the use of neural networks to evaluate groundwater levels in fractured media. *J. Hydrol.* 307:92.

Ma, T.S.; Sophocleous, M. & Yu, Y.S. (1999). Geostatistical applications in groundwater modeling in south-central Kansas. *J. Hydrol. Eng.* 16:57.

Maier, HR. & Dandy, GC. (1998). Understanding the behavior and optimizing the performance of back-propagation neural network: an empirical study. *Environ. Model. Softw.* 13:179.

Maier, HR. & Dandy, GC. (2000). Neural network for the prediction and forecasting water resources variables: a review of modeling issues and applications. *Environ. Model. Softw.* 15:101.

Matheron, G. (1963). Principles of geostatistics. *Economic Geol.* 58:1246.

Myers, D.E. (1982). Matrix formulation of Cokriging . *Math. Geol.* 14:249.

McCulloch, W.S. & Pitts, W. (1943). A logical calculus of the ideas immanent in nervous activity. *Bulletin Math. Biophys.* 5: 115.

Nash, J.E. & Sutcliffe, J.V. (1970). River flow forecasting through conceptual models: Part I. A conceptual models discussion of principles. *J. Hydrol.* 10:282.

Neuman, SP. & Jacobsen, EA. (1984). Analysis of non-intrinsic spatial variability by residual kriging with application to regional groundwater level. *Math. Geol.* 16:499.

Nourani, V. & Mano, A. (2007). Semi-disrtibuted flood runoff model at the sub continental scale for southwestern Iran. *Hydrol. Process.* 21:3137.

Nourani, V.; Mogaddam, A.A. & Nadiri A. (2008). An ANN-based model for spatiotemporal groundwater level forecasting. *Hydrol. Process.* 22:5054.

Nourani, V.; Alami, M.T. & Aminfar, M.H., (2009). A combined neural-wavelete model for prediction of Ligvanchai watershed precipitation. *Eng. Appl. Artif. Intell.* 22:466.

Nourani, V.; Ejlali, RG. & Alami, MT. (2010).Spatiotemporal groundwater level forecasting in coastal aquifers by hybrid Artificial Neural Network-Geostatisics model: A case study. *Environ. Eng. Sci.*28:217.

Qu, ZY.; Chen, YX. & Shi, HB. (2004). Structure and algorithm of BP network for underground hydrology forecasting. *J. Water Resour.* 2:88.

Rumelhart, DE. & McClelland , JL. (1986). Parallel distributed processing : Explorations in the microstructure of cognition, I and II. MIT Pess, Cambridge.

Sahoo, GB.; Raya, C. & Wadeb, HF. (2005). Pesticide prediction in groundwater in North Carolina domestic wells using artificial neural network. *Ecol. Model.* 183:29.

Ta'any, R.; Tahboub, A. & Saffarini, G. (2009). "Geostatistical analysis of spatiotemporal variability of groundwater level fluctuations in Amman-Zarqa basin, Jordan: a case study". *Environ. Geol.* 57:525.

Tayfur, G.; Swiatek, D.; Wita, A. & Singh, VP. (2005). Case study : finite element method and artificial neural network models for flow through Jeziorsko earth dam in Poland. *J. Hydraulic Eng.* 131:431.

Tayfur, G. & Singh, VP. (2006). ANN and fuzzy logic models for simulating event-based rainfall-runoff. *J. Hydraulic Eng.* 132:1321.

Toth, E.; Brath, A. & Montanari, A. (2000). Comparsion of short-term rainfall prediction models for real-time flood forecasting. *J. Hydrol.* 239:132.

Zhang, B. L. & Dong, Z. Y. (2001). An adaptive neural-wavelet model for short term load forecasting. *Elect. Pow. Sys. Res.*59:121.

Permissions

The contributors of this book come from diverse backgrounds, making this book a truly international effort. This book will bring forth new frontiers with its revolutionizing research information and detailed analysis of the nascent developments around the world.

We would like to thank Purna Nayak, for lending his expertise to make the book truly unique. He has played a crucial role in the development of this book. Without his invaluable contribution this book wouldn't have been possible. He has made vital efforts to compile up to date information on the varied aspects of this subject to make this book a valuable addition to the collection of many professionals and students.

This book was conceptualized with the vision of imparting up-to-date information and advanced data in this field. To ensure the same, a matchless editorial board was set up. Every individual on the board went through rigorous rounds of assessment to prove their worth. After which they invested a large part of their time researching and compiling the most relevant data for our readers. Conferences and sessions were held from time to time between the editorial board and the contributing authors to present the data in the most comprehensible form. The editorial team has worked tirelessly to provide valuable and valid information to help people across the globe.

Every chapter published in this book has been scrutinized by our experts. Their significance has been extensively debated. The topics covered herein carry significant findings which will fuel the growth of the discipline. They may even be implemented as practical applications or may be referred to as a beginning point for another development. Chapters in this book were first published by InTech; hereby published with permission under the Creative Commons Attribution License or equivalent.

The editorial board has been involved in producing this book since its inception. They have spent rigorous hours researching and exploring the diverse topics which have resulted in the successful publishing of this book. They have passed on their knowledge of decades through this book. To expedite this challenging task, the publisher supported the team at every step. A small team of assistant editors was also appointed to further simplify the editing procedure and attain best results for the readers.

Our editorial team has been hand-picked from every corner of the world. Their multi-ethnicity adds dynamic inputs to the discussions which result in innovative outcomes. These outcomes are then further discussed with the researchers and contributors who give their valuable feedback and opinion regarding the same. The feedback is then collaborated with the researches and they are edited in a comprehensive manner to aid the understanding of the subject.

Apart from the editorial board, the designing team has also invested a significant amount of their time in understanding the subject and creating the most relevant covers. They scrutinized every image to scout for the most suitable representation of the subject and create an appropriate cover for the book.

The publishing team has been involved in this book since its early stages. They were actively engaged in every process, be it collecting the data, connecting with the contributors or procuring relevant information. The team has been an ardent support to the editorial, designing and production team. Their endless efforts to recruit the best for this project, has resulted in the accomplishment of this book. They are a veteran in the field of academics and their pool of knowledge is as vast as their experience in printing. Their expertise and guidance has proved useful at every step. Their uncompromising quality standards have made this book an exceptional effort. Their encouragement from time to time has been an inspiration for everyone.

The publisher and the editorial board hope that this book will prove to be a valuable piece of knowledge for researchers, students, practitioners and scholars across the globe.

List of Contributors

Matjaž Glavan and Marina Pintar
University of Ljubljana, Biotechnical Faculty, Department of Agronomy, Chair for Agrometeorology, Agricultural Land Management, Economics and Rural Development, Slovenia

Michael P. Strager
Division of Resource Management, West Virginia University, USA

Raymond Abudu Kasei
University for Development Studies, Ghana

P.C. Nayak
Deltaic Regional Centre, National Institute of Hydrology, Kakinada, India

K.P. Sudheer
Dept of Civil Engineering, Indian Institute of Technology Madras, India

S.K. Jain
NEEPCO, Department of Water Resources Development and Management, Indian Institute of Technology, Roorkee, India

Oscar Delgado-González, Fernando Marván-Gargollo, Adán Mejía-Trejo and Eduardo Gil-Silva
Instituto de Investigaciones Oceanológicas, Universidad Autónoma de Baja California, Ensenada, Baja California, México

Luc Descroix and Okechukwu Amogu
IRD / UJF-Grenoble 1 / CNRS / G-INP, LTHE UMR 5564, LTHE, Laboratoire d'études des Transferts en Hydrologie et Environnement, France

Dorota Miroslaw-Swiatek
Warsaw University of Live Sciences – SGGW, Poland

Paul Bradley and Celeste Journey
U.S. Geological Survey, USA

Frappart Frédéric and Ramillien Guillaume
Université de Toulouse, OMP-GET, CNRS, IRD, France

Arshad Ashraf
National Agricultural Research Center, Islamabad

Zulfiqar Ahmad
Department of Earth Sciences, Quaid-i-Azam University, Islamabad, Pakistan

Mohsen Masihi
Department of Chemical and Petroleum Engineering, Sharif University of Technology, Iran

Peter R. King
Earth Science and Engineering Department, Imperial College London, UK

Alon Rimmer
Israel Oceanographic and Limnological Research Ltd., The Yigal Alon Kinneret Limnological Laboratory, Israel

Andreas Hartmann
Institute of Hydrology, Freiburg University, Germany

Vahid Nourani and Reza Goli Ejlali
University of Tabriz, Iran